ALGEBRA
WITH
APPLICATIONS
Technique and THOUGHT

Revised Edition

ALGEBRA
WITH
APPLICATIONS
Technique and THOUGHT

Revised Edition

William J. Adams

Mathematics Department, Pace University

with illustrations by
Ramunė B. Adams

Library of Congress Control Number: 2007908899
ISBN: Hardcover   978-1-4363-0179-4
                 Softcover   978-1-4363-0178-7

This book is available on the web at webpage.pace.edu/wadams

This book was printed in the United States of America.

To order additional copies of this book, contact:
Xlibris Corporation
1-888-795-4274
www.Xlibris.com
Orders@Xlibris.com
40316

To Onutė Adams

CONTENTS

Chapter 3

Chapter 4

Chapter 5

Chapter 6

Chapter 7

Chapter 8

Chapter 9

Chapter 10

Chapter 11

To Students Studying
Algebra from this Book

Those who found their earlier study of algebra less than a satisfying experience (and perhaps some who found it agreeable) might at this point be ready to ask whether to force us to study more of this stuff should be viewed as unconstitutional in that it may be a violation of the constitutional provision against cruel and unusual punishment.

If I had agreed with this view I would not have written this book, for I strongly oppose infliction of cruel and unusual punishment in any form.

Every undertaking has its tedious and boring patches—learning the alphabet, learning to read, studying piano scales and the like come to mind as examples for some of us—and in the study of mathematics such topics as arithmetic, fractions, factoring, solving equations and the like might come to mind.

Yet, if we persist (and this includes me), the door to experience great treasures becomes open to us—literature, the thinking and insights of our fellow humans, theater, music. The study of algebra opens the door to great treasures as well. An indication of some of these treasures is found in this book in terms of applications that provide us with insight into the world we find ourselves in and ideas from mathematics itself.

Beauty, it is often said, is in the eyes of the beholder, to which we may add what is interesting. I should like to encourage you to persist in moving beyond what you might view as the tedious and boring patches of algebra so that the door to a rich harvest of real world applications and mathematical ideas will open up to you.

Algebra and mathematics more generally are great triumphs of the human intellect and, like fine art, music, and literature, bring us together as human beings interested in partaking of the best of the human spirit.

W.J.A.

INTRODUCTION

Still another book on algebra; why? What's the point? At the very least an explanation is in order.

Many of the books that I am familiar with do a good job in their explanation of algebra procedures and technique but fall short, in my view, in their attention to fundamental concepts and ideas. There is also the issue of addressing what may be termed "serious" applications to give students a sense of the importance of algebra for the study of the real world. The books I am familiar with give little if any attention to this dimension. The simplistic and contrived applications they present, are, I would submit, counterproductive.

My objective in this book is to present an exposition of algebra that balances technique with thought along with a spectrum of reality grounded applications whose intent is to show off the power of algebra in particular and, by extension, mathematics in general as a tool for studying the real world.

Much attention is given to developing the mathematical modeling perspective to make clear what mathematics can and cannot do for us. A point continually stressed is that reaching a mathematical answer to a problem is not the end of the story. It is in a sense the end of a chapter, but the next chapter is concerned with questions of whether and how the mathematical answer should be implemented.

Self-tests have been included to help students see the unity of algebra and get blocks of material under better control.

As to which topics are taken up and the order pursued, this will, needless to say, depend on course requirements and the instructor's judgment.

Answers to the odd-numbered exercises and self-tests are provided in the last part of the book.

The *Companion to Algebra* provides answers to even-numbered exercises and, for selected problems, details on how the answers were obtained.

To make this book widely available I have put it on the web at webpage. pace.edu/wadams.

I am greatly indebted to my daughter Ramunė for preparing the illustrations.

W.J.A.

PERSPECTIVE ON THE CHAPTERS AND EXERCISES

Chapter 1:   The World of Real-Numbers

Chapter 1 provides the foundation for "standard" tools of algebra and the backdrop against which properties of the complex-number system (ch. 10) and matrices (ch. 11) are discussed. The issue of division by zero is given appropriate attention.

Chapter 2:   Equations and Inequalities: A First Look

The building of algebra technique initiated in chapter 1 is continued. The development of a "serious" application to the world of finance is initiated (sec. 2.3); the growth of the seeds planted here is continued in section 9.2.

Chapter 3:   Functions

The pivotal concept of function is introduced and its dimensions are indicated through discussion of a number of examples. Much attention is paid to the domain of definition of functions arising in applied settings. The concept mathematical model function for a function that arises in an applied setting is introduced.

Chapter 4:   Linear Structures

This chapter sets the foundation for topics discussed in chapters 6 (Linear Program Models) and 11 (Life in the Land of Matrices). section 4.2 on the elimination of a variable technique for solving systems of two-variable linear equations is important for development of the corner point method for solving two-variable linear programs taken up in section 6.3. The tableau method for solving systems of linear equations (which is the Gauss-Jordan elimination process with a refinement that enhances its flexibility) developed in sections 4.4 and

4.5 opens the way for discussion of significant applications of larger systems of linear equations. It also serves as a foundation for discussion of matrix inversion (sec. 11.3).

Section 4.3 presents an application of systems of two-variable linear equations to determining a regression line. It is interesting and shows off a "serious" application of systems of two-variable linear equations, but may be omitted without loss of continuity. Section 4.6 takes up two significant (and self-contained) problems from accounting (Service Charge Allocation and Income Consolidation) which lead to systems of linear equations. They are solved by the tableau method and later addressed by matrix methods (sec. 11.5), at which point an advantage of the matrix approach becomes clear. While interesting, this development may, if necessary, be omitted without loss of continuity. The application of the tableau method to systems of linear equations with infinitely many solutions (sec. 4.7) is interesting, but may also be omitted without loss of continuity.

Chapter 5:   More on Mathematical Modeling

This chapter provides the foundation for discussion of what mathematical technique (algebra technique, in particular) can do for us, and its limitations. Two vehicles are employed to introduce the nature of mathematical modeling: the problem of determining the financial cost of smoking and determining the time it takes to travel to a vacation spot. Mathematical methods (addition, multiplication, division) are kept simple so that the main features of mathematical modeling stand out.

The payoff of the mathematical modeling point of view introduced here is most fully demonstrated in chapter 6.

Chapter 6:   Linear Program Models

Chapter 6 effects a synthesis of the algebra technique developed in chapter 4 and the modeling point of view introduced in chapter 5. The Austin Company's production scheduling problem, introduced in section 6.2, serves as a vehicle for carrying out this synthesis. Two linear program models with different solutions emerge for this problem, setting the stage for discussion of which solution is best, should be implemented, and why. The corner point method for solving linear programs is developed and a variety of situations leading to linear program models is examined.

Chapter 7:   Further Food for Thought

Consideration of the advertising media selection problem from marketing (sec. 7.1) provides insight into what makes a linear program model successful or not, and the limitations of this approach.

A problem leading to a linear program model with inconsistent constraints is employed to illustrate that not all linear program models have solutions (sec. 7.2).

What can we do if a problem requires a solution in integers, but the linear program model set up for it does not yield a solution in integers? This question is addressed in sec. 7.3. Math technique discussed in sec. 4.1 is put to use here.

Chapter 8:   Topics in Algebra with Consideration of Reality Concerns

Further development of algebra technique, including consideration of quadratic equations and functions which are applied to questions arising in economics.

Chapter 9:   Math Modeling for Money

Consideration of arithmetic and geometric progressions, with the main applied dimension being to money matters. This applied dimension picks up on seeds planted in section 2.3.

Chapter 10:   The World of Complex-Numbers

Development of the complex-number system and examination of its properties against the backdrop provided by the real-number system's properties considered in chapter 1. The problem of determining roots of real-numbers and complex-numbers is addressed.

Chapter 11:   Life in the Land of Matrices

Development of matrix algebra and examination of its properties against the backdrop provided by the real-number system's properties considered in chapter 1. Matrix inversion and its application to solving systems of linear equations are taken up, with a return to service charge allocation and income consolidation problems first considered in section 4.6.

Leontief input-output models for economies are discussed with a reaffirmation of the mathematical modeling spirit stressing the importance of obtaining realistic input and final demand matrices for such models to be realistic descriptions of economies.

Exercises

In the spirit of the show 20-questions, the following twenty questions are among my favorites in being thought provoking and insightful.

1. Ch. 1, Sec. 1.4, 41/46
2. Ch. 1, Sec. 1.4, 41/47
3. Ch. 1, Sec. 1.4, 41/48
4. Ch. 1, Sec. 1.4, 41/49
5. Ch. 1, Sec. 1.6, 65/64
6. Self-Test 3, 151/9
7. Ch. 4, Sec. 4.7, 210/7
8. Ch. 5, Sec. 5.1, 226/2
9. Ch. 6, Sec. 6.5, 263/2
10. Ch. 6, Sec. 6.6, 269/3
11. Self-Test 2, 294/1
12. Self-Test 4, 301/4
13. Ch. 7, Sec. 7.2, 220/27
14. Ch. 8, Sec. 8.2, 322/5
15. Ch. 8, Sec. 8.2, 327/9
16. Self-Test 2, 350/2
17. Self-Test 1, 386/8
18. Self-Test 2, 387/10
19. Self-Test 1, 426/11
20. Self-Test 1, 303/12

CHAPTER 1

The World of
Real-Numbers

1.1 THE REAL-NUMBER SYSTEM

Since algebraic operations and developments are dependent on the number system adopted as a foundation, let us begin our study of algebraic principles and applications by reviewing and summarizing basic data about numbers and their mathematical life. The real-number system, as it is called, underlies a wide range of developments and applications and will serve as the foundation for the topics considered in this book, that is, for the most part.

Words and Wordings

It is important for us to keep in mind that there is no particular significance to the word real in real number system or real number. Real is not an adjective modifying the noun number. It is not that real numbers are more real in the sense of real-world applications than others, with the suggestion that other numbers are less so.

Real number system as well as real numbers are single entities in their own right. Perhaps the best way to make this clear is to write them realnumber and realnumber system rather than the usual way, real number and real number system. As a compromise I will settle for real-number and real-number system.

The real-number system is a collection of mathematical objects, called real-numbers, which acquire mathematical life by virtue of certain fundamental principles, or rules, that we adopt. The situation is somewhat similar to a game, like chess, for example. The chess system, or game, is a collection of objects, called chess pieces, which acquire life by virtue of the rules of the game, that is, the principles that are adopted to define allowable moves for the pieces and the way in which they may interact. A word about representation by symbols is perhaps in order at this point. The number two, for example, is represented by such symbols as 2, $1+1$, $3-1$, $\dfrac{4}{2}$, and so on. We should be careful not to equate the symbol used to express a number and the number itself. They are not the same, just as the signature of a person is not the same as the person, although a person's signature can be used to identify or represent him under certain circumstances. When it is necessary to be ultracareful about this distinction, the symbol used to represent a number is called a **numeral.**

Our working experience with numbers has provided us all with some familiarity with the principles that govern the real-number system. However, to establish a common ground of understanding and avoid certain errors that have become very common, we shall explicitly state and illustrate many of these principles. The real-number system includes such numbers as:

$$-27, \ -\pi, \ -\sqrt{2}, \ -1, \ -\frac{2}{3}, \ 0, \ \frac{1}{2}, \ \frac{1}{\sqrt{3}}, \ \sqrt{3}, \ \pi, \ 87.4$$

It is worthy of note that positive numbers, $\dfrac{1}{2}$, 1, $\sqrt{3}$, and 4, for example, are sometimes expressed as $+\dfrac{1}{2}$, $+1$, $+\sqrt{3}$, $+4$ The plus sign, +, used here does not express the operation of addition, but is rather part of the symbolism for the numbers themselves. Similarly, the minus sign, $-$, used in expressing such numbers as $-\dfrac{1}{2}$, -1, and -2, is part of the symbolism used for these numbers.

Within the real-number system, numbers of various kinds are identified and named. The numbers

$$1, 2, 3, 4, 5, \ldots$$

which are used in the counting process, are called **natural numbers.** The natural numbers, together with

$$-1, \ -2, \ -3, \ -4, \ -5, \ldots$$

and zero, are called **integers.** Since $1, 2, 3, 4, 5, \ldots$ are greater than 0, they are also called **positive integers;** $-1, \ -2, \ -3, \ -4, \ -5, \ldots$ are less than 0, and for this reason are called **negative integers.** A real-number is said to be a **rational-number** if it can be expressed as the ratio of two integers, where the denominator is not zero. (For reasons we shall say more about later, division by zero is not defined, and thus denominators with zero must be excluded.) For example,

$$-10, \quad \frac{1}{2}, \quad -\frac{2}{3}, \quad 0, \quad 1, \quad \frac{25}{7}, \quad 5$$

are rational-numbers. The integers are included among the rational-numbers since any integer can be expressed as the ratio of the integer itself and one. For example:

$$-10 = \frac{-10}{1}, \quad 1 = \frac{1}{1}, \quad 0 = \frac{0}{1}, \quad 5 = \frac{5}{1}$$

A real-number that cannot be expressed as the ratio of two integers is said to be an **irrational-number.** $\sqrt{2}$, π, and $-\sqrt{3}$ illustrate irrational-real-numbers. This classification for real-numbers is shown in Figure 1.1.

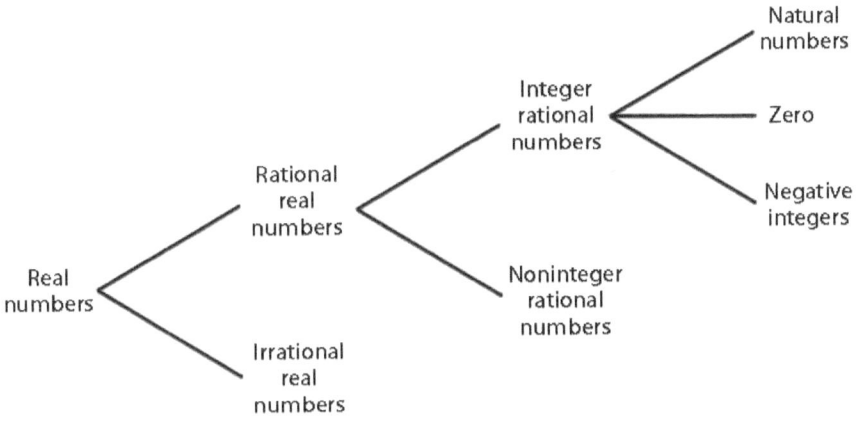

Figure 1.1

1.2   PROPERTIES OF ADDITION AND MULTIPLICATION

Addition and multiplication are primary operations on real-numbers. Most, if not all, of the following basic properties of these operations are familiar to us from experience.

Closure property of addition and multiplication. Whenever two real-numbers are added or multiplied, we obtain a real-number as the result. That is, performing the operations of addition and multiplication leaves us within the real-number system. We are not taken outside this system. Formally stated, we have the following:

> **A1.** If a and b are any numbers in the real-number system R, then the sum of a and b, denoted by $a+b$, is in R.
>
> **M1.** If a and b are any numbers in the real-number system R, then the product of a and b, denoted by $a \cdot b$ or ab, is in R.

Commutative property of addition and multiplication. The order in which two real-numbers are added or multiplied does not affect the result obtained. Such a property is called a commutative property. Addition and multiplication of real-numbers are commutative operations. Formally stated, we have the following:

> **A2.** If a and b are any numbers in the real-number system R, then
>
> $$a+b=b+a.$$
>
> **M2.** If a and b are any numbers in the real-number system R, then
> $$ab=ba.$$

Associative property of addition and multiplication. Parentheses, brackets, and the like, we recall, are used in algebra to group together whatever terms are within them. Thus $2+(3+4)$ means that 2 is to be added to the sum of 3 and 4, yielding $2+7$ or 9, whereas $(2+3)+4$ means that the sum of 2 and 3, or 5, is to be added to 4, yielding 9. Similarly, $2 \cdot (3 \cdot 4)$ yields $2 \cdot (12) = 24$, whereas $(2 \cdot 3) \cdot 4$ yields the same end result via the route $(6) \cdot 4 = 24$. That such is the case in general is the content of the associative property of addition and multiplication of real-numbers.

> **A3.** If a, b, and c are any numbers in the real-number system R, then
>
> $$a+(b+c)=(a+b)+c.$$
>
> **M3.** If a, b, and c are any numbers in the real-number system R, then
>
> $$a(bc)=(ab)c.$$

The properties we have surveyed apply no matter how many terms are involved in a sum or product. In total they say that the sum, or product, or any number of real numbers is a real-number, and the order in which they are added, or multiplied, is immaterial.

The numbers zero and one. The following are the basic properties of the numbers zero and one.

> **A4.** There is an unique real-number, called **zero** and denoted by 0, with the property that $a + 0 = 0 + a$, where a is any real-number.
>
> **M4.** There is an unique real-number, **different from zero**, called **one** and denoted by 1, with the property that $a \cdot 1 = 1 \cdot a = a$, where a is any real-number.

Additive and multiplicative inverses of a number. Corresponding to the numbers 2 and $-\frac{1}{3}$, for example, are the numbers -2 and $\frac{1}{3}$ with the property:

$$2 + (-2) = (-2) + 2 = 0$$

$$(-\frac{1}{3}) + \frac{1}{3} = \frac{1}{3} + (-\frac{1}{3}) = 0$$

-2 is called the **additive inverse of**, or **negative of**, 2, and vice versa. $\frac{1}{3}$, is called the **additive inverse of**, or **negative of**, $-\frac{1}{3}$, and vice versa

Also, corresponding to the number 2 is the number $\frac{1}{2}$ with the property:

$$2(\frac{1}{2}) = \frac{1}{2}(2) = 1$$

Corresponding to $-\frac{1}{3}$ is -3 with the property:

$$(-\frac{1}{3})(-3) = (-3)(-\frac{1}{3}) = 1$$

$\dfrac{1}{2}$ is called the **multiplicative inverse, or reciprocal of,** 2, and vice versa; -3 is called the multiplicative inverse, or reciprocal of, $-\dfrac{1}{3}$, and vice versa. More generally, every real-number has a unique additive inverse, and every **nonzero** real-number has an unique multiplicative inverse. Formally stated, we have the following:

> **A5.** If a is any real-number, then there is an unique real-number x, called the **additive inverse of a,** or **negative of a,** with the property that $a + x = x + a = 0$.
>
> **M5.** If a is any nonzero real-number, then there is an unique real-number y, called the **multiplicative inverse of a,** or **reciprocal of a,** with the property that $ay = ya = 1$.

The negative of a (denoted by x in A5) is usually symbolized by prefixing the minus sign, $-$, before a, yielding $-a$. Thus, in terms of this notation, the negative of -3, which is 3 [since $(-3) + 3 = 3 + (-3) = 0$], is denoted by $-(-3)$. We have, then, that $-(-3) - 3$. More generally, if a is any real-number, then:

$$-(-a) = a$$

The multiplicative inverse of a (denoted by y in M5) is often represented by the symbol $1/a$ or a^{-1}. Note that since the product of any number y and 0 is 0,

$$y \cdot 0 = 0 \cdot y = 0 :$$

> **0 cannot have a multiplicative inverse. Thus 1/0 does not exist.**

Distributive property. We know that $2(3+4) = 2(7) = 14$, and that $2 \cdot 3 + 2 \cdot 4 = 6 + 8 = 14$. Thus $2(3+4) = 2 \cdot 3 + 2 \cdot 4$. That such is the case in general for all real numbers is the content of the distributive property of multiplication over addition, more simply called the distributive property.

D. If a, b, and c are any real-numbers, then $a(b+c)=ab+ac$.

That is, the multiplier a is distributed between the terms b and c. In reading from left to right, we speak of multiplying out or removing the parentheses. In reading from right to left, we speak of taking out the common factor a from $ab+ac$, thus leaving the product of a and the other factor $b+c$.

More generally, the distributive property holds for any number of terms.

$$a(b+c+d+\cdots+z)=ab+ac+ad+\cdots+az$$

To illustrate, consider the sum:

$$3x+2x+4x$$

By taking out the common factor x, the distributive property yields:

$$3x+2x+4x = x(3+2+4)$$
$$= x(9)$$
$$= 9x$$

From this example we see how the distributive property allows us to combine a sum of like terms into one term.

As another example consider:

$$3xy+3xz+3x$$

By taking out the common factor $3x$ and observing what is left over in each term in our sum, we obtain:

$$3xy+3xz+3x = 3x(y+z+1)$$

In general, using the distributive property to take out a factor common to all members of a sum of terms and expressing the sum as a product of the term taken out and the remaining sum is called **factoring**. The role of factoring in simplifying algebraic fractions is considered in section 1.5.

To illustrate the role of the distributive property in removing parentheses, consider $-1(x+3)$:

$$-1(x+3) = (-1)x + (-1)3$$
$$= -x + (-3)$$

As another example, consider $2a(x+2y+3z)$. Multiplying each term in the sum $x+2y+3z$ by $2a$ yields:

$$2a(x+2y+3z) = 2ax + 2a(2y) + 2a(3z)$$
$$= 2ax + 4ay + 6az$$

From our experience with multiplication we recognize the correctness of such statements as

$$(2)(-4) = -8$$
$$(-2)(4) = -8$$
$$(-2)(-4) = 8$$

These examples illustrate a general result called the **rule of signs** for multiplication. If a and b are positive real-numbers, then:

$a(-b) = -(ab)$	(1.1)
$(-a)b = -(ab)$	(1.2)
$(-a)(-b) = ab$	(1.3)

In colloquial language, the product of two numbers with unlike signs is negative; the product of two numbers with like signs is positive.

If $a = 1$, then from (1.2) we obtain:

$$(-1)b = -b$$

Thus, for example,

$$-(x+3) = (-1)(x+3)$$
$$= (-1)x + (-1)3$$
$$= -x + (-3).$$

We also have:

$$-(a + 2b + 4c + d) = (-1)(a + 2b + 4c + d)$$
$$= (-1)a + (-)2b + (-1)4c + (-1)d$$
$$= -a + (-2b) + (-4c) + (-d)$$

In colloquial language, a minus sign in front of terms grouped by parentheses changes the sign of each term in the group.

My colleague Leo Greenfest refers to it as the disguised -1 and recommends that when you see a minus sign before a group of terms within parentheses replace it, mentally if not physically, by $-1\cdot$ This helps us to keep in mind that it's multiplying through by -1 that changes the sign of each term in the group. Thank you Leo; an excellent idea.

The following are two other important properties of real-number multiplication.

M6. The product of two nonzero real-numbers is nonzero. That is, if $a \neq 0$ and $b \neq 0$, then $ab \neq 0$.

This implies that if the product of a and b is zero, then a, b, or both a and b must be zero. The only way zero can be obtained from a product is for at least one of the members of the product to be zero.

> **M7. Cancellation Property.** If $ax = ay$, and $a \neq 0$, then $x = y$.

For example, if

$$3xy = 3xz$$

and $x \neq 0$, then the common factor $3x$ can be canceled, yielding:

$$y = z$$

EXERCISES

Find the following sums.

1. $-2 + (-4)$
2. $3 + (-2)$
3. $4 + -(-2)$
4. $2 + (-2)4$
5. $-3 + (-1)(-2)$
6. $3(-1) + (-2)$
7. $(-2)(-3) + 4$
8. $-(-3) + (-5)$
9. $-3 + -(-4)$
10. $-(-1) + (-2)(-1) + (-1)3$
11. $(-1)(-4) + -(-3) + 3(-2)$
12. $(-1)2 + (-2)(-3) + -(-2)$

Simplify by removing parentheses or combining terms.

13. $3a + 5a + 7a$
14. $2xy + 3xy + 4xy$
15. $4wz + (-2)wz + 3wz$
16. $3x^2 + 4x^2 + 2x^2$
17. $6ab + (-3)ab + 4ab + (-5)ab$
18. $2ab + 4ab + ab$
19. $5xyz + (2)xyz + 7xyz$
20. $3(2y + 3a + 4z)$
21. $-2(2x + 3y + 4z)$
22. $-1(4w + 2v + 3z)$
23. $-3(3x + 3) + 4x$
24. $-(2a + 4b + 6c)$
25. $-(2n + 5m) + 4n + (-3)m$
26. $-(3x + 2y + 3) + 2x + 4$
27. $-(xy + 2a) + 4xy + 3a$
28. $-(3x^2 + 4y) + 4x^2 + 3y$

Simplify by factoring.

29. $2x + 4y$
30. $9n + 9m$
31. $xy + xz$
32. $3ab + 6ac$
33. $4xyz + 8xyw$
34. $3x^2y + 6x^2z$
35. $2x + 4x^2$
36. $2ab + 4ac + 8ad$
37. $a^2y + a^2z + a^2w$
38. $2mnp + 4mnv + 10mnw$
39. $4wk^2 + (-3)k^2 + ak^2$
40. $3xmy + 2my + 5xm$

41. $3abc + 6ab + 3ac$ 42. $4vw^2 + 2xw^2 + 8w^2$

43. $2xy + 3xz + x$ 44. $3rst + 6rs + 12rsk$

45. Is $-x + 4$ the negative of $x - 4$? Explain.

46. Is $xy + 3$ the negative of $-xy - 2$? Explain.

47. Is -2 the multiplicative inverse of 2? Explain.

48. Does 0 have an additive inverse? Explain.

49. Does 1 have a multiplicative inverse? Explain.

50. What are the additive inverses of 6, -1, 1\4, $\sqrt{2}$, and π?

51. What are the multiplicative inverses of 6, -1, 1\4, $\sqrt{2}$, and π?

1.3   PROPERTIES OF AN ORDER RELATION

One of the basic properties of the real-number system is that there is a relation < (read less than) which permits us to compare numbers for size. If a and b are real numbers, we write

$$a < b$$

to signify that **a is less than b**. Another way of saying the same thing is to write

$$b > a$$

which is read "**b is greater than a**." Thus $0 < 1$, $1 < 3$, $-\dfrac{1}{2} < 0$, $-2 < -1$.

These relations are also expressed by writing $1 > 0$, $3 > 1$, $0 > -\dfrac{1}{2}$, $-1 > -2$.

The concepts positive number and negative number are defined by comparing the number in question with 0. If $a > 0$, then a is said to be **positive**; if $a < 0$, a is said to be **negative**. Thus $-\dfrac{1}{2}$ is negative since $-\dfrac{1}{2} < 0$; 1 is positive since $1 > 0$; 0 itself is neither positive nor negative.

We write

$$a \leq b$$

and, equivalently,

$b \geq a$ to mean that **a is less than or equal to b**, and, equivalently, **b is greater than or equal to a**. Thus $1 \leq 2$ is correct since $1 < 2$; $1 \leq 1$ is correct since $1 = 1$.

Relations of the ≤ and ≥ type, which contain an equality component, are also referred to as inequalities.

Geometrically, real-numbers are identified with points on a straight line. We choose a straight line, a unit of distance, and an initial point of reference called the **origin.** To the origin we assign the number zero. By marking off the unit of length in both directions from the origin, we assign positive integers to marked-off points in one direction (by convention, to the right of the origin) and negative integers to marked-off points in the other direction. By following through in terms of the chosen unit of length, a real-number is attached to each point on the number line, and each point on the number line has attached to it one number (see Figure 1.2).

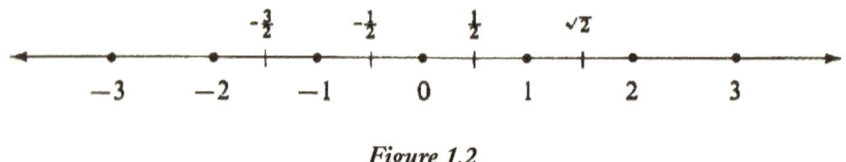

Figure 1.2

Geometrically, in terms of our number line, to say that $a < b$ is to say that a is to the left of b; $b > a$ means that b is to the right of a. Thus $-4 < -3$ since -4 is to the left of -3, and $-\dfrac{1}{2} > -1$ since $-\dfrac{1}{2}$ is to the right of -1.

The addition operation also has a simple geometric interpretation in terms of the number line. Standing at the origin, interpret each number as a movement in a direction. $+3$, for example, means move three units in the positive direction (to the right) on our number line; -1 means move one unit in the negative direction (to the left); 0 means stay where you are. The addition operation + is interpreted as "and then." Thus $(+3) + (-1)$ means move three units to the right and then one unit to the left, leaving us at $+2$ (see Figure 1.3).

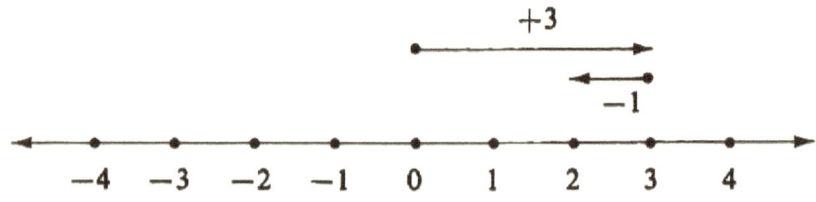

Figure 1.3

$(-3)+(-2)$ means move three units to the left and then move two units to the left, leaving us at -5 (see Figure 1.4).

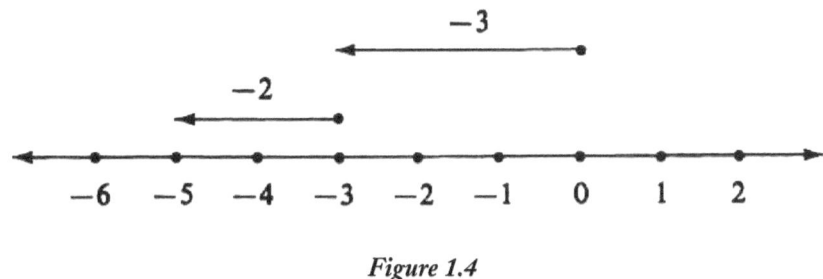

Figure 1.4

The concept of **negative of a number** should not be confused with the concept of **negative number**; they are not the same. "Negative of" means "additive inverse of." A "negative number," on the other hand, is a number that is less than zero. The negative of -2 is 2, which is positive; the negative of 0 is 0, which is neither positive nor negative.

Formally speaking, the following conditions give life to an order relation $<$. Without them there is no order relation.

O1. For any two a and b in the real-number system one and only one of the relations $a < b$, $a = b$, $a > b$ holds.

O2. If $a < b$, then $a + c < b + c$. Adding the same real-number to both sides of $a < b$ preserves the sense of $<$.

O3. If $a < b$ and $c > 0$, then $ac < bc$; if $c < 0$, then $ac > bc$. Multiplying both sides of $a < b$ by a positive number preserves its sense; multiplying by a negative number reverses its sense.

04. If $a < b$ and $b < c$, then $a < c$.

1.4   SUBTRACTION AND DIVISION

Subtraction is defined in terms of addition in the following way.

If a and b are any two real-numbers, then the **difference** $a-b$ is defined by

$$a-b=c$$

where c is such that $b+c=a$. Let us note that

$$c=a+(-b)$$

since

$$b+c=b+a+(-b)=a.$$

Thus the difference $a-b$ is given by

$$a-b=a+(-b).$$

That is, to *subtract b* from *a* means to add the negative of *b* (additive inverse of *b*) to *a*. For example,

$$3-2=3+(-2)=1$$
$$-4-3=-4+(-3)=-7$$
$$-5-(-4)=-5+-(-4)$$
$$=-5+4$$
$$=-1.$$

In connection with this last example, recall that $-(-4)$ expresses the additive inverse of -4, that is, the number that if added to -4 yields 0, which, of course, is 4 itself. In colloquial language such a result is often described by saying that "two minus signs make a plus sign."

The distributive property discussed in section 1.2 can be extended to include subtraction. Thus we have:

$$a(b-c+d-e+\cdots+z)=ab-ac+ad-ae+\cdots+az$$

For example,

$$-3(x-2y+1)=(-3)x-(-3)2y+(-3)1$$
$$=-3x+6y-3.$$

Division is defined in terms of multiplication in the following way.

If a and b are any real-numbers, where $b \neq 0$, then the quotient $a \div b$ is defined by:

$$a \div b = a \cdot \left(\frac{1}{b}\right) = a \cdot b^{-1}$$

That is, to *divide* a by b means to multiply a by the multiplicative inverse (reciprocal) of b. The quotient $a \div b$ is also expressed by the fraction symbol $\frac{a}{b}$ or a/b. For example:

$$3 \div 2 = 3 \cdot (\frac{1}{2}) = \frac{3}{2}$$

$$5 \div (-3) = 5 \cdot (-\frac{1}{3}) = -\frac{5}{3}$$

$$6 \div \frac{1}{3} = 6 \cdot (\frac{1}{3})^{-1} = 6 \cdot 3 = 18$$

$$0 \div 4 = 0 \cdot (4^{-1}) = 0$$

Note that the last example illustrates that **division into zero is permitted.** However, **division by zero is not permitted.** For any real-number a, $a \div 0$ is not defined because the multiplicative inverse of 0 does not exist.

Although this analysis shows that division by zero cannot be defined in terms of the multiplicative inverse of zero, might it not be possible to define division by zero in some other way? The answer is no; no definition of division by zero that is compatible with the structure of the real-number system as a whole is possible. Any proposed definition for division by zero must lead to a contradiction within the real-number system.

As an indication of this mathematical fact, let us suppose that division by zero has been defined in some way and consider the following argument (which requires some recall of elementary algebra).

Let x and y denote nonzero numbers such that:

$$x = y$$

Multiplying both sides of this equation by y yields:

$$xy = y^2$$

Subtracting x^2 from both sides yields:

$$xy - x^2 = y^2 - x^2$$

By factoring we obtain:

$$xy - x^2 = x(y - x), \quad y^2 - x^2 = (y - x)(y + x)$$

Thus we have:

$$x(y - x) = (y - x)(y + x)$$

Dividing both sides by $y - x$ (which is 0 since $x = y$) yields:

$$x = y + x$$

Since $x = y$, by substituting x for y in the above we obtain

$$x = x + x$$

or

$$x = 2x.$$

Dividing both sides by x yields:

$$1 = 2$$

But then

$$0 = 1$$
$$3 = 2 + 1 = 1 + 1 = 2$$
$$4 = 3 + 1 = 2 + 1 = 3$$

and so on. That is, $0 = 1 = 2 = 3 = 4$, and so on, which is in contradiction to the structure of the real-number system which requires, at the very least, that 0 and 1 not be the same (see M4, sec. 1.2). The source of this inconsistency is the step in which we divided by $y - x = 0$ (under the **assumption** that division by zero could be defined in some way).

If $1 = 2$, then we may conclude that $\$1 = \$2$. If a person can be found who accepts this, then he, logically speaking, should be prepared to give you two dollars for every one dollar that you give him. Whether he would be prepared to do so emotionally speaking, is another matter.

Division by zero does not usually announce itself with trumpets blaring, and when division by an algebraic expression arises it must be ascertained whether division by zero could arise under certain circumstances. A spectacular case in point arose in connection with Albert Einstein's theory of relativity. At one point in his work Einstein found that certain of his equations yielded peculiar results; as a consequence he turned in another direction, a direction that subsequently had to be abandoned. These peculiarities were later explained in 1922 when the Russian physicist Alexander Friedman found that in his derivation Einstein had divided an equation by a quantity that, Friedman showed, could be zero under certain circumstances.

This discovery and the refinement in Einstein's analysis that it brought about laid the foundation for the mathematical theory of the expanding universe.

While noting that division by zero is not defined, we should again observe that there is nothing wrong with dividing *into* zero as long as we are dividing by a nonzero number. For example,

$$0 \div 3 = 0 \cdot \left(\frac{1}{3}\right) = 0$$

$$0 \div c = 0 \cdot \left(\frac{1}{c}\right) = 0$$

where $c \neq 0$.

EXERCISES

Find the following values.

1. $2 - 4$
2. $4 - 6$
3. $-3 - 4$
4. $-2 - 5$
5. $-1 - 6$
6. $3 - (-2)$
7. $5 - (-3)$
8. $-3 - (-4)$
9. $-(-2) - (-1)$
10. $-(-3) - (-7)$
11. $-3 - (-5) + 2$
12. $-3 + 2 - 4$
13. $2 - (-3) - 3$
14. $-2(-3 + 2 - 4)$
15. $-3(4 - 2 - (-1))$

16. $-1(-3 + 4 - (-2))$
17. $-2 \div 4$
18. $-3 \div \frac{1}{3}$

19. $4 \div -\frac{1}{2}$
20. $\frac{1}{3} \div 2$
21. $\frac{1}{4} \div -2$

22. $\frac{1}{2} \div -4$
23. $\frac{1}{5} \div 0$
24. $0 \div \frac{1}{2}$

25. $\frac{1}{8} \div -2$

Simplify by removing parentheses or combining terms.

26. $-2(3a - 4b + 3c)$
27. $3a - 3(4a - 2b - 3c)$
28. $-(3a + 4b - 2c)$
29. $2a - 4b - (a - 2b + 4c)$

30. $-2(x-2y+3z)-3(2x+3y-2z)$
31. $-(-3x+2y-z)-(4x-4y-2z)$
32. $-3(-2x-4y-z)-(-x+2y-4z)$
33. $3ab+2ab-4ab$ 34. $-6xy+2xy-3xy$
35. $-3x^2+4x^2-x^2$ 36. $-3ab+4a-2ab-2a$
37. $-3x+2xy+4y+4x-3xy+2y-x$

Simplifying by factoring.

38. $3xy-4xz-5x$ 39. $4a^2b-3bx-b$
40. $4mn-5mn^2+3mnp$ 41. $2xy^2-4y^2+6zy^2$

42. $\dfrac{1}{2}a-2ab$ 43. $gh-3gh^2$

44. $\dfrac{1}{3}xy+6xyz$ 45. $3x^2y^3-2xy^2+xy$

"Did you know that $0/0=5$?" John asked his friends Jim and Burt. "It follows from the following mathematical analysis," he exclaimed.

Consider the ratio:

$$y=\frac{x^2+x-6}{x^2-3x+2}$$

For $x=2$, calculation yields $y=0/0$. But the numerator and denominator of the preceding ratio can be simplified by factoring. This gives us:

$$y=\frac{(x-2)(x+3)}{(x-2)(x-1)}$$

Cancellation of $(x-2)$ from the numerator and denominator of the preceding allows us to write y in the following simpler form.

$$y=\frac{x+3}{x-1}$$

For $x = 2$, calculation yields $y = 5$.

Since $y = 0/0$ and 5 for $x = 2$, it follows that $0/0 = 5$.

46. Do you agree with John's analysis? Explain.

47. Do you agree with Jim's conclusion that $0/0 = 1$ and his reasoning? Explain.

48. Do you agree with Burt's conclusion that $0/0 = 0$, and his reasoning? Explain.

49. Do you agree with John's conclusion that "what this means is 5, 0, and 1 are equal"?

1.5   FRACTIONS

$$-\frac{103}{4}, \qquad -\frac{1}{2}, \qquad \frac{2}{3}, \qquad \frac{\sqrt{3}}{2}, \qquad \frac{\pi}{5}$$

The above are examples of numerical fractions;

$$\frac{x}{2}, \qquad \frac{x-1}{y+1}, \qquad \frac{x^2 + 3x + 4}{3x - 1}$$

illustrate algebraic fractions, which become numerical fractions when the variables involved are given specific numerical values. Of course it is understood that the variables in such expressions cannot take on values that make the denominator zero.

To begin our survey of basic properties of fractions, let us first recall that

$$\frac{a}{b} = \frac{ax}{bx}$$

That is, a **common factor** may be introduced into, or removed from, the numerator and denominator of a fraction without changing its value. When the need is to simplify a fraction by removing, or canceling, common factors, the process is often called **cancellation.** For example,

$$\frac{2}{4} = \frac{1 \cdot \cancel{2}}{2 \cdot \cancel{2}} = \frac{1}{2}$$

Cancellation of the common factor 2 yields $\dfrac{1}{2}$.

$$\frac{2(\cancel{x-1})}{3(\cancel{x-1})} = \frac{2}{3}$$

Cancellation of the common factor $(x-1)$ yields $\dfrac{2}{3}$, assuming that $x-1 \neq 0$.

Danger, be on the alert!

$$\frac{2\cancel{x}}{3+\cancel{x}} = \frac{2}{4}$$

is an incorrect cancellation; x is not a common factor of both numerator and denominator.

To be sure, x is a factor of the numerator since we have the product $2x$. But x is not a factor of the denominator since we do not have the product $3x$, but rather the sum $3+x$. Thus the cancellation principle is not applicable in this situation.

Incorrect simplifications of the kind just illustrated have gained enormous popularity over the years and one must be on continuous alert against them.

To simplify algebraic fractions by cancellation, the numerator and denominator components must be expressed as products of factors. To express these components as products of factors, we often employ the distributive property. The following examples illustrate.

EXAMPLE 1

$$\frac{5x-10}{3x-6}$$

By factoring the numerator and denominator, we obtain

$$\frac{5x-10}{3x-6} = \frac{5(x-2)}{3(x-2)}$$

Canceling the common factor $(x-2)$, where $x-2 \neq 0$, yields

$$\frac{5x-10}{3x-6} = \frac{5\cancel{(x-2)}}{3\cancel{(x-2)}} = \frac{5}{3}$$

EXAMPLE 2

$$\frac{2xy+3x}{6y+9}$$

By factoring numerator and denominator, we obtain:

$$\frac{2xy+3x}{6y+9} = \frac{x\cancel{(2y+3)}}{3\cancel{(2y+3)}}$$

Canceling the common factor $(2y+3)$, where $2y+3 \neq 0$, yields:

$$\frac{2xy+3x}{6y+9} = \frac{x}{3}$$

EXAMPLE 3

$$\frac{x-2}{2-x}$$

At first sight it seems that no simplification is possible. But let us observe that $2-x$ is the negative of $x-2$ (since the sum of $2-x$ and $x-2$ is zero). The quotient of two quantities that are negatives of each other is -1. This we establish by factoring -1 from the numerator and following through as shown.

$$\frac{x-2}{2-x} = \frac{-1(-x+2)}{2-x} = \frac{-1\overset{1}{\cancel{(2-x)}}}{\underset{1}{\cancel{(2-x)}}} = -1$$

It is assumed that $2-x \neq 0$; that is, $x \neq 2$.

EXERCISES

Specify whether the cancellation is correct.

1. $\dfrac{x+1}{3y+1} = \dfrac{x+1}{3y+1}$

2. $\dfrac{\overset{1}{\cancel{4}}-3}{\cancel{4}+4} = \dfrac{-2}{5}$

3. $\dfrac{\overset{1}{\cancel{2}}}{x+\cancel{2}_1} = \dfrac{1}{x+1}$

4. $\dfrac{2x(x+y)}{3x(x-y)} = \dfrac{2(x+y)}{3(x-y)}$

5. $\dfrac{4\,\overset{1}{\cancel{x+y}}}{\underset{1}{\cancel{x+y}}} = 4$

6. $\dfrac{\overset{1}{\cancel{2}}+(x+y)}{\underset{1}{\cancel{2}}} = 1+x+y$

7. $\dfrac{\overset{1}{\cancel{a}}-\overset{1}{\cancel{b}}}{\underset{1}{\cancel{a}}+\underset{1}{\cancel{b}}}=\dfrac{1-1}{1+1}=\dfrac{0}{2}=0$ 8. $\dfrac{2\,\overset{1}{\cancel{a+3}}}{\underset{1}{\cancel{a+3}}}=2$

Simplify, if possible, by canceling common factors.

9. $\dfrac{9}{27}$ 10. $\dfrac{8}{24}$ 11. $\dfrac{-9}{-33}$

12. $\dfrac{2ax}{3ay}$ 13. $\dfrac{3abc}{4b}$ 14. $\dfrac{3x+1}{x+1}$

15. $\dfrac{4(a+b)}{a+b}$ 16. $\dfrac{3}{x+3}$ 17. $\dfrac{2+(x+y)}{x+y}$

18. $\dfrac{3a(x+y)}{4x(x+y)}$ 19. $\dfrac{3(x-y)(x+y)}{2(x-y)}$ 20. $\dfrac{-4(a+2b)}{-2(a+2b)}$

21. $\dfrac{2a+b}{a+b}$ 22. $\dfrac{x+y+1}{y+1}$ 23. $\dfrac{3-x}{x-3}$

24. $\dfrac{x-5}{x-5}$ 25. $\dfrac{2x+1}{2x}$

Simplify, if possible, by factoring and canceling common factors.

26. $\dfrac{2x+10}{x+5}$ 27. $\dfrac{2x-4}{x-2}$ 28. $\dfrac{2a-10}{5-a}$

29. $\dfrac{xy+y}{y(x+1)}$ 30. $\dfrac{2a+3ab}{2a(2+3ab)}$ 31. $\dfrac{4x-16y}{x-4y}$

32. $\dfrac{3xy+4xyz}{xy}$ 33. $\dfrac{3x-15}{10-2x}$ 34. $\dfrac{3xy+x}{-3y-1}$

35. $\dfrac{4m+3n}{4mx^2+3nx^2}$ 36. $\dfrac{yz-3ys}{3yz-9ys}$ 37. $\dfrac{4cd-c}{4de-e}$

38. $\dfrac{5x-5y}{x^2-xy}$ 39. $\dfrac{3ab-abc}{9-3c}$

40. For what x are the fractions stated in 23-27 undefined? Explain.

Rules of Sign

In Section 1.2 we noted that $(-1)b = -b$. For fractions this implies that:

$$\frac{-a}{-b} = \frac{\cancel{(-1)}a}{\cancel{(-1)}b} = \frac{a}{b}$$

By multiplying numerator and denominator of $-a/b$ by -1, we obtain:

$$\frac{-a}{b} = \frac{(-1)(-a)}{(-1)b} = \frac{a}{-b}$$

Thus we have the following rules of sign:

$$\frac{-a}{-b} = \frac{a}{b}$$

$$\frac{-a}{b} = \frac{a}{-b}$$

$\dfrac{-a}{b}$ and $\dfrac{a}{-b}$ are equal to the negative (additive inverse) of $\dfrac{a}{b}$, as we now show. The negative of $\dfrac{a}{b}$, denoted by $-\left(\dfrac{a}{b}\right)$ (with the minus sign in front of the fraction), is that number which yields zero when added to $\dfrac{a}{b}$. We must show that:

$$-\frac{a}{b} = \frac{-a}{b}$$

(In colloquial language, the minus sign in front of a fraction can be moved to the numerator of the fraction.) To establish this result we must show that the sum of

$\dfrac{a}{b}$ and $\dfrac{-a}{b}$ is zero. No request could be simpler.

$$\frac{a}{b}+\frac{-a}{b}=\frac{1}{b}\cdot a+\frac{1}{b}\cdot(-a)=\frac{1}{b}(a+(-a))=\frac{1}{b}\cdot 0=0$$

Since

$$-\frac{a}{b}=\frac{-a}{b}\quad\text{and}\quad\frac{-a}{b}=\frac{a}{-b}$$

we have the following rule of signs:

$$-\frac{a}{b}=\frac{-a}{b}=\frac{a}{-b}$$

That is, the minus sign in front of a fraction can be shifted from the front of the fraction to the numerator of the fraction or to the denominator of the fraction, and vice versa.

Test for Equality

$\dfrac{a}{b}=\dfrac{c}{d}$ if and only if $ad=bc$. Passing from $\dfrac{a}{b}=\dfrac{c}{d}$ to $ad=bc$ is sometimes

called cross multiplication. Thus, for example, $\dfrac{7}{9}\neq\dfrac{13}{15}$ since $15(7)=105$, whereas

$13(9)=117$; $\dfrac{6}{12}=\dfrac{14}{28}$ since $28(6)=168$ and $14(12)=168$.

Multiplication of Fractions

As we recall, to multiply fractions, multiply numerators and denominators. That is:

$$\frac{a}{b}\cdot\frac{c}{d}=\frac{ac}{bd}$$

EXAMPLE 4

$$\frac{1}{2} \cdot \frac{3}{4} = \frac{3}{8}$$

EXAMPLE 5

$$\frac{2x}{3y} \cdot \frac{y}{z} = \frac{2x \, \cancel{y}^{1}}{3 \, \cancel{y}_{1} z} = \frac{2x}{3z}$$

Alternatively,

$$\frac{2x}{3 \, \cancel{y}_{1}} \cdot \frac{\cancel{y}^{1}}{z} = \frac{2x}{3z}, \text{ where } y \neq 0$$

In multiplying fractions, common factors can be canceled after numerators and denominators are multiplied, as was first done, or before numerators and denominators are multiplied, as is illustrated by the second method.

EXAMPLE 6

$$\frac{a-1}{b} \cdot \frac{2}{b(a-1)} = \frac{(\cancel{a-1})^{1} 2}{b^{2}(\cancel{a-1})_{1}} = \frac{2}{b^{2}}$$

where $a - 1 \neq 0$.

Division of Fractions

$$\frac{a}{b} \div \frac{c}{d} = \frac{a}{b} \cdot \frac{d}{c} = \frac{ad}{bc}$$

That is, **to divide fractions we invert the divisor and multiply.** For example,

$$\frac{1}{2} \div \frac{2}{3} = \frac{1}{2} \cdot \frac{3}{2} = \frac{3}{4}.$$

This procedure for finding the quotient of two fractions follows from the definition of division of real-numbers. By definition:

$$\frac{a}{b} \div \frac{c}{d} = \frac{a}{b} \cdot \left(\text{reciprocal of } \frac{c}{d} \right)$$

The reciprocal of $\frac{c}{d}$ is $\frac{d}{c}$ $\left(\text{since } \frac{c}{d} \cdot \frac{d}{c} = 1 \right)$. Thus we have:

$$\frac{a}{b} \div \frac{c}{d} = \frac{a}{b} \cdot \frac{d}{c} = \frac{ad}{bc}$$

We should also note that sometimes the quotient $\frac{a}{b} \div \frac{c}{d}$ is written:

$$\frac{\dfrac{a}{b}}{\dfrac{c}{d}} \text{ or } \frac{a}{b} / \frac{c}{d}$$

To avoid ambiguity, we make the division line longer or heavier than the fraction lines of the component fractions. To illustrate the ambiguity that arises when we are not careful, let us note that $\dfrac{\frac{1}{3}}{2}$ could mean $\dfrac{1}{\frac{3}{2}}$, which equals $\dfrac{1}{6}$, or $\dfrac{\frac{1}{3}}{2}$, which equals $\dfrac{2}{3}$.

EXAMPLE 7

Find $\dfrac{1}{6} \div \dfrac{2}{3}$ and simplify.

$$\frac{1}{6} \div \frac{2}{3} = \frac{1}{6} \cdot \frac{3}{2} = \frac{1}{4}$$

EXAMPLE 8

Find $\dfrac{3a}{b-1} \div \dfrac{2a}{b}$ and simplify.

$$\frac{3a}{b-1} \div \frac{2a}{b} = \frac{3\cancel{a}^{\,1}}{b-1} \cdot \frac{b}{2\cancel{a}_{1}} = \frac{3b}{2(b-1)}, \quad a \neq 0, \quad b \neq 0, \quad b \neq 1$$

EXAMPLE 9

Find $\dfrac{4a}{a-1} \div 3a$ and simplify.

$$\frac{4a}{a-1} \div 3a = \frac{4\cancel{a}^{\,1}}{a-1} \cdot \frac{1}{3\cancel{a}_{1}} = \frac{4}{3(a-1)}, \quad a \neq 0, \quad a \neq 1$$

EXERCISES

Determine if the given fractions are equal.

41. $\dfrac{6}{21}, \dfrac{8}{24}$

42. $\dfrac{7}{12}, \dfrac{9}{15}$

43. $\dfrac{14}{23}, \dfrac{17}{25}$

44. $\dfrac{3}{18}, \dfrac{5}{30}$

45. $\dfrac{6}{21}, \dfrac{14}{49}$

46. $\dfrac{6}{16}, \dfrac{8}{25}$

Express the product in simplest terms.

47. $\dfrac{2}{3} \cdot \dfrac{6}{7}$

48. $\dfrac{1}{4} \cdot \dfrac{8}{7}$

49. $\dfrac{2}{7} \cdot \dfrac{14}{5}$

50. $\dfrac{2a}{3b} \cdot \dfrac{5b}{3c}$

51. $\dfrac{3x}{4y} \cdot \dfrac{8y-1}{x}$

52. $\dfrac{xy}{z} \cdot \dfrac{x}{y}$

53. $\dfrac{a}{4} \cdot \dfrac{16}{ab}$

57. $\dfrac{4x}{y} \cdot \dfrac{3y}{x^2}$

58. $\dfrac{1}{x-5} \cdot \dfrac{5-x}{3}$

59. $\dfrac{3x-2}{x} \cdot \dfrac{1}{6x-4}$

60. $\dfrac{5a}{a-1} \cdot \dfrac{2a-2}{5}$

61. $\dfrac{3}{2-x} \cdot \dfrac{x-2}{4}$

62. $\dfrac{4a+1}{b} \cdot \dfrac{2a}{4ab+b}$

63. $\dfrac{4}{3-n} \cdot \dfrac{2n-6}{n}$

64. $\dfrac{3}{v-3} \cdot \dfrac{vx-3x}{x}$

Divide and express the quotient in simplest terms.

65. $\dfrac{1}{2} \div \dfrac{3}{4}$

66. $-\dfrac{1}{6} \div \dfrac{2}{3}$

67. $\dfrac{3}{5} \div -\dfrac{1}{4}$

68. $3 \div \dfrac{1}{2}$

69. $\dfrac{1}{4} \div -2$

70. $\dfrac{2}{3} \div -6$

71. $\dfrac{2a}{b} \div \dfrac{3}{ab}$

72. $\dfrac{3x}{y} \div \dfrac{x}{y-1}$

73. $\dfrac{4}{a} \div \dfrac{2b}{a}$

74. $\dfrac{x-1}{2y} \div \dfrac{3}{x-1}$

75. $\dfrac{4}{x-2} \div \dfrac{2x}{x-2}$

76. $\dfrac{a}{b-b} \div \dfrac{a+b}{a-b}$

77. $\dfrac{1}{x-2} \div \dfrac{2}{2-x}$

78. $\dfrac{2}{a+4} \div \dfrac{1}{2a+8}$

79. $\dfrac{2}{3-a} \div \dfrac{2a}{a-3}$

80. $\dfrac{n}{m+n} \div \dfrac{2m}{mn+n^2}$

81. $\dfrac{1}{2x-y} \div \dfrac{2}{y-2x}$

82. $\dfrac{y}{2y-z} \div \dfrac{3z}{6y-3z}$

Addition of Fractions

As we recall, **to add fractions with the same denominator, add their numerators and retain the common denominator.** That is:

$$\dfrac{a}{c} + \dfrac{b}{c} = \dfrac{a+b}{c}$$

For example:

$$\frac{1}{3}+\frac{4}{3}=\frac{1+4}{3}=\frac{5}{3}$$

$$\frac{x}{2}+\frac{x-1}{2}=\frac{x+x-1}{2}=\frac{2x-1}{2}$$

$$\frac{2x}{x-1}+\frac{x+1}{x-1}=\frac{2x+x+1}{x-1}=\frac{3x+1}{x-1}, \quad x \neq 1$$

To add fractions with different denominators, $\frac{1}{3}$ and $\frac{2}{5}$ for example, find a common denominator, express each fraction as an equivalent fraction with this common denominator, add numerators, and retain the common denominator. $3(5) = 15$ is a common denominator for $\frac{1}{3}$ and $\frac{2}{5}$. Expressing $\frac{1}{3}$ and $\frac{2}{5}$ in terms of the common denominator 15 and adding yields:

$$\frac{1}{3}=\frac{1}{3}\cdot\frac{5}{5}=\frac{5}{15}, \qquad \frac{2}{5}=\frac{2}{5}\cdot\frac{3}{3}=\frac{6}{15}$$

$$\frac{1}{3}+\frac{2}{5}=\frac{5}{15}+\frac{6}{15}=\frac{11}{15}$$

EXAMPLE 10

Find $\frac{1}{2x}+\frac{2}{3x}$ and simplify.

6x is a common denominator. Multiplying numerator and denominator of $\frac{1}{2x}$ by 3 and numerator and denominator of $\frac{2}{3x}$ by 2 and adding yields:

$$\frac{1}{2x}+\frac{2}{3x}=\frac{3}{3(2x)}+\frac{2(2)}{2(3x)}$$

$$=\frac{3}{6x}+\frac{4}{6x}$$

$$=\frac{7}{6x}, \quad x\neq 0$$

BEWARE: You **cannot** add fractions by adding their numerators and denominators.

$$\frac{1}{3}+\frac{2}{5}\neq\frac{3}{8}$$

$$\frac{1}{2x}+\frac{2}{3x}\neq\frac{3}{5x}$$

This incorrect way of proceeding has gained in popularity in recent years.

EXAMPLE 11

Find $\dfrac{2}{a}+\dfrac{1}{a-1}$ and simplify.

We obtain the common denominator $a(a-1)$ by multiplying numerator

and denominator of $\dfrac{2}{a}$ by $(a-1)$ and multiplying numerator and denominator

of $\dfrac{1}{a-1}$ by a. We obtain:

$$\frac{2}{a}+\frac{1}{a-1}=\frac{2(a-1)}{a(a-1)}+\frac{a}{a(a-1)}$$

$$=\frac{2(a-1)+a}{a(a-1)}$$

$$=\frac{3a-2}{a(a-1)}, \quad a\neq 0,\ \ 1$$

EXERCISES

Add the given fractions and simplify.

83. $\dfrac{2}{3} + \dfrac{7}{3}$

84. $\dfrac{5}{12} + \dfrac{8}{12}$

85. $\dfrac{4}{13} + \dfrac{7}{13}$

86. $\dfrac{1}{2} + \dfrac{2}{3}$

87. $\dfrac{2}{5} + \dfrac{1}{8}$

88. $\dfrac{3}{2} + \dfrac{7}{3}$

89. $\dfrac{3}{4} + \dfrac{2}{7}$

90. $\dfrac{4}{5} + \dfrac{1}{2}$

91. $\dfrac{1}{3} + \dfrac{5}{4}$

92. $\dfrac{2}{x} + \dfrac{3}{x}$

93. $\dfrac{x}{y} + \dfrac{x-1}{2y}$

94. $\dfrac{2x}{5} + \dfrac{4x}{5}$

95. $\dfrac{x-1}{3} + \dfrac{1-2x}{3}$

96. $\dfrac{a}{2b} + \dfrac{a-2}{2b}$

97. $\dfrac{3x-1}{3y} + \dfrac{1-x}{3y}$

98. $\dfrac{4x}{x-1} + \dfrac{-4}{x-1}$

99. $\dfrac{3}{x+2} + \dfrac{x-1}{x+2}$

100. $\dfrac{a+b}{a-b} + \dfrac{b-a}{a-b}$

101. $\dfrac{5x+6}{2x+1} + \dfrac{3x+2}{2x+1}$

102. $\dfrac{1}{2x} + \dfrac{2}{3x}$

103. $\dfrac{4}{5m} + \dfrac{2}{3m}$

104. $\dfrac{x}{7} + \dfrac{x-1}{3}$

105. $\dfrac{y+1}{3} + \dfrac{2y-2}{5}$

106. $\dfrac{m+4}{2} + \dfrac{2m-3}{3}$

107. $\dfrac{x+y}{4} + \dfrac{2x-y}{3}$

108. $\dfrac{x+1}{2x} + \dfrac{2x}{3x}$

109. $\dfrac{1}{x+1} + \dfrac{1}{x}$

110. $\dfrac{1}{m-n} + \dfrac{2}{m+n}$

111. $\dfrac{3}{x} + \dfrac{2}{x-2}$

112. $\dfrac{5}{x-1} + 6$

113. $\dfrac{4}{x+2} + 3$

114. $\dfrac{3}{xy} + \dfrac{2}{ty}$

115. $4 + \dfrac{3}{t+3}$

116. $\dfrac{3}{x-2} + \dfrac{1}{x+1}$

117. $\dfrac{3}{x-2} - x$

118. $\dfrac{x}{1-x} + \dfrac{2x}{x-1}$

119. $\dfrac{x}{y}+\dfrac{y}{x-1}$ 120. $\dfrac{4}{2x}+\dfrac{3x-1}{3}$ 121. $\dfrac{x}{2x-3}+\dfrac{4}{x-1}$

Subtraction of Fractions

Subtraction of fractions is a special case of real-number subtraction. The difference $\dfrac{a}{b}-\dfrac{b}{c}$ is defined by:

$$\frac{a}{b}-\frac{b}{c}=\frac{a}{c}+\left(\frac{-b}{c}\right)$$

That is, to subtract $\dfrac{b}{c}$ from $\dfrac{a}{c}$ means to add the negative of $\dfrac{b}{c}$ to $\dfrac{a}{c}$. Since

$-\dfrac{b}{c}=\dfrac{-b}{c}$ (see p. 47), we have:

$$\frac{a}{c}-\frac{b}{c}=\frac{a}{c}+\frac{-b}{c}=\frac{a-b}{c}$$

For example:

$$\frac{3}{5}-\frac{1}{5}=\frac{3}{5}+\frac{-1}{5}=\frac{3-1}{5}=\frac{2}{5}$$

$$\frac{x}{2}-\frac{x-1}{2}=\frac{x}{2}-\frac{-(x-1)}{2}$$

$$=\frac{x}{2}+\frac{-x+1}{2}$$

$$=\frac{x-x+1}{2}$$

$$=\frac{1}{2}$$

To subtract fractions with different denominators, express each fraction in terms of a common denominator and then subtract.

EXAMPLE 12

Find $\dfrac{1}{2} - \dfrac{1}{5}$ and simplify.

$$\frac{1}{2} - \frac{1}{5} = \frac{5}{10} - \frac{2}{10} = \frac{5-2}{10} = \frac{3}{10}$$

EXAMPLE 13

Find

$$\frac{x}{4} - \frac{x+2}{2}$$

and simplify.

To obtain a common denominator of 4, we multiply numerator and denominator of $\dfrac{x+2}{2}$ by 2. We have:

$$\frac{x}{4} - \frac{x+2}{2} = \frac{x}{4} - \frac{2(x+2)}{4}$$

$$= \frac{x}{4} + \frac{-2(x+2)}{4}$$

$$= \frac{x}{4} + \frac{-2x-4}{4}$$

$$= \frac{x-2x-4}{4}$$

$$= \frac{-x-4}{4}$$

EXAMPLE 14

Find

$$\frac{1}{x-2} - \frac{1}{x+1}$$

and simplify.

We obtain the common denominator $(x-2)(x+1)$ by multiplying numerator and denominator of $\dfrac{1}{x-2}$ by $(x+1)$ and multiplying numerator and denominator of $\dfrac{1}{x+1}$ by $(x-2)$. This yields:

$$\frac{1}{x-2} - \frac{1}{x+1} = \frac{x+1}{(x-2)(x+1)} - \frac{(x-2)}{(x-2)(x+1)}$$

$$= \frac{x+1}{(x-2)(x+1)} + \frac{-(x-2)}{(x-2)(x+1)}$$

$$= \frac{x+1}{(x-2)(x+1)} + \frac{-x+2}{(x-2)(x+1)}$$

$$= \frac{x+1-x+2}{(x-2)(x+1)}$$

$$= \frac{3}{(x-2)(x+1)}, \quad x \neq -1, \ 2$$

EXERCISES

Add or subtract the given fractions and simplify.

122. $\dfrac{1}{2} - \dfrac{1}{3}$

123. $\dfrac{2}{5} - \dfrac{4}{7}$

124. $\dfrac{2}{3} - \dfrac{1}{5}$

125. $\dfrac{1}{4} - \dfrac{1}{6}$

126. $\dfrac{7}{3} - \dfrac{2}{4}$

127. $\dfrac{3}{4} - \dfrac{9}{2}$

128. $\dfrac{2a}{3} - a$

129. $\dfrac{4}{x} - \dfrac{2-x}{3x}$

130. $\dfrac{3x}{4} - \dfrac{2}{2x}$

131. $\dfrac{1}{x+1} - \dfrac{2}{3}$

132. $\dfrac{1}{a-1} - \dfrac{2}{b}$

133. $\dfrac{1}{2x} - \dfrac{x-1}{x}$

134. $\dfrac{2x-3}{x} - 4$

135. $\dfrac{1}{3x+4} - 2$

136. $\dfrac{5x}{x-1} - \dfrac{3}{2}$

137. $\dfrac{1}{x} - \dfrac{2}{3x} + \dfrac{4}{2x}$

138. $\dfrac{1}{2x-1} - \dfrac{1}{2x}$

139. $\dfrac{a+b}{2} - \dfrac{a-b}{3}$

140. $\dfrac{x+1}{2x} - \dfrac{2}{x} + 4$

141. $\dfrac{3b-1}{2} - \dfrac{b+1}{5}$

142. $\dfrac{1}{m+n} - \dfrac{1}{m-n}$

143. Refer to Examples 5, 6, 8-11, 13, 14 on pages 48 through 57. For what values of the variable(s) are the expressions considered undefined? Explain.

1.6   POWERS, ROOTS, AND RADICALS

If n is a positive integer, then x^n stands for the product $x \cdot x \cdots \cdots x$ with n factors. n, the number of factors in $x \cdot x \cdots \cdots x = x^n$, is called the **exponent** of x^n and x^n itself is called the **n^{th} power of x.** Thus:

$$x^1 = x, \qquad \text{the exponent of } x^1 \text{ is 1}$$
$$x^2 = x \cdot x, \qquad \text{the exponent of } x^2 \text{ is 2}$$
$$x^3 = x \cdot x \cdot x, \qquad \text{the exponent of } x^3 \text{ is 3}$$

$$(-2)^4 = (-2)(-2)(-2)(-2) = 16$$

In particular, x^2 is called the **square of x,** and x^3 is called the **cube of x.** Exponents are extended to include zero and negative integers as follows. If $x \neq 0$, then x^0 is defined by:

$$x^0 = 1$$

If $x \neq 0$ and n is a positive integer, then x^{-n} **is defined by:**

$$x^{-n} = \frac{1}{x^n}$$

For example:

$$8^0 = 1, \qquad x^{-2} = \frac{1}{x^2} \qquad 0^0 \text{ is}$$
$$\qquad\qquad\qquad\qquad\qquad\qquad \text{not}$$
$$x^{-3} = \frac{1}{x^3}, \qquad 3^{-2} = \frac{1}{3^2} = \frac{1}{9} \qquad \text{defined}$$

$3^{-2} = \dfrac{1}{9}$ serves to illustrate that a number raised to a negative exponent need not necessarily be negative.

Let us also note that:

$$\frac{1}{x^{-n}} = \frac{1}{\frac{1}{x^n}} = 1 \cdot \frac{x^n}{1} = x^n$$

Thus a **negative exponent** of a term in the **numerator** throws the term with corresponding positive exponent into the denominator, whereas a **negative exponent** of a term in the **denominator** throws the term with corresponding positive exponent into the numerator.

As further illustrations, we have:

$$\frac{1}{x^{-3}} = x^3, \qquad \frac{y^2}{x^{-4}} = y^2 x^4, \qquad \frac{2x^{-3}}{3y^{-2}} = \frac{2y^2}{3x^3}$$

The definitions of zero and negative exponents, which perhaps seem strange at first glance, are natural in terms of the properties that positive, negative, and zero exponents have. The following examples illustrate one of these basic properties.

$$x^2 \cdot x^3 = x^{2+3} = x^5, \quad x^2 \cdot x^{-4} = x^{2-4} = x^{-2}, \quad x^3 \cdot x^0 = x^{3+0} = x^3$$

We verify these relations as follows:

$$x^2 \cdot x^3 = \underbrace{(x \cdot x)(x \cdot x \cdot x)}_{5 \text{ factors}} = x^5$$

$$x^2 \cdot x^{-4} = \frac{x^2}{x^4} = \frac{x \cdot x}{x \cdot x \cdot x \cdot x} = \frac{1}{x^2} = x^{-2}$$

$$x^3 \cdot x^0 = x^3 \cdot 1 = x^3$$

More generally,

$$x^n \cdot x^m = x^{n+m}$$

where n and m are positive, negative, or zero. Thus we have

$$x^4 \cdot x = x^4 \cdot x^1 = x^5 \qquad\qquad (1+v)(1+v) = (1+v)^2$$
$$x^{-3} \cdot x^{-2} = x^{-3} \qquad\qquad (1+v)^2(1+v) = (1+v)^3$$
$$x^6 \cdot x^{-2} = x^4 \qquad\qquad (1+v)^3(1+v)^{-1} = (1+v)^2$$

The division counterpart of the above result is the following. If x is not zero, then

$$\frac{x^n}{x^m} = x^{n-m}$$

where n and m are positive, negative, or zero. For example,

$$\frac{x^4}{x^2} = x^{4-2} = x^2 \qquad\qquad \frac{x^3}{x^{-2}} = x^{3-(-2)} = x^{3+2} = x^5$$

$$\frac{x^{-2}}{x^3} = x^{-2-3} = x^{-5} \qquad\qquad \frac{x^{-3}}{x^{-4}} = x^{-3-(-4)} = x^1 = x$$

The following are three other basic properties of exponents:

$$(x^n)^m = x^{nm}$$
$$(xy)^n = x^n y^n$$
$$\left(\frac{x}{y}\right)^n = \frac{x^n}{y^n}$$

For example:

$$(x^2)^3 = x^{2(3)} = x^6 \qquad\qquad (x^{-3})^2 = x^{(-3)2} = x^{-6}$$
$$(xy)^5 = x^5 y^5 \qquad\qquad (xy)^{-3} = x^{-3} y^{-3}$$
$$\left(\frac{x}{y}\right)^2 = \frac{x^2}{y^2} \qquad\qquad \left(\frac{x}{y}\right)^{-3} = \frac{x^{-3}}{y^{-3}}$$

In summary, then, we have the following basic definitions and properties of exponents.

Definition. If n is a positive integer, then

$$x^n = \underbrace{x \cdot x \cdot x \cdots \cdot x}_{n \text{ factors}}$$

$$x^{-n} = \frac{1}{x^n}$$

$$x^0 = 1, \quad \text{where } x \neq 0$$

Properties:

$$\frac{1}{x^{-n}} = x^n \qquad\qquad x^n x^m = x^{n+m}$$

$$\frac{x^n}{x^m} = x^{n-m} \qquad\qquad (x^n)^m = x^{nm}$$

$$(xy)^n = x^n y^n \qquad\qquad \left(\frac{x}{y}\right)^n = \frac{x^n}{y^n}$$

The following examples further illustrate how these properties are used to simplify expressions involving exponents.

$$(x^2)^0 = 1$$

$$(2x^{-3})^2 = 2^2(x^{-3})^2 = 4x^{-6} = \frac{4}{x^6}$$

$$(3x^{-2})^{-2} = 3^{-2}(x^{-2})^{-2} = 3x^{-2}x^4 = \frac{x^4}{3^2} = \frac{1}{9}x^4$$

$$(x^2 y^{-3})^3 = (x^2)^3 (y^{-3})^3 = x^6 y^{-9} = \frac{x^6}{y^9}$$

$$\frac{3x^2 y^{-3}}{2x^{-3} y^2} = \frac{3}{2} x^{2-(-3)} y^{-3-2} = \frac{3}{2} x^5 y^{-5} = \frac{3x^5}{2y^5}$$

$$\left(\frac{x}{y}\right)^{-3} = \frac{x^{-3}}{y^{-3}} = \frac{y^3}{x^3}$$

$$\left(\frac{x^{-1}}{y^2}\right)^3 = \frac{(x^{-1})^3}{(y^2)^3} = \frac{x^{-3}}{y^6} = \frac{1}{x^3 y^6}$$

EXERCISES

Evaluate each of the following.

1. $(-2)^5$

2. $(-3)^{-2}$

3. $(-2)^{-3}$

4. $(\frac{1}{2})^0$

5. $3(4^{-2})$

6. $(4^0)(3^{-2})$

7. $\dfrac{(-2)^{-3}}{3^{-2}}$

8. $\dfrac{(3^2)^{-2}}{2^{-3}}$

9. $(5^0)(10^3)$

10. $\dfrac{(-2)^{-4}}{5^{-2}}$

11. $(-6)^{-2}(3^{-2})$

12. $[(-2)^{-3}]^{-2}$

13. $[(-4)^0]^{-3}$

14. $\dfrac{4^{-2}}{[(-3)^{-1}]^{-3}}$

15. $\dfrac{(-4)^{-3}}{(-5)^{-2}}$

Simplify each of the following by using properties of exponents. Remove zero and negative exponents.

16. $3x^{-2}y$

17. $4x^{-1}y^{-2}$

18. $\dfrac{3a^{-1}}{b^{-2}}$

19. $(x^{-2})^{-3}$

20. $\dfrac{x^{-2}}{(y^{-2})^3}$

21. $(4m^{-3})^0$

22. $(1+a)^3(1+a)$

23. $(a+b)^2(a+b)^{-2}$

24. $(m+2n)^3(m+2n)^{-1}$

25. $(2x^3y^4)(3x^2y^{-2})$

26. $\dfrac{(2x^2)x^{-3}y}{4x^2y^2}$

27. $(5x^{-2})^2(3x^4)^{-1}$

28. $2x^2(3x^4+4x^{-2})$

29. $x^{-3}(2x^3+x^2+1)$

30. $\dfrac{2y^4z^3}{3y^{-2}z^4}$

31. $(2ab^2)^{-3}$

32. $(4a^2b^3)^{-2}$

33. $(2a^{-1}b^2)^{-3}$

34. $(2xy)^3(3x^2y^{-1})^2$

35. $\dfrac{4x^2y^{-3}}{3x^{-2}y^2}$

36. $\left(\dfrac{x^2}{y^3}\right)^3 \cdot \dfrac{x^{-2}}{y^{-1}}$

37. $2a^2\left(4a^3+\dfrac{2}{a^3}\right)$

38. $\dfrac{(2x^3)^{-3}}{(4x^2)^{-2}}$

39. $\dfrac{(3x^2y^{-1})^3}{(2xy^2)^{-2}}$

40. $5x^2\left(3x^{-3} + \dfrac{2x^2}{x^{-3}}\right)$

41. $x^{-2}(x^4 + x^3)$

42. $\dfrac{2x^3 + x^4}{x^{-2}}$

43. $\dfrac{(5mn)^{-3}}{1 + 2m^2n^3}$

44. $\left(\dfrac{a}{b}\right)^{-2}(a^2b)^{-3}$

45. $\dfrac{3v^2 + 2v^{-3}}{v^2}$

46. $\dfrac{(2a^2b^{-2})^3}{(a^{-1}b^{-3})^2}$

47. $\dfrac{2y^{-2}(y^3 + y^2 + 1)}{y}$

48. $\dfrac{(z^2)^{-3} + (z^4)^{-2})}{z^3}$

Roots and Radicals

Let us recall that if $x^2 = c$, then x is called a **square root of c**. Thus -3 and 3 are square roots of 9, since $(-3)^2 = 9$ and $3^2 = 9$. More generally, if n is a positive integer greater than 1 and $x^n = c$, then x is called an **nth root of c**. In particular, if $x^3 = c$, then x is called a **cube root of c**. Thus 4 is a cube root of 64, since $4^3 = 64$; -2 and 2 are fourth roots of 16, since $(-2)^4 = 16$ and $2^4 = 16$.

From these examples it is clear that a number c may have more than one nth root. One of these, the principal nth root, is singled out in the following way.

> **Definition.** If c is **positive**, the **principal nth root of c** is the **positive** n^{th} root of c; if c is **negative** and n is odd (that is, $n = 3, 5, 7, 9$, etc.), the **principal nth root of c** is the **negative** nth root of c.

Thus, for example, the principal square root of 25 is 5, since 5 is positive and $5^2 = 25$; the principal cube root of -8 is -2 since -2 is negative, $n = 3$ is odd, and $(-2)^3 = -8$.

> The principal nth root of c is denoted by the radical sign $\sqrt[n]{c}$ The number c under the radical sign is called the **radicand**, and the number n, which indicates the root to be taken and is written above the radical sign, is called the **index** of the radical. By definition, then, $\sqrt[n]{c}$ designates the principal nth root of c and has the property that:
> $$\sqrt[n]{c^n} = c$$

When the principal square root of c, denoted by $\sqrt[2]{c}$, is involved, it is customary to omit the index 2 on the radical sign and write $\sqrt{c}$. In terms of radical notation, we have:

$$\sqrt{25} = 5, \qquad \sqrt[3]{-8} = -2, \qquad \sqrt[4]{16} = 2$$

Note $\sqrt{25} \neq -5$ since $\sqrt{25}$ designates the principal square root of 25, which is positive 5. The negative square root of 25, -5, is designated by $-\sqrt{25}$.

The number 25 has two square roots, -5 and 5. This observation prompts more general questions. How many n^{th} roots does a real-number have and how can we determine them? We address these questions in Section 10.8.

EXERCISES

Determine the value of each of the following.

49. $\sqrt{64}$

50. $\sqrt[3]{-27}$

51. $\sqrt[3]{-64}$

52. $\sqrt{49}$

53. $\sqrt[5]{-32}$

54. $\sqrt[6]{64}$

55. $\sqrt[4]{81}$

56. $\sqrt[3]{-125}$

57. $(\sqrt{5})^2$

58. $(\sqrt[3]{-9})^3$

59. $(\sqrt[5]{-32})^2$

60. $\dfrac{1}{(\sqrt{25})^{-3}}$

61. $(\sqrt[4]{6})^4$

62. $\dfrac{1}{\sqrt[4]{81}}$

63. $\dfrac{1}{\sqrt[4]{16}} \cdot \dfrac{1}{(\sqrt[3]{125})^{-2}}$

64. Does $\sqrt{(-2)^2} = -2$? Explain.

Computation with Radicals

Only a comparatively small number of radicals can be determined directly from the definition of radical itself. Thus we have such results as $\sqrt{4} = 2$, $\sqrt[3]{-8} = -2$, $\sqrt[4]{16} = 2$. But how are such radicals as $\sqrt{2}$, $\sqrt{3}$, and $\sqrt[3]{10}$ to be determined? How does one work with numbers expressed in terms of radicals? "Simple, use a calculator," you might say. Indeed the calculator is a most valuable computational ally, but we should be careful not to allow ourselves to become its slave so that without its use the simplest of calculations becomes an impossible task. Therefore, basic principles that aid calculation never become obsolete. Square root computations, in particular, arise in many situations, and square root tables have been constructed to facilitate such calculations. A primitive square root table is given in Appendix Table 1 (p. 430), and from this table we see, for example, that $\sqrt{2} \simeq 1.414$ and $\sqrt{3} \simeq 1.732$. (The symbol $\simeq$ means "is approximately equal to.")

The calculation of radicals (with or without a calculator) and the simplification of expressions involving radicals are facilitated by certain basic properties of radicals, two of which we now state and illustrate. In many situations it suffices to express a numerical result in radical form and leave the explicit calculation of the radical to be undertaken when necessary.

> **Product and Quotient Theorems for Radicals.** If a and b are positive real-numbers, then:
>
> $$\sqrt[n]{ab} = \sqrt[n]{a} \cdot \sqrt[n]{b}, \qquad \sqrt[n]{\frac{a}{b}} = \frac{\sqrt[n]{a}}{\sqrt[n]{b}}$$

That is, the nth root of a product (or quotient) of positive numbers is equal to the product (or quotient) of the nth roots of the component parts. For example:

$$\sqrt{2} \cdot \sqrt{8} = \sqrt{2(8)} = \sqrt{16} = 4$$

$$\sqrt{200} = \sqrt{100(2)} = \sqrt{100}\sqrt{2} = 10\sqrt{2}$$

$$\sqrt[3]{\frac{1}{8}} = \frac{\sqrt[3]{1}}{\sqrt[3]{8}} = \frac{1}{2}$$

$$3\sqrt{50} = 3[\sqrt{25(2)}] = 3(\sqrt{25}\sqrt{2}) = 3(5\sqrt{2}) = 15\sqrt{2}$$

$$4\sqrt[3]{16} = 4[\sqrt[3]{8(2)}] = 4(\sqrt[3]{8}\sqrt[3]{2}) = 4(2\sqrt[3]{2}) = 8\sqrt[3]{2}$$

$$3\sqrt{4x^2 y^5} = 3\sqrt{4}\sqrt{x^2}\sqrt{y^4 y} = 3(2)x\sqrt{y^4}\sqrt{y} = 6xy^2\sqrt{y}$$

Appendix Table 1 allows us to approximate square roots of numbers up to 150, but by expressing 200, for example, as the product of the perfect square 100 and 2 and using the product theorem for radicals, we can bring $\sqrt{200}$ within range of our table. (Any number that is the square of an integer is said to be a **perfect square.** Since $100 = 10^2$, 100 is a perfect square.)

$$\sqrt{200} = 10\sqrt{2} \approx 10(1.414) = 14.14$$

Thus:

$$\sqrt{200} \approx 14.14$$

> **DANGER, be on the alert! The product and quotient theorems for radicals cannot be extended to sums and differences.** In general:
>
> $$\sqrt[n]{a+b} \neq \sqrt[n]{a} + \sqrt[n]{b}, \quad \sqrt[n]{a-b} \neq \sqrt[n]{a} - \sqrt[n]{b}$$

To illustrate, compare $\sqrt{2} = \sqrt{1+1}$ with $\sqrt{1} + \sqrt{1}$.

$$\sqrt{1+1} = \sqrt{2} \approx 1.414$$
$$\sqrt{1} + \sqrt{1} = 1 + 1 = 2$$

EXAMPLE 1   TEMPTATION MUST BE RESISTED

Simplify, to the extent possible, $\sqrt{9x^2 + 81y^2}$.

For some reason this sort of situation brings out the mathematical Mr. Hyde in many of us. Suggested simplifications include $3x + 9y$, which may be obtained by overlooking the afore danger sign and arguing thusly:

$$\sqrt{9x^2 + 81y^2} = \sqrt{9x^2} + \sqrt{81y^2}$$
$$= 3x + 9y$$

> If this were correct, then squaring $3x + 9y$ should give us $9x^2 + 81y^2$. It does not; we obtain $9x^2 + 54xy + 81y^2$.

The only possibility of simplification rests on expressing $9x^2 + 81y^2$ as a product and using the product theorem for radicals. The pickings are thin.

$$\sqrt{9x^2 + 81y^2} = \sqrt{9(x^2 + 9y^2)}$$
$$= \sqrt{9}\sqrt{x^2 + 9y^2}$$
$$= 3\sqrt{x^2 + 9y^2}$$

EXERCISES

Simplify the following by using the product and quotient theorems for radicals.

65. $\sqrt{48}$ 66. $\sqrt{72}$ 67. $\sqrt{54}$

68. $\sqrt{300}$ 69. $\sqrt{124}$ 70. $4\sqrt{50}$

71. $3\sqrt{175}$ 72. $\frac{1}{2}\sqrt{180}$ 73. $-\frac{1}{2}\sqrt{28}$

74. $\sqrt{\dfrac{12}{5}}$ 75. $\sqrt{\dfrac{100}{3}}$ 76. $\sqrt{\dfrac{98}{5}}$

77. $2\sqrt{\dfrac{45}{3}}$ 78. $4\sqrt{\dfrac{27}{4}}$ 79. $\frac{1}{4}\sqrt{80}$

80. $\sqrt[3]{250}$ 81. $\sqrt[4]{48}$ 82. $\sqrt[3]{180}$

83. $\sqrt{4a^2b}$ 84. $\sqrt{9a^4c}$ 85. $\frac{1}{2}\sqrt{16x^3y}$

86. $\sqrt{x^2 - 4}$ 87. $\sqrt{4x^2 + 16y^2}$ 88. $\sqrt{x^2 + 2x + 1}$

Radicals and Fractional Exponents

Earlier in this section we defined a concept of exponent for integer values and observed that the following properties hold:

1. $\dfrac{1}{x^{-n}} = x^n$.

2. $\dfrac{x^n}{x^m} = x^{n-m}$.

3. $(xy)^n = x^n y^n$.

4. $x^n x^m = x^{n+m}$.

5. $(x^n)^m = x^{nm}$.

6. $\left(\dfrac{x}{y}\right)^n = \dfrac{x^n}{y^n}$.

We now turn our attention to an extension of the concept of exponent to rational numbers. In making such an extension it is natural to preserve properties 1 through 6, which hold for integral exponents, since it is these very properties that make the exponent concept useful.

To obtain a clue about how to proceed, consider $x^{1/2}$, which has not been defined and is therefore a condidate for definition. If property 5 is to hold for fractional exponents, then $(x^{1/2})^2$ must yield $x^{(1/2)\cdot 2} = x^1 = x$. Thus $x^{1/2}$ will have to stand for a quantity which is squared yields x. This leads us to define:

$$x^{1/2} = \sqrt{x}$$

where $\sqrt{x}$ is the principal square root of x. This definition also makes sense from the point of view of property 4 since we obtain

$$x^{1/2} \cdot x^{1/2} = x^{1/2+1/2} = x^1$$

which corresonds to the result $\sqrt{x} \cdot \sqrt{x} = x$.

More generally, the definition

$$x^{1/q} = \sqrt[q]{x} ,$$

ensures that property 5 holds for rational exponents of the form $1/q$. We obtain

$$(x^{1/q})^q = x^{(1/q)\cdot q} = x$$

which corresponds to the meaning of $\sqrt[q]{x}$, that $(\sqrt[q]{x})^q = x$. Furthermore, we define $x^{-1/q}$ by:

$$x^{-1/q} = \frac{1}{x^{1/q}}$$

Thus, for example:

$$9^{1/2} = \sqrt{9} = 3 \qquad\qquad (-8)^{1/3} = \sqrt[3]{-8} = -2$$

$$(81)^{-1/4} = \frac{1}{(81)^{1/4}} = \frac{1}{3} \qquad (-32)^{-1/5} = \frac{1}{(-32)^{1/5}} = -\frac{1}{2}$$

Finally, if property 5 is to hold when $x^{1/q}$ is raised to the power p and when x^p is raised to the power $1/q$, we must have

$$(\sqrt[q]{x})^p = (x^{-1/q})^p = x^{1/q \cdot p} = x^{p/q}$$

and:

$$\sqrt[q]{x^p} = (x^p)^{1/q} = x^{p \cdot (1/q)} = x^{p/q}$$

This leads us to the following definition of $x^{p/q}$: If p and q are nonnegative integers and $q \neq 0$, $x^{p/q}$ and $x^{-p/q}$ are defined by:

$$\boxed{\; x^{p/q} = (\sqrt[q]{x})^p = (\sqrt[q]{x^p}, \qquad x^{-p/q} = \frac{1}{x^{p/q}} \;}$$

To avoid unpleasant surprises we also require that x be nonnegative in defining $x^{p/q}$ and positive for $x^{-p/q}$.

For example, we have:

$$9^{3/2} = (\sqrt{9})^3 = 3^3 = 27 \qquad\qquad 8)^{2/3} = (\sqrt[3]{8})^2 = 2^2 = 4$$

$$(81)^{-3/4} = \frac{1}{(81)^{3/4}} = \frac{1}{(\sqrt[4]{81})^3} = \frac{1}{3^3} = \frac{1}{27}$$

It can be proved that rational exponents satisfy conditions 1 through 6, and that any rational exponent can be replaced by an equivalent rational number.

Thus, for example, $x^{2/4} = x^{1/2}$. The following examples further illustrate the application of properties 1 through 6 to the simplification of expressions involving radicals or rational exponents.

$$\sqrt{x} \cdot \sqrt[3]{x} = x^{1/2} \cdot x^{1/3} = x^{1/2+1/3} = x^{5/6} = \sqrt[6]{x^5}$$
$$(3^{1/2})^4 = 3^{(1/2) \cdot 4} = 3^2 = 9 \qquad (8x^6)^{1/3} = 8^{1/3}(x^6)^{1/3} = 2x^2$$

EXERCISES

Evaluate each of the following.

89. $25^{1/2}$ 90. $49^{1/2}$ 91. $125^{1/3}$

92. $64^{-1/2}$ 93. $216^{-1/3}$ 94. $100^{-3/2}$

95. $8^{-1/3}$ 96. $36^{-2/4}$ 97. $27^{2/3}$

98. $8^{-4/3}$ 99. $3^{5/2} \cdot 3^{1/2}$ 100. $(4^3)^{-1/3}$

101. $(2^8)^{-1/2}$ 102. $2^{1/4} \cdot 2^{-3/4}$ 103. $6^{1/3} \cdot 6^{5/3}$

104. $\dfrac{32^{1/2}}{32^{1/3}}$ 105. $\dfrac{4^{-1/2}}{2^{-6}}$ 106. $3^{-1/2} \cdot 3^{3/2}$

Simplify each of the following.

107. $(9x^4)^{1/2}$ 108. $(x^2 x^{1/2})^4$ 109. $\dfrac{x^{1/2}}{x^{-1/2}}$

110. $\dfrac{x}{x^{1/2}}$ 111. $\left(\dfrac{x^{1/2}}{x^3}\right)^2$ 112. $(x^{-1/4} y^{1/2})^4$

113. $\sqrt{x}(\sqrt{x})^5$ 114. $\dfrac{(\sqrt[3]{x})^2}{\sqrt[3]{x^5}}$ 115. $\dfrac{(\sqrt[3]{x})^5}{\sqrt[10]{x}}$

CHAPTER 2

Equations and Inequalities:
A First Look

Equations and inequalities arise in a wide variety of situations, and their study is a basic concern of this book. In this chapter we begin this study by surveying basic concepts and properties of equations and inequalities. We also turn to the world of finance to illustrate how equation relationships are derived and applied. In later chapters the foundation developed here is built upon in considering a number of equation and inequality structures and their applications.

2.1   EQUATIONS

$$2x - 1 = 5 \tag{2.1}$$

$$\frac{a}{b} = \frac{ax}{bx} \tag{2.2}$$

$$3(x + 2) = 3x + 6 \tag{2.3}$$

$$2x + y = 8 \tag{2.4}$$

$$xy + 2z = 10 \tag{2.5}$$

$$\frac{1}{x+1} = 0 \tag{2.6}$$

and

$$x^2 = 16 \tag{2.7}$$

are equations. As these examples illustrate, an **equation** is a statement of equality between two algebraic expressions. Some equations involve one variable; others involve more than one variable. Two kinds of equations, identities and conditional equations, should be distinguished.

An **identity** is an equation that becomes a mathematically correct statement for all allowable values of the variables involved. By **allowable values** of the variables we mean values for which the expressions involved are meaningful. The equation

$$\frac{a}{b} = \frac{ax}{bx} \tag{2.2}$$

is an identity, where b and x are unequal to zero. Zero is not an allowable value for b and x since division by zero is meaningless. From the distributive property of real-numbers it follows that the equation

$$3(x+2) = 3x + 6 \tag{2.3}$$

is an identity that holds for all values of x.

An equation that does not become a mathematically correct statement for all allowable values of the variables involved is called a **conditional equation** or, more simply, an **equation.** Equations (2.1) and (2.4) through (2.7) are conditional equations. The equation

$$2x - 1 = 5 \tag{2.1}$$

becomes a correct statement for one allowable value of x, that is, $x = 3$. For this reason 3 is called a solution, or root, of $2x - 1 = 5$, and is said to satisfy this equation. More generally, a number is said to be a **solution,** or **root,** of a one-variable equation if the equation becomes a correct statement when the number is substituted for the variable in the equation. Such a number is said to **satisfy** the equation. To **solve** an equation in one variable means to find all its solutions.

The equation

$$\frac{1}{x+1} = 0 \tag{2.6}$$

has no solutions. By inspection we can see that there is no value which yields a correct statement when substituted for x in (2.6).

$$2x + y = 8 \tag{2.4}$$

illustrates a conditional equation in two variables, x and y. A **solution** of an equation in two variables, x and y, is any ordered pair of numbers that yields a correct statement when substituted for x and y, respectively, in the equation. For example, (3, 2) is a solution of $2x + y = 8$, since substitution of 3 for x and 2 for y yields $2(3) + 2 = 8$, which is correct. Here too we say that (3, 2) satisfies $2x + y = 8$. The ordered pair (2, 3), on the other hand, is not a solution of $2x + y = 8$, since substitution of 2 for x and 3 for y yields $2(2) + 3 = 8$, which is an incorrect statement. We say that (2, 3) does not satisfy $2x + y = 8$. The equation $2x + y = 8$ has many solutions; for example, as can easily be verified,

$(1, 6)$, $(\frac{1}{2}, 7)$, $(-2, 12)$ and $(6, -4)$ are solutions.

The equation

$$xy + 2z = 10 \tag{2.5}$$

illustrates a conditional equation in three variables, x, y, and z. A **solution** of an equation in three variables, x, y, and z, is any ordered triple of numbers that yields a correct statement when substituted for x, y, and z, respectively, in the equation. The ordered triple (1, 2, 4) is a solution of (satisfies) $xy + 2z = 10$, since substitution of 1 for x, 2 for y, and 4 for z yields $1(2) + 2(4) = 10$, which is a correct statement. The notion of solution of an equation in any number of variables is defined in an analogous way.

Operations for Solving Equations

Two equations are said to be **equivalent** if they have the same solutions.

$$2x - 1 = 5 \quad \text{and} \quad 2x = 6$$

are equivalent equations since both have the same solution, $x = 3$. The equations

$$2x + y = 8 \quad \text{and} \quad y = 8 - 2x$$

are equivalent since any ordered pair of numbers that satisfies one also satisfies the other.

Of basic importance to solving equations are operations that enable us to pass from an equation to an equivalent equation, that is, operations which leave unaltered the solutions of an equation. The following **operations of equivalence** lead to equivalent equations.

---

1. **The same constant may be added to, or subtracted from, each side of an equation.**

For example, by adding 1 to both sides of $2x - 1 = 5$, we obtain the equivalent equation $2x = 6$.

---

2. **A term may be transferred from one side of an equation to the other side, provided that its sign is changed. This operation is called transposition.**

For example, transposing $2x$ from the left side of $2x + y = 8$ yields the equivalent equation $y = 8 - 2x$. Transposing $2x$ is the same as adding $-2x$ to both sides of $2x + y = 8$, and such an operation has no effect on solutions.

---

3. **Both sides of an equation may be multiplied or divided by the same nonzero constant.**

For example, by dividing both sides of $2x = 6$ by 2 we obtain the equivalent equation and solution $x = 3$. The equation

$2x - 1 = 5$ can be solved by first employing operation 1 and adding 1 to both sides of $2x - 1 = 5$ to obtain the equivalent equation $2x = 6$, and then employing operation 3 and dividing both sides of $2x = 6$ by 2 to obtain the **equivalent equation** and solution $x = 3$. It is a wise precaution to substitute the value obtained into the original equation and verify that it is a solution. Substituting 3 for x in $2x - 1 = 5$ yields $2(3) - 1 = 5$, which is a correct statement.

EXAMPLE 1

Solve $2x + 3 = x - 2$.

To solve for x we must isolate x; to isolate x we bring all terms involving x to one side of the equation and all other terms to the other side. We are starting with:

$$2x+3 = x-2$$

Transposing x yields:

$$x+3 = -2$$

Transposing 3 yields:

$$x = -5$$

Substituting $x = -5$ into the original equation $2x+3 = x-2$ yields $2(-5)+3 = -5-2$, which is correct since we have -7 in both cases. Thus -5 is the solution of the given equation.

EXAMPLE 2

Determine three solutions of $2x+3y = 6$.

Solutions to such an equation in two variables can be obtained by giving values to one of the variables and determining the corresponding values of the other variable. If we give x the value 0, we obtain:

$$2(0)+3y = 6, \qquad 3y = 6$$

Dividing both sides of $3y = 6$ by 3 yields $y = 2$. Thus $(0, 2)$ is a solution.

If we give y the value 0, we obtain:

$$2x+3(0) = 6, \qquad 2x = 6$$

Dividing both sides of $2x = 6$ by 2 yields $x = 3$. Thus $(3, 0)$ is a solution.

If we give x the value 1, we obtain:

$$2(1)+3y = 6, \qquad 2+3y = 6$$

Transposing 2 yields:

$$3y = 4$$

Dividing both sides of $3y = 4$ by 3 yields $y = \frac{4}{3}$. Thus $(1, \frac{4}{3})$ is a solution.

As a precaution, we return to the original equation $2x + 3y = 6$ and verify that $(0, 2)$, $(3, 0)$, and $(1, \frac{4}{3})$ are solutions. Substituting 0 for x and 2 for y in $2x + 3y = 6$ yields $2(0) + 3(2) = 6$, which is correct, thus verifying that $(0, 2)$ is a solution. In a similar way we verify that $(3, 0)$ and $(1, \frac{4}{3})$ are solutions.

It often happens that an equation arises for which it is desirable to express one variable in terms of the others. The following two examples illustrate.

EXAMPLE 3

Given $xy + 2z = 10$, express z in terms of x and y.

Transposing xy yields:

$$2z = 10 - xy$$

By dividing both sides of $2z = 10 - xy$ by 2, we obtain:

$$z = \frac{10 - xy}{2}$$

EXAMPLE 4

The area A of a rectangular region with length x and width w is $A = xw$. Express w in terms of A and x.

Dividing both sides of $A = xw$ by x yields:

$$\frac{A}{x} = w$$

x

w

Another principle that is useful in connection with solving equations is the following:

---

4. If $x^2 = y^2$, then $x = \pm y$; if $x = \pm y$, then $x^2 = y^2$.

For example, $x^2 = 16$ implies that $x = -\sqrt{16} = -4$ and $x = \sqrt{16} = 4$.

EXERCISES

Solve the following equations and check.

1. $x - 2 = 4$

2. $x + 3 = 5$

3. $3x = 12$

4. $\frac{1}{3}x = 9$

5. $\frac{x}{2} = 7$

6. $2x + 3 = 15$

7. $3x + 4 = 8$

8. $5x - 7 = 22$

9. $\frac{1}{2}x - 3 = 8$

10. $3 + 2x = 4x - 1$

11. $2 - x = 3x + 6$

12. $5 - x = 10 + x$

13. $2(x - 1) = 4$

14. $3(x - 2) = 12$

15. $2(3x - 4) = 4$

Determine three solutions for each of the following equations and check.

16. $5x + 2y = 10$

17. $3x - y = 6$

18. $2x - 3y = 12$

19. $x - 2y = 4$

20. $3x - 2y = 18$

21. $7x - 2y = 14$

By giving values to two of the variables and determining the corresponding value of the third variable, find three solutions to each of the following equations.

22. $2x + y + z = 4$ 23. $3x - 2y + z = 12$ 24. $4x - y + 5z = 20$

Express y in terms of x for each of the following equations.

25. $5x + 2y = 10$ 26. $2x - 3y = 12$ 27. $x^2 + y^2 = 25$

Express z in terms of x and y for each of the following equations.

28. $4x - y + 5z = 20$ 29. $\dfrac{1}{x} - 3y + z = 10$ 30. $3xz + 2y = 4$

31. $4xy - 3z = 12$ 32. $2xyz + x = 10$ 33. $\dfrac{4}{z} + xy = 10$

34. The perimeter p of a rectangular region with length x and width w is $p = 2x + 2w$. Express w in terms of p and x.

35. The average A of three value x, y, and z is $A = (x + y + z)/3$. Express z in terms of x, y, and A.

36. The area A of a circle with radius r is $A = \pi r^2$. Express r in terms of A.

37. The circumference C of a circle with radius r is $C = 2\pi r$. Express r in terms of C.

2.2 GRAPHS

To begin, let us recall the structure of a rectangular coordinate system. We start with two lines, called **coordinate axes,** that are drawn at right angles to each other. It is customary to call the horizontal line the x-axis and the vertical line the y-axis, although other notations are sometimes used in connection with applications. The plane determined by these coordinate axes is called the **coordinate plane,** or **xy-plane,** and the point O of intersection of these coordinate axes is called the **origin** (see Figure 2.1).

The coordinate axes divide the coordinate plane into four regions called **quadrants,** which are numbered I, II, III and IV, as shown in Figure 2.1. A unit of length is chosen for each axis, and in terms of these units of length a number is assigned to each point on each axis. Often the same unit of length is used on both axes, but in some applications it is more appropriate to use different units of length.

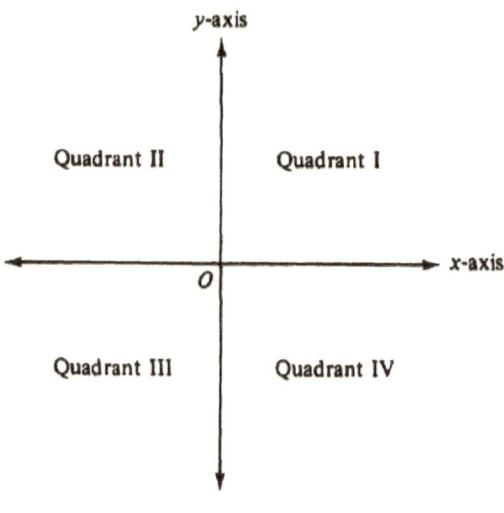

Figure 2.1

Thus a grid is established, as shown in Figure 2.2. Each point in the *xy*-plane can now be labeled with an ordered pair of numbers, with the first number, called the **x-coordinate,** or **abscissa**, identifying the point on the *xy*-grid with respect to the *x*-axis and the second number, called the **y-coordinate**, or **ordinate**, identifying the point on the *xy*-grid with respect to the *y*-axis. This ordered pair of numbers is called the **coordinates** of the point.

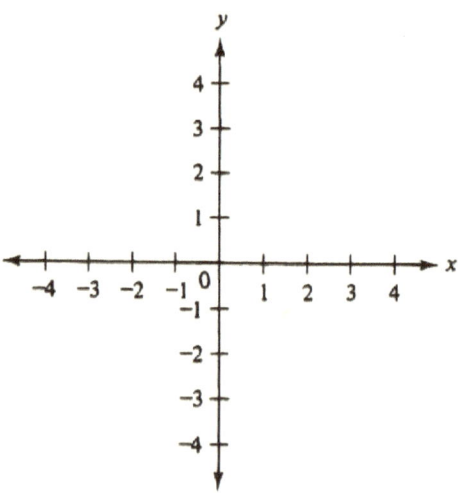

Figure 2.2

Thus the point P, shown in Figure 2.3, which is three units to the right of the origin along the x-axis and one unit up, has coordinates $(3, 1)$; the point Q, shown in Figure 2.3,

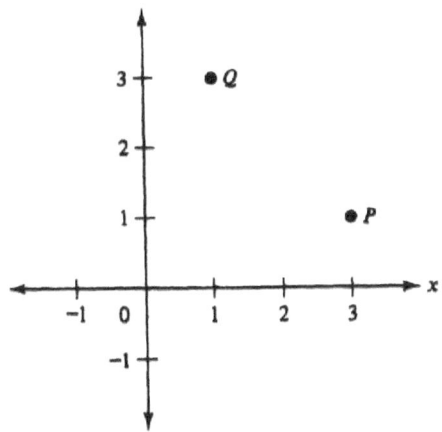

Figure 2.3

which is one unit to the right of the origin along the x-axis and three units up, has coordinates $(1, 3)$. If coordinates, such as $(-1, 2)$, are specified, the point R determined is found by going one unit to the left of the origin along the x-axis and two units up (see Figure 2.4).

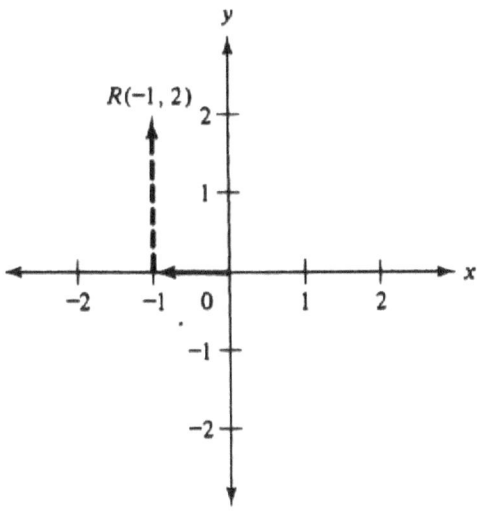

Figure 2.4

Marking a point with given coordinates is called **plotting** the point. Note that every point on the x-axis has y-coordinate 0, while every point on the y-axis has x-coordinate 0.

Since there is a one-to-one correspondence between points in the xy-plane and ordered pairs of real-numbers—each point is tagged with one coordinate pair and each coordinate pair singles out one point—the terms "point" and "coordinates of a point" are used interchangeably to simplify language usage. Thus, for example, it is acceptable practice to say, "consider the point (2, 4)," rather than "consider the point with coordinates (2, 4)."

EXERCISES

1. Plot the points whose coordinates are (2, 3), (3, 2), $(\frac{1}{2}, 3)$, $(-1, -2)$, and $(-\frac{1}{2}, -1)$.

2. Plot the points whose coordinates are $(-1, -3)$, (3, 0), (0, 3), $(-\frac{1}{2}, -\frac{1}{2})$, $(-1, -5)$, $(-1, 0)$, and $(0, -1)$.

The concept of coordinate system owes its importance to the fact that it permits us to unite the worlds of geometry and algebra and attack algebraic problems by means of geometric methods and geometric problems by means of algebraic methods. Both disciplines have benefited immeasurably from this mutual transfusion of ideas and methods.

The geometric counterpart of an equation in two variables is its graph. The **graph** of an equation in two variables is the collection of all those points, and only those points, whose coordinates satisfy the equation. A crude, but often satisfactory approach to sketching the graph of an equation is to determine a number of solutions of the equation, plot the points corresponding to these solutions on a rectangular coordinate system, and join these points by a smooth curve.

This technique must be used with caution, since it is based on the **assumption** that joining the points plotted with a smooth curve gives an accurate picture of the other points on the graph. Such an **assumption** is not always correct; moreover, there are generally many ways of joining points by a smooth curve. The following examples illustrate the graphing of equations.

EXAMPLE 1

Sketch the graph of $2x + y = 2$.

To obtain solutions of this equation, we substitute values of our choice for one variable in the equation, or its equivalent, calculate the corresponding values of the other variable, and tabulate the results. To simplify calculations we first express y in terms of x by transposing $2x$. This yields

$$y = 2 - 2x$$

Substituting -2, -1, 0, 1, 2, 3, and 4 for x and calculating y yields the results shown in Table 2.1. Plotting these points and connecting them gives us the line shown in Figure 2.5. To indicate that the graph is the entire line, and not a line segment or some other component, arrowheads are used as shown.

Table 2.1

x	-2	-1	0	1	2	3	4
y	6	4	2	0	-2	-4	-6

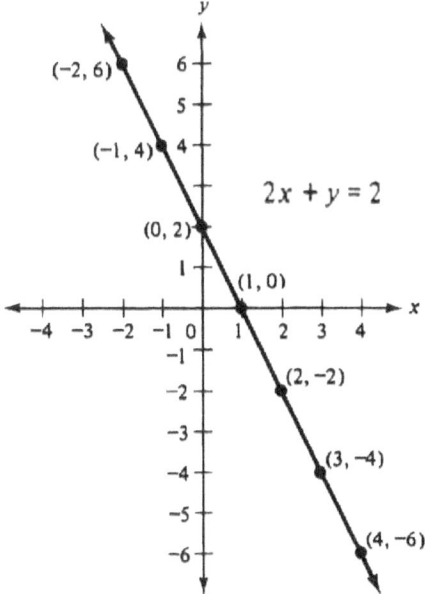

Figure 2.5

The equation $2x + y = 2$ illustrates a **linear equation** in two variables, whose general structure is

$$Ax + By = C$$

where A, B, and C are constants, with not both A and B equal to zero. It is not coincidental that the term "linear" is used here and that $2x + y = 2$ represents a line. The term linear suggests a connection with the line, and this is indeed the case.

The graph of every linear equation in two variables

$$Ax + By = C$$

is a line; moreover, every line is the graph of a linear equation in two variables.

Because of this intimate connection between two-variable linear equations and lines, it is acceptable practice to simplify language and say, for example, "consider the line $2x + y = 2$," rather than "consider the line that is the graph of $2x + y = 2$."

Since a line is determined by two points, to plot the graph of a linear equation it suffices to plot two points.

EXAMPLE 2

Sketch the graph of $2x - 3y = 6$.

Since this is a linear equation, it suffices to determine two solutions to obtain its graph. If we give x the value 0, y is -2; if we give y the value 0, x is 3. Thus $(0, -2)$ and $(3, 0)$ determine the graph of $2x - 3y = 6$, which is shown in Figure 2.6.

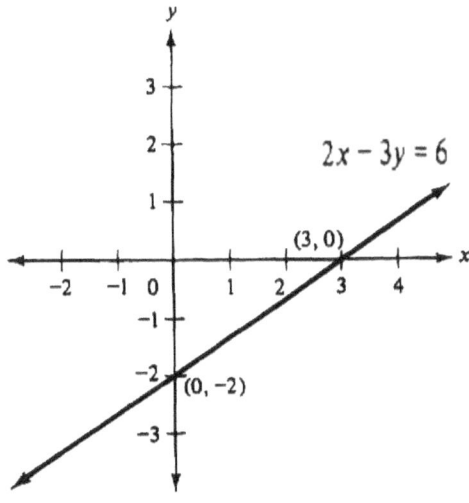

Figure 2.6

EXAMPLE 3

Sketch the graph of $x^2 + y^2 = 25$.

To simplify the determination of solutions of this equation, we first express y in terms of x. Transposing x^2 yields

$$y^2 = 25 - x^2$$
$$y = \pm\sqrt{25 - x^2}$$

If $x = 0$, $y = 5$ and -5; thus $(0,5)$ and $(0,-5)$ are solutions; if $x = -5$, $y = 0$; if $x = 5$, $y = 0$; if $x = -4$, $y = 3$ and -3; if $x = 4$, $y = 3$ and -3; if $x = -3$, $y = 4$ and -4; if $x = 3$, $y = 4$ and -4. In summary we have Table 2.2. Plotting these points and connecting them yields the circle shown in Figure 2.7.

Table 2.2

x	−5	−4	−4	−3	−3	0	0	3	3	4	4	5
y	0	3	−3	4	−4	5	−5	4	−4	3	−3	0

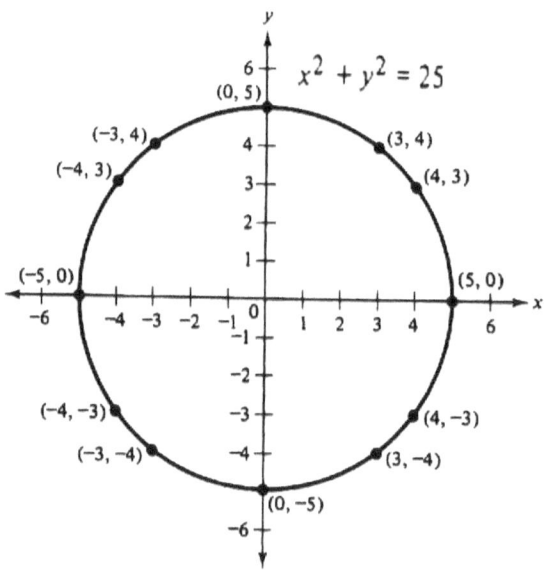

Figure 2.7

EXAMPLE 4

Sketch the graph of $xy = 1$.

Note that x and y can be given any values except 0. We first express y in terms of x by dividing both sides of $xy = 1$ by x, thus obtaining

$$y = \frac{1}{x}$$

If $x = 1$, $y = 1$; if $x = 5$, $y = \frac{1}{5}$; if $x = 10$, $y = \frac{1}{10}$; if $x = 100$, $y = \frac{1}{100}$. In general, as x takes on larger and larger positive values, $y = 1/x$ remains positive but gets closer and closer to zero, as shown in Figure 2.8(a).

Still giving positive values to x, but taking them closer and closer to zero, such as $\frac{1}{2}$, $\frac{1}{5}$, $\frac{1}{10}$, and $\frac{1}{100}$, yields 2, 5, 10, and 100 as respective values for y. As x takes on positive values closer and closer to zero, $y = 1/x$ gets larger and larger. In summary, we have Figure 2.8(b).

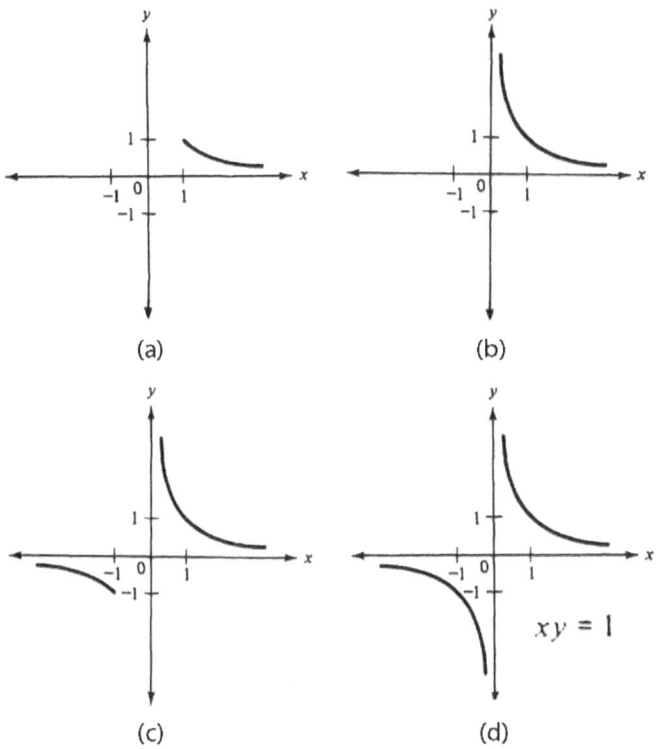

Figure 2.8

Now give negative values to x. If $x=-1$, $y=-1$; if $x=-5$, $y=-\dfrac{1}{5}$; if $x=-10$, $y=-\dfrac{1}{10}$; if $x=-100$, $y=-\dfrac{1}{100}$. In general, as x takes on smaller and smaller negative values, the corresponding values of $y=1/x$ are negative but get closer and closer to zero, as shown in Figure 2.8(c).

Giving negative values to x that are getting closer and closer to zero, such as $-\dfrac{1}{2}$, $-\dfrac{1}{5}$, $-\dfrac{1}{10}$, and $-\dfrac{1}{100}$, yields -2, -5, -10, and -100 as respective values for y. As x takes on negative values that get closer and closer to zero, $y=1/x$ gets smaller and smaller in a negative sense.

The complete graph of $xy=1$, or equivalently $y=1/x$, is shown in Figure 2.8(d). Observe that this graph consists of two separate pieces. Any attempt to connect all points on the graph by one continuous smooth curve would be disastrous.

EXERCISES

Sketch the graphs of the following equations.

3. $5x + 2y = 10$ 4. $3x - y = 6$ 5. $2x - 3y = 12$

6. $y = 2$ 7. $x = 3$ 8. $x - y = 5$

9. $y = -1$ 10. $x - 2y = 4$ 11. $x = -2$

12. $-x + 4y = 0$ 13. $x + 1.2y = 12$ 14. $-x + 3y = 0$

15. $y = x^2$ 16. $y = -x^2$ 17. $y = x^2 + 1$

18. $y = -x^2 + 1$ 19. $y = \dfrac{1}{x^2}$ 20. $y = -\dfrac{1}{x^2}$

2.3 \ A FIRST LOOK AT THE WORLD OF FINANCE

We now turn to the world of finance and show how the algebraic methods discussed can be used to develop basic equations that underlie simple and compound interest situations.

The fee charged for the use of borrowed money is called **interest**. It is an amount that is stated in terms of some monetary unit (dollars, cents, pounds, etc.). Thus if Mr. Needy borrows $500 from Mr. Plenty with the understanding that $550 is to be paid back one year later, the interest is $50. The time period at the end of which the amount borrowed, called the **principal**, and the interest owed are to be paid is called the **interest period**. In the Needy-Plenty transaction the interest period is one year. The sum, principal plus interest, to be repaid at the end of the interest period is called the **amount**. In the Needy-Plenty transaction the amount is $550. The ratio of the interest charged during the interest period to the sum owed at the beginning of the interest period is called the **interest rate**. Interest rates are usually stated in percent per annum, but for computation purposes the decimal form must be used.

Although it is incorrect to do so, in colloquial language the terms interest and interest rate are used interchangeably. In the Needy-Plenty transaction the interest rate is

$$\frac{50}{500} = 0.10 \text{ or } 10\% \text{ per annum}$$

More generally, if P is the principal and c is the interest due at the end of one interest period, then the **interest rate** i per interest period is expressed by

$$i = \frac{c}{P} \tag{2.8}$$

Multiplying both sides of Equation (2.8) by P yields c in terms of i and P:

$$c = Pi \tag{2.9}$$

In the world of finance, two basic interest structures, called simple interest and compound interest, are distinguished. Simple interest is interest that is proportional to the length of time of the loan; that is, if c is the interest due at the end of the interest period and n is the number of interest periods, then the **simple interest** I owed at the end of the n interest periods is expressed by

$$I = cn \tag{2.10}$$

From Equation (2.9) we have $c = iP$ Substituting iP for c in (2.10) yields the interest equation

$$I = Pin \tag{2.11}$$

The amount $A = P + I$ is given by

$$A = P + Pin$$

Factoring P yields the **amount equation**

$$A = P(1 + in) \tag{2.12}$$

EXAMPLE 1

Janet Arman borrowed \$800 for $2\frac{1}{2}$ years at simple interest of 9 percent per annum. Find the interest and the amount.

The rate 9 percent per annum tells us that the length of an interest period is 1 year. Since the loan is for $2\frac{1}{2}$ years, $n = 2.5$. Also, $P = \$800$ and $i = 0.09$. Thus from Equation (2.11) the interest is

$$I = 800(0.09)(2.5) = 180 \text{ or } \$180$$

and the amount is:

$$A = 800 + 180 = 980 \text{ or } \$980$$

EXAMPLE 2

$1000 is borrowed for 6 months at simple interest of 10 percent per annum. Find the amount.

$P = \$1000$, $i = 0.10$, and $n = \frac{1}{2}$. From Equation (2.12) we have

$$A = 1000[1 + (0.10)\frac{1}{2}] = 1050 \text{ or } \$1050$$

EXAMPLE 3

A car loan of $1200 is to be repaid in three installments of $400 plus simple interest at the rate of 9 percent per annum on the principal outstanding during each payment period. Find the total interest and the total amount due at the end of the year.

Three payments are to be made. The interest due with the first payment (due after 4 months) is:

$$1200(0.09)\frac{1}{3} = 36 \text{ or } \$36$$

The interest due with the second payment is:

$$800(0.09)\frac{1}{3} = 24 \text{ or } \$24$$

The interest due with the third payment is:

$$400(0.09)\frac{1}{3} = 12 \text{ or } \$12$$

Thus the total interest to be paid is $36 + $24 + $12 = 72, and the total amount is $1200 + $72 = 1272.

Sometimes confused with simple interest is the concept of simple discount. Let us suppose that Ms. Borrower borrows $100 from a lender for 1 year at 10 percent with the understanding that she is to receive $90 now and to repay $100 a year from now. Then the 10 percent is called a **simple discount rate**.

The actual interest rate is more than 10 percent, for the lender receives $10 interest plus the principal invested at the end of 1 year, but the principal invested is only $90. If we let i denote the interest rate, then we have:

$$i(90) = 10$$

$$i = \frac{10}{90} = 0.1111 \text{ or } 11.11\%$$

EXAMPLE 4

Jeff Jackson borrows $300 for 1 year at simple discount rate of 8 percent per annum. How much does he receive? What interest rate is actually involved?

The amount he receives now is $300 minus 8 percent of $300.

$$300 - 0.08(300) = 300 - 24 \text{ or } $276$$

Letting i denote the interest rate, we have:

$$i(276) = 24$$

$$i = \frac{24}{276} = 0.0869 \text{ or } 8.69\%$$

EXERCISES

Find the simple interest and amount on the following loans.

1. $2000 at 6 percent per annum for 2 years.

2. $5000 at 8 percent per annum for 3 years.

3. $6000 at 9 percent per annum for $3\frac{1}{2}$ years.

4. A car loan of $8000 is to be repaid in quarterly installments of $2000 plus simple interest at the rate of 10 percent per annum on the principal outstanding during each payment period. Find the total interest and total amount at the end of the year.

5. A home improvement loan of $1800 is to be repaid in three installments of $600 plus simple interest at the rate of 12 percent per annum on the principal outstanding during each payment period. Find the total interest and total amount due at the end of the year.

Compound Interest

If Mr. Needy sought to extend the loan for another year, Mr. Plenty might well adopt the point of view that, since the amount due at the end of the year is $550 (consisting of the $500 principal and $50 interest), in extending the loan for another year it is appropriate to charge 10 percent interest on the $550 owed rather than on the $500 originally borrowed. Thus the interest that would have to be paid over the second year is:

$$550(0.10) = 55 \text{ or } \$55$$

The total interest for the 2 years would be $\$50 + \$55 = \$105$, and the total amount would be $605.

In general, if the interest due is added to the principal at stated intervals of time and itself earns interest thereafter, then the sum by which the original principal has been increased at the end of any time is called **compound interest**. The time interval between successive additions of interest to principal is called the **interest period** or **conversion period**, and at the end of each conversion period the new principal, consisting of the original principal plus the compound interest, is called the **compound amount**.

To obtain an equation for the compound amount, let us suppose that an initial amount of money P is invested at compound interest at the rate i per interest period. Then at the end of the first interest period the compound amount is:

$$A_1 = P + Pi = P(1+i) \tag{2.13}$$

At the end of the second interest period the new amount A_2 is A_1 plus the interest $A_1 i$ obtained from A_1:

$$A_2 = A_1 + A_1 i = A_1(1+i)$$

Replacing A_1 by $P(1+i)$ [see (2.13)] yields:

$$A_2 = P(1+i)(1+i) = P(1+i)^2 \qquad (2.14)$$

At the end of the third interest period the new amount A_3 is A_2 plus the interest $A_2 i$ obtained from A_2:

$$A_3 = A_2 + A_2 i = A_2(1+i)$$

Replacing A_2 by $P(1+i)^2$ [see (2.14)] yields:

$$A_3 = P(1+i)^2(1+i) = P(1+i)^3 \qquad (2.15)$$

At the end of the fourth interest period the new amount A_4 is A_3 plus the interest $A_3 i$ obtained from A_3:

$$A_4 = A_3 + A_3 i = A_3(1+i)$$

Replacing A_3 by $P(1+i)^3$ [see (2.15)] yields:

$$A_4 = P(1+i)^3(1+i) = P(1+i)^4$$

More generally, at the end of the nth interest period the **compound amount** A is given by the equation:

$$A = P(1+i)^n \qquad (2.16)$$

EXAMPLE 5

$500 is invested at 2 percent per month compounded monthly. What will it grow to in 18 months?

Since $P = \$500$, $i = 0.02$, and $n = 18$, we have:

$$A = 500(1.02)^{18}$$

$(1.02)^{18} = 1.42825$ (see Appendix Table 2). Thus:

$$A = 500(1.42825) = 714.13 \text{ or } \$714.13$$

In the world of finance an interest rate stated as 6 percent per annum compounded twice a year (semiannually) means that 3 percent interest is added every 6 months to the amount accumulated.

More generally, the rate r **per annum compounded m times a year** envisions the year being divided into m interest periods of equal length, with interest at the rate of $i = r/m$ being added to the amount accumulated at the end of each period. The rate r per annum compounded m times a year is called a **nominal rate.** Returning to Equation (2.16). $A = P(1+i)^n$, and replacing i by r/m yields

$$A = P\left(1 + \frac{r}{m}\right)^n$$

as the compound amount after n interest periods. It is useful to express n in terms of years. Since the number of interest periods in 1 year is m, the number of periods n in x years is $n = mx$. Thus the equation

$$A = P\left(1 + \frac{r}{m}\right)^{mx} \tag{2.17}$$

expresses the **compound amount after x years when principal P is invested at the rate r per annum compounded m times a year.** x can assume nonnegative integer values ($x = 0, 1, 2$, etc.) and fractional values consistent with m ($x = 1/m$, $2/m$, etc.).

EXAMPLE 6

What will $1000 grow to in 5 years if invested at 8 percent per annum compounded quarterly?

$P = \$1000$, $r = 0.08$, $m = 4$, $x = 5$, and $mx = 20$. Thus

$$A = 1000\left(1 + \frac{0.08}{4}\right)^{20} = 1000(1.02)^{20} = 1000(1.48595) = 1485.95 \text{ or } \$1485.95$$

EXAMPLE 7

What principal should be invested at 9 percent per annum compounded three times a year if $5000 is to become available in 10 years?

$A = \$5000$, $r = 0.09$, $m = 3$, $x = 10$, and $mx = 30$. Substituting these values into Equation (2.17) yields:

$$5000 = P(1.03)^{30}$$

Dividing both sides by $(1.03)^{30}$ gives us:

$$P = \frac{5000}{(1.03)^{30}} = 5000(1.03)^{-30}$$

Appendix Table 3 yields $(1.03)^{30} = 0.41199$ Thus:

$$P = 5000(0.41199) = 2059.95 \text{ or } \$2059.95$$

More generally, solving for P by dividing both sides of Equation (2.17) by $[1 + (r/m)]^{mx}$ yields:

$$P = \frac{A}{\left(1 + \dfrac{r}{m}\right)^{mx}}$$

From the definition of negative exponent we have:

$$P = A\left(1 + \frac{r}{m}\right)^{-mx} \tag{2.18}$$

Equation (2.18) expresses the sum P that must be initially invested if amount A is to become available mx years in the future, where the interest rate is r per annum compounded m times a year. In this setting P is called the **present value** of A.

Nominal and Effective Interest Rates

Your cousin Jack the broker calls and says: "Don't make a move! I've got the investment for you; it brings a return of 12% per annum compounded six times a year." "It sounds good Jack," you reply, "but what interest would I have at the end of the year?" "The corresponding effective rate is 12.6%," fires back Jack. How did he get that and what does it mean, you think.

We turn to these questions after considering their more general counterparts.

For a given nominal rate r per annum compounded m times a year, the corresponding **effective rate** v is the rate which compounded annually yields the same interest at the end of the year. The amount yielded by effective rate v at the end of the year is $P(1+v)$. Since $x=1$ the amount yielded at the end of the year by the nominal rate r per annum compounded m times a year is $P[1+(r/m)]^m$. Since the amounts attained at the end of the year are the same under both arrangements, we have:

$$P(1+v) = P\left(1+\frac{r}{m}\right)^m$$

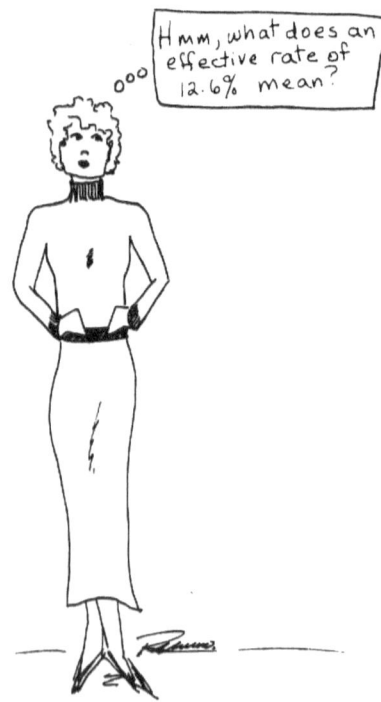

Dividing both sides by P yields:

$$1+v = \left(1+\frac{r}{m}\right)^{m}$$

Thus:

$$v = \left(1+\frac{r}{m}\right)^{m} - 1$$

EXAMPLE 8

Find the effective rate corresponding to a nominal rate of 12 percent per annum compounded six times a year.

$r = 0.12$ and $m = 6$. Thus:

$$v = (1.02)^{6} - 1$$

From Appendix Table 2, $(1.02)^{6} = 1.12616 \approx 1.126$. Thus:

$$v = 1.126 - 1 = 0.126 \text{ or } 12.6\%$$

Therefore, 12.6 percent compounded annually yields the same interest at the end of the year as 12 percent per annum compounded six times a year. $1000 invested at 12 percent per annum compounded six times a year yields a return of $(1000)(0.126) = 126$ or $126 at the end of the year.

EXERCISES

6. Find the amount on deposit at the end of 3 years if $2000 is invested at 12 percent per annum compounded (a) three times a year; (b) four times a year; (c) six times a year.

7. Find the amount on deposit at the end of 4 years if $3000 is invested at 8 percent per annum compounded (a) semiannually; (b) quarterly.

8. The Andrius Company expects that $5000 will be needed in 5 years to meet the cost of equipment replacement. How much should be initially invested at 12 percent per annum compounded quarterly to meet this cost?

9. The Rodgers family wishes to have $8000 available in 6 years for the education of their children. How much should be initially invested at 8 percent per annum compounded quarterly to meet this expense?

10. An appliance was purchased by paying $100 down and agreeing to pay $200 in 9 months. If money is worth 12 percent per annum compounded quarterly, what was the selling price of the appliance?

11. Arnold List borrows $1000 for 1 year at a simple discount rate of 6 percent per annum, (a) How much does he receive? (b) What interest rate is actually involved?

12. Rosalie Formann borrows $2000 for 6 months at a simple discount rate of 8 percent per annum. (a) How much does she receive? (b) What interest rate is actually involved?

13. Peter Thompson plans to borrow some money at a simple discount rate of 6 percent per annum for 1 year. If he wishes to receive $1200 in cash, how much must he borrow?

14. Show that, if r is the simple interest rate when money is borrowed for 1 year at a simple discount rate of d, then $r = d/(1-d)$ and $d = r/(1+r)$.

15. Determine the effective rate corresponding to a nominal rate of 8 percent per annum compounded 8 times a year. Interpret the result obtained.

16. Determine the effective rate corresponding to a nominal rate of 12 percent per annum compounded 4 times a year. Interpret the result obtained.

17. Which investment opportunity yields the larger return, 12 percent per annum compounded three times a year or 11 percent per annum compounded 11 times a year? How much interest would be generated at the end of the year under these plans if $1000 were invested?

2.4   INEQUALITIES

Inequalities come in a variety of forms. Any statement involving one of the symbols > (greater than), ≥ (greater than or equal to), < (less than), or ≤ (less than or equal to) is termed an **inequality**. Inequalities of the < type and > type are sometimes called **strict inequalities**. Two kinds of inequalities, absolute inequalities and conditional inequalities, are to be distinguished. An inequality that involves numbers only or holds for all permissible values of the variables involved is called an **absolute inequality**.

$$-3 < -2, \qquad 8 > \frac{1}{2}, \quad \text{and} \quad x^2 \geq 0$$

are absolute inequalities. An inequality that does not hold for all permissible values of the variables involved is called a **conditional inequality**, or more simply, an **inequality**.

$$2x + 3 < 7, \qquad x^2 \leq 16, \quad \text{and} \quad 2x + 3y \geq 6$$

are conditional inequalities.

When 1 is substituted for x in the inequality

$$2x + 3 < 7$$

we obtain $2(1) + 3 < 7$, or $5 < 7$, which is a correct statement. The number 1 or any other number that yields a correct statement when substituted for x in $2x + 3 < 7$ (such as $\frac{3}{2}, \frac{1}{2}, 0, -4$) is said to be a solution of $2x + 3 < 7$. The number 1 and the other solutions of $2x + 3 < 7$ are said to **satisfy** this inequality. The number 2 is not a solution of (does not satisfy) $2x + 3 < 7$ since substitution of 2 in $2x + 3 < 7$ yields $2(2) + 3 < 7$, or $7 < 7$, which is an incorrect statement.

More generally, a number is said to be a **solution** of a one-variable inequality if a correct statement is obtained when the number is substituted for the variable of the inequality.

A **solution** of an inequality in two variables, x and y, let us say, is any ordered pair of numbers that yields a correct statement when substituted for x and y, respectively. For example, $(3, 0)$ is a solution of

$$2x+3y \geq 6$$

since substitution of 3 for x and 0 for y yields $2(3)+3(0) \geq 6$, or $6 \geq 6$, which is a correct statement. $(3, 0)$ is said to satisfy $2x+3y \geq 6$. $(0, 1)$ is not a solution of **(does not satisfy)** $2x+3y \geq 6$, since substitution of 0 for x and 1 for y yields $2(0)+3(1) \geq 6$, or $3 \geq 6$, which is an incorrect statement.

A **solution** of an inequality in three variables, x, y, and z, is any ordered triple of numbers that yields a correct statement when substituted for x, y, and z, respectively. Thus $(-1,3,4)$ is a solution of

$$x^2y+z<10$$

since substitution of -1 for x, 3 for y, and 4 for z yields $(-1)^2(3)+4<10$, or $7<10$, which is a correct statement. Here too we say that $(-1,3,4)$ satisfies $x^2y+z<10$. The ordered triple of numbers $(2,4,-1)$ does not satisfy (is not a solution of) $x^2y+z<10$, since substitution of 2 for x, 4 for y, and -1 for z yields $(2)^2(4)-1<10$, or $15<10$, which is an incorrect statement.

The concept of solution of an inequality in any number of variables is defined in an analogous way.

EXERCISES

Determine which of $(0, 2)$, $(3,-2)$, $(-4,0)$, $(\frac{1}{2},5)$, $(2, 3)$, *and* $(6, 2)$ *satisfy the following inequalities.*

1. $3x+2y \leq 6$ 2. $2x-y^2<4$ 3. $2x+xy \geq 10$
4. $2x-3y<4$ 5. $5x^2-y>6$ 6. $2x^2-y^2 \leq 4$

Determine which of $(2, 3, 1)$, $(1,3,-1)$, $(4, 2, 1)$, $(3,-1,4)$, $(\frac{1}{2},\frac{1}{3},-3)$, *and* $(-2,3,1)$ *satisfy the following inequalities.*

7. $3x+2y-z \leq 8$ 8. $2x+5y-3z \leq 10$ 9. $3x^2y-2z>6$

10. $x^2+2y^2-z \leq 12$ 11. $\dfrac{2x^2-y^2}{z} \leq 6$ 12. $4x-2y+3z \leq 8$

Operations of Equivalence

Two inequalities are said to be **equivalent** if they have the same solutions. The following operations of equivalence lead to equivalent inequalities:

> 1. The same constant may be added to, or subtracted from, both sides of an inequality; the direction of the inequality is maintained. That is, **if $a \leq b$ and c is any number, then** $a + c \leq b + c$ **(and** $a - c \leq b - c$).

For example, adding 5 to both sides of $2x - 5 \leq 3$ yields the equivalent inequality $2x \leq 8$.

> 2. A term may be transferred from one side of an inequality to the other side, provided that its sign is changed; the direction of the inequality is maintained. This operation is called **transposition.**

For example, transposing $2x$ from the left side to the right side of $2x + y \leq 8$ yields the equivalent inequality $y \leq 8 - 2x$.

> 3. Both sides of an inequality may be multiplied or divided by the same nonzero constant; multiplying or dividing by a **positive** constant **maintains** the direction of the inequality; multiplying or dividing by a **negative** constant **reverses** the direction of the inequality.
>
>   That is, **if $a \leq b$ and c is positive, then** $ac \leq bc$ **and** $\dfrac{a}{c} \leq \dfrac{b}{c}$.
>
>   **If $a \leq b$ and c is negative, then** $ac \geq bc$ **and** $\dfrac{a}{c} \geq \dfrac{b}{c}$.

For example, multiplying both sides of $2 < 3$ by -1 yields $-2 > -3$. The direction of the inequality is reversed since -1 is negative. Dividing both sides of $2x \leq 8$ by 2 yields the equivalent inequality $x \leq 4$. The direction of the original inequality is maintained since 2 is positive.

These operations of equivalence for inequalities are analogous to the ones cited for equations, with one difference. In multiplying or dividing both sides of an inequality by a value one must pay particular attention to the sign of the value. Positive values maintain the direction of an inequality; negative values reverse the direction of an inequality.

EXAMPLE 1

Determine the solutions of $10 - 3x \geq 7$.

Subtracting 10 from both sides of $10 - 3x \geq 7$ yields:

$$-3x \geq -3$$

By dividing both sides of $-3x \geq -3$ by -3 (which is negative), we obtain:

$$x \leq 1$$

Since the operations employed lead to equivalent inequalities, $10 - 3x \geq 7$ is equivalent to $x \leq 1$. Every number that is less than or equal to 1 is a solution of $10 - 3x \geq 7$, and every solution of $10 - 3x \geq 7$ is less than or equal to 1.

We may represent the solutions described by $x \leq 1$ by emphasizing on the x-axis all points less than or equal to 1 as shown in Figure 2.9.

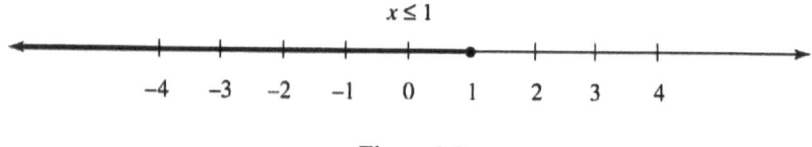

Figure 2.9

This gives us a half-line with end point 1. Since 1 is part of the graph we indicate it by means of a heavy dot.

If the solutions were given by $x < 1$ so that 1 is not part of the graph, we would indicate it by means of an open circle as shown in Figure 2.10.

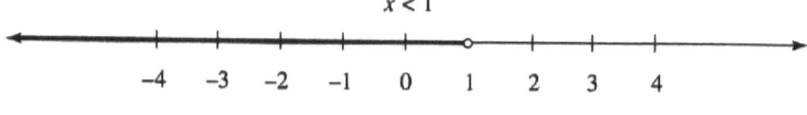

Figure 2.10

EXAMPLE 2

Show that $y \geq \dfrac{1}{5}(x+y)$ is equivalent to $-x+4y \geq 0$.

Multiplying both sides of $y \geq \dfrac{1}{5}(x+y)$ by 5 yields:

$$5y \geq x+y$$

By transposing $x+y$ we obtain:

$$-x+4y \geq 0$$

EXAMPLE 3

Show that $200 - x - y \geq 0$ is equivalent to $x+y \leq 200$.

Subtracting 200 from both sides of $200 - x - y \geq 0$ yields:

$$-x-y \geq -200$$

Multiplying both sides of $-x-y \geq -200$ by -1 (which is negative) gives us:

$$x+y \leq 200$$

EXAMPLE 4

Show that $2(x-4) > 3x+2$ is equivalent to $x < -10$.

$$2(x-4) > 3x+2$$

Since $2(x-4) = 2x-8$, we have:

$$2x-8 > 3x+2$$

Transposing $3x$ yields:

$$-x-8 > 2$$

Adding 8 to both sides gives us:

$$-x > 10$$

Multiplying both sides by -1 yields:

$$x < -10$$

EXERCISES

13. Show that $100 - x \geq 0$ is equivalent to $x \leq 100$.
14. Show that $300 - y \geq 0$ is equivalent to $y \leq 300$.
15. Show that $1000 - x - y \geq 0$ is equivalent to $x + y \leq 1000$.
16. Show that $2x + y < 6$ is equivalent to $y < 6 - 2x$.
17. Show that $x \leq 2$ is equivalent to $3x - 4 \leq 2$.
18. Show that $2(x - 3y) < 8$ is equivalent to $x < 4 + 3y$.
19. Show that $3(x + 2y) < 2x + 4$ is equivalent to $x < 4 - 6y$.
20. Show that $-2(x + 3) < 4x + 12$ is equivalent to $x > -3$.

Find the solutions of the following inequalities in one variable. Sketch their graphs.

21. $3x + 6 > 4$	22. $-2x \geq 5$	23. $5x + 3 \leq 13$
24. $3y + 2 < 11$	25. $3 - 4y > -5$	26. $2 - 3y < 17$
27. $2(x + 4) < x - 7$	28. $-3(2x + 2) < 6x + 5$	
29. $2(x - 4) < 3(2x - 6)$	30. $4(3 - 2x) \geq 7 - 3x$	
31. $3(4 - 2y) \leq y - 2$	32. $-2(3 - y) \geq y + 1$	

2.5   GRAPHS OF TWO-VARIABLE INEQUALITIES

The **graph** of an inequality in two variables is the collection of all those points, and only those points, whose coordinates satisfy the inequality. In a number of situations the geometric view of an inequality provided by its graph is most useful. We shall find graphs of particular importance in our discussion of basic linear programming. In this connection especially, graphs of linear inequalities occupy the spotlight.

A two-variable **linear equation**, we recall, is of the form

$$Ax + By = C$$

where A and B are not both zero. A **two-variable linear inequality** is an inequality that can be obtained from $Ax + By = C$ by replacing the equality condition (=) by any one of the inequality conditions >, ≥, <, or ≤. The inequality $2x - 3y \le 6$, for example, is a two-variable linear inequality; it can be obtained from the linear equation $2x - 3y = 6$ by replacing = by ≤.

The graph of $2x - 3y = 6$ is a line, termed a **boundary line,** that divides the rest of the coordinate plane into two components (see Figure 2.11).

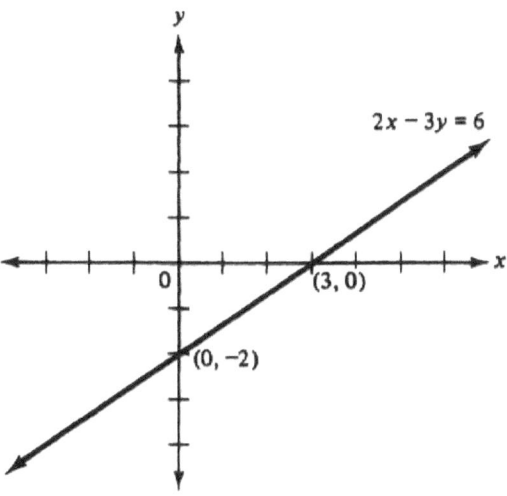

Figure 2.11

All points on one side of this boundary line satisfy $2x - 3y < 6$, while all points on the other side satisfy $2x - 3y > 6$.

To find which side corresponds to which inequality, it suffices to choose a **test point on one side** of $2x - 3y = 6$ and substitute it into $2x - 3y < 6$. If this test point satisfies $2x - 3y < 6$, then all points on the same side as this test point satisfy $2x - 3y < 6$, and all points on the other side of the boundary line satisfy $2x - 3y > 6$. If the test point does not satisfy $2x - 3y < 6$, then it satisfies $2x - 3y > 6$, as do all other points on the same side as the test point. The points on the other side of the boundary line satisfy $2x - 3y < 6$. It's an heads you win tails you win situation.

To illustrate, let us take (0, 0) as our **test point.** (Any other point not on the boundary line $2x - 3y = 6$ would do just as well.) Substituting (0, 0) into

$2x-3y<6$ yields $0<6$, which is a correct statement. Thus (0, 0) satisfies $2x-3y<6$, and all points on the same side as (0, 0) (above $2x-3y=6$) satisfy $2x-3y<6$. The inequality $2x-3y>6$ is satisfied by all points below $2x-3y=6$ (see Figure 2.12).

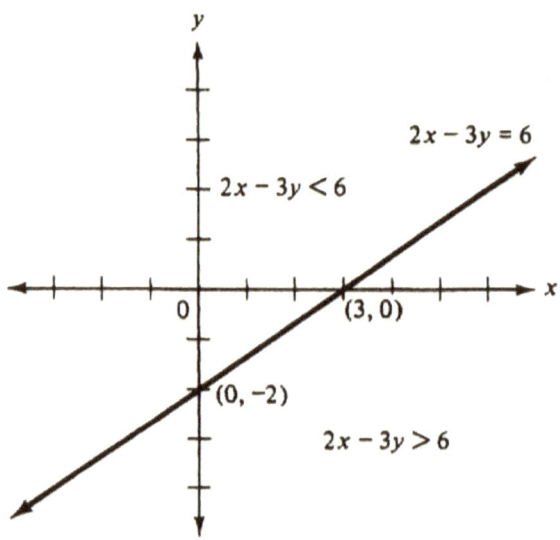

Figure 2.12

More generally, the graph of $Ax+By=C$ is a line that divides the rest of the coordinate plane into two components. All points on one side of $Ax+By=C$ satisfy $Ax+By<C$, while all points on the other side satisfy $Ax+By>C$.

Examples 1 and 2 further illustrate the graphing of linear inequalities.

EXAMPLE 1

Sketch the graph of $5x+2y\geq10$.

We first graph the boundary line $5x+2y=10$. For $x=0$, $y=5$; for $y=0$, $x=2$. Thus $5x+2y=10$ passes through (0, 5) and (2, 0) (see Figure 2.13).

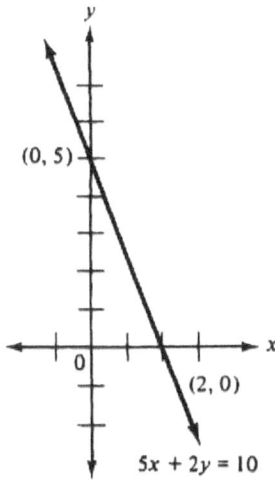

Figure 2.13

We next choose a test point not on $5x+2y=10$, (0, 0), for example, and determine if it satisfies $5x+2y>10$. Substituting 0 for x and 0 for y in $5x+2y>10$ yields $0>10$, which is an incorrect statement. Since (0, 0) does not satisfy $5x+2y>10$, the points that do satisfy $5x+2y>10$ are on the other side of (above) the boundary line $5x+2y=10$.

The graph of $5x+2y\geq10$ consists of all points on and above the boundary line $5x+2y=10$. This is indicated visually by drawing $5x+2y=10$ as a solid line and shading the region above it (see Figure 2.14).

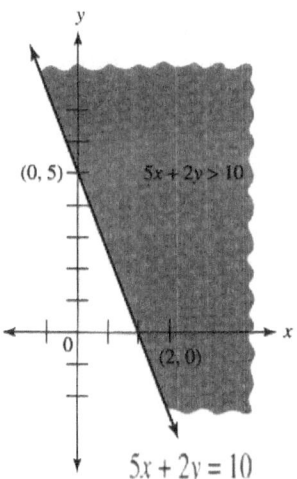

Figure 2.14

EXAMPLE 2

Sketch the graph of $x - 2y < 4$.

We first graph the boundary line $x - 2y = 4$. For $x = 0$, $y = -2$; for $y = 0$, $x = 4$. Thus $x - 2y = 4$ passes through $(0, -2)$ and $(4, 0)$. Since this boundary line is not part of the graph of $x - 2y < 4$ (this inequality is a strict inequality), we draw a dashed line as shown in Figure 2.15.

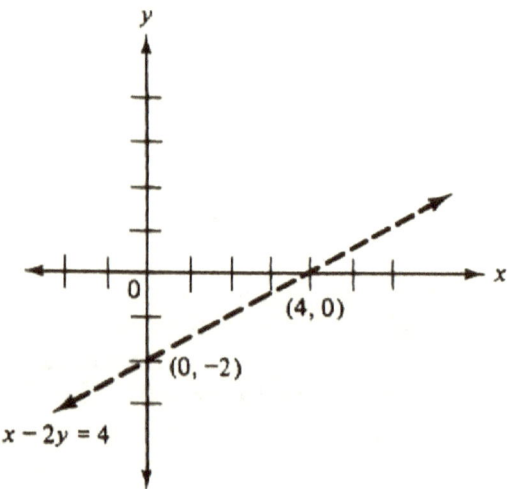

Figure 2.15

We next choose a test point not on $x - 2y = 4$, $(0, 0)$, for example, and determine if it satisfies $x - 2y < 4$. Substituting 0 for x and 0 for y in $x - 2y < 4$ yields $0 < 4$, which is a correct statement. Since $(0, 0)$ satisfies our inequality and is above the boundary line $x - 2y = 4$, the graph of $x - 2y < 4$ consists of all points above this boundary line (see Figure 2.16).

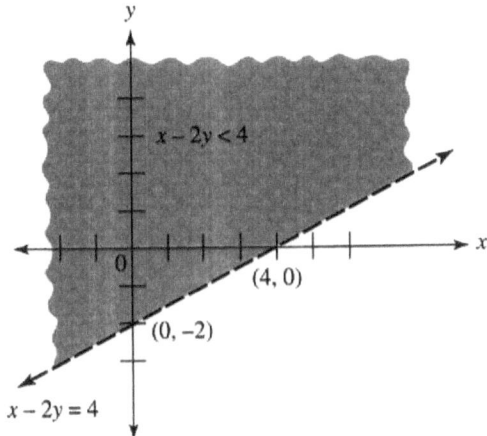

Figure 2.16

To illustrate a nonlinear situation, let us note that the circle defined by $x^2 + y^2 = 25$ (see Section 2.2, Example 3) divides the rest of the coordinate plane into two regions. In the interior of the circle $x^2 + y^2 < 25$ is satisfied; outside the circle $x^2 + y^2 > 25$ is satisfied (see Figure 2.17).

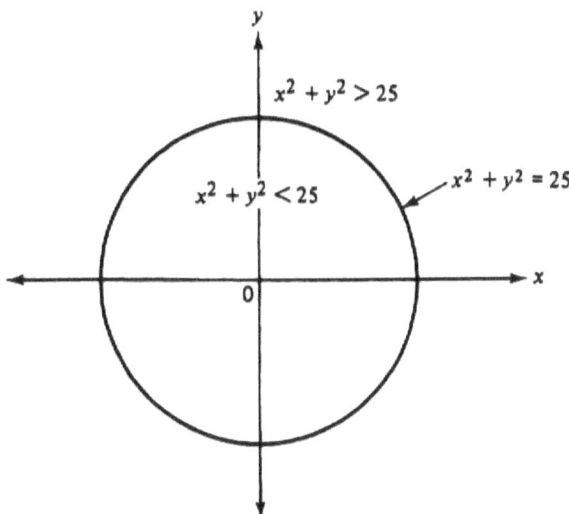

Figure 2.17

EXERCISES

Sketch the graphs of the following inequalities.

1. $x + y \geq 3$

2. $x + 2y \leq 6$

3. $x - y \geq 4$

4. $2x + y < 8$

5. $2x - y < 1$

6. $x + 3y < 6$

7. $x \geq 0$

8. $y \geq 0$

9. $3x - y \leq 3$

10. $4x - 3y > 12$

11. $x + y \geq 0$

12. $x - y < 0$

13. $-x + 4y \geq 0$

14. $4x + 5y \geq 75$

15. $2x + y \geq 23$

16. $2x + 3y \geq 120$

17. $3x + 2y \leq 150$

18. $-x + 3y \geq 0$

19. $x + 1.2y \leq 12$

20. $5x + 3y \geq 95$

21. $-3x + 10y \leq 0$

22. $x^2 + y^2 > 16$

23. $x^2 + y^2 < 9$

24. $x^2 + y^2 \leq 4$

CHAPTER 3

Functions

3.1 FUNCTIONS OF ONE VARIABLE

The idea of function is suggested by essentially the following background. Let us assume that two factors connected with some process under study have been singled out for consideration. Moreover, these factors can be quantified and expressed in terms of variables x and y. Also, the process gives rise to a rule of association such that, whenever a value for x is specified, exactly one value for y is determined by means of this rule. Such a structure is called a function. In summary, then, a **function,** or **function of one variable,** has the following structure.

There are two variables x and y, say, and a rule with the property that for each value of x the rule determines exactly one value for y. Here x is envisioned as taking on values with corresponding values of y being determined. To distinguish these different roles, x is called an **independent variable** and y is called a **dependent variable.** If their roles were to be reversed, with y taking on values and corresponding values of x being determined, then y would be the independent variable and x would be the dependent variable. If x is the independent variable and y is the dependent variable, then we say that **y is a function of x.** The term "function of one variable" is due to the presence of one independent variable. The set of values that can be taken on by the independent variable is called the **domain of definition,** or **domain,** of the function and the independent variable. When the domain of definition is not explicitly stated, it is understood to be the set of all real-numbers for which the rule makes mathematical sense.

If the function arises from some applied setting, economics, for example, then the domain, when not explicitly stated, is understood to be all numbers for which the rule makes economic as well as mathematical sense.

Often, but not always, the rule of the function can be expressed in terms of a single algebraic expression. For example, the rule

$$y = x^3 - x + 10$$

defines y as a function of x. The domain, which is not explicitly stated, is understood to be the set of all real-numbers, since this rule can be applied to all real numbers.

For $x = 2$,	$y = (2)^3 - 2 + 10 = 16$
For $x = -1$,	$y = (-1)^3 - (-1) + 10 = 10$

The rule

$$y = \sqrt{x}$$

defines y as a function of x. Recall that $\sqrt{x}$ expresses the principal, or nonnegative, square root of x. The domain, which is not explicitly stated, is understood to be all nonnegative real-numbers, since we are operating within the real-number system and $\sqrt{x}$ does not yield a real-number for negative x.

The practice of using letters such as f, g, h, F, G, H, and so on, to designate either the rule of the function or the function itself has become widespread in recent years. In terms of this practice, the notation $f(x)$, read "f of x" or "f at x," is used to express the value assigned by the function f to x. $f(x)$, in other words, is another symbol for the dependent variable and, in fact, we often see the statement $y = f(x)$ written as an expression of this fact. The notation $f(x)$ is also often used to denote the function itself, as well as the value assigned by the function to x. Thus, for example, the rule of the function

$$y = x^3 - x + 10$$

is expressed in terms of the $f(x)$ notation by

$$f(x) = x^3 - x + 10$$

The value assigned by this function to 2, for example, is expressed by $f(2)$; the value assigned to -1 is expressed by $f(-1)$. We have:

$$f(2) - (2)^3 - 2 + 10 = 16$$
$$f(-1) = (-1)^3 - (-1) + 10 = 10$$

Two functions f and g are said to be **equal** if (1) they have the same domain of definition, and (2) the rules of f and g assign the same value to each number in their common domain of definition. Thus, for example, the functions f and g defined by

$$f(x) = x^3 - x + 10$$

and

$$g(x) = x^3 - x + 10, \qquad\qquad \text{where } x \geq 0$$

are not equal since they have different domains of definition. $f(-1) = 10$, whereas $g(-1)$ is not defined.

The following examples further illustrate the nature of a function.

EXAMPLE 1   *THE CONSTANT FUNCTION*

Let f denote the function that assigns the value 3 to each real-number. That is,

$$f(x) = 3$$

Thus $f(-4) = 3$, $f(0) = 3$, $f(\frac{1}{2}) = 3$, $f(\sqrt{2}) = 3$, $f(10) = 3$, and so on. Since the dependent variable takes on one value, the value 3 in this case, such a function is called a **constant function.**

EXAMPLE 2   *A FUNCTION WITH A TWO-PIECE RULE*

Let f denote the function defined on the reals by

$$f(x) = \begin{cases} 2x+1, & \text{for } x \leq 1 \\ x-2, & \text{for } x > 1. \end{cases}$$

This is **one function** with **one rule** and **one domain of definition.** The unusual feature is that the rule behaves differently on different parts of the domain of definition according to whether $x \leq 1$ or $x > 1$. Thus $f(1) = 2(1) + 1 = 3$, $f(-2) = 2(-2) + 1 = -3$, but $f(4) = 4 - 2 = 2$.

An analogy which is helpful to a point is to think of a function in terms of a production process which converts raw materials into finished products.

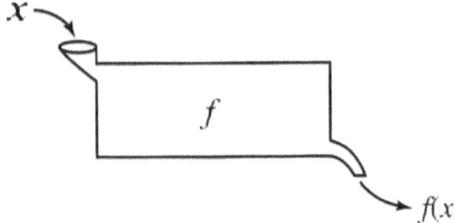

The x values in the domain of definition correspond to raw materials to be used. Only these raw materials may be fed to the machine, which corresponds to the rule of the function, for the production of finished products—the $f(x)$ values. The machine, or rule f, may be a very simple one, as in the case of a constant function which yields the same finished product, $f(x) = 3$, for example, for all raw materials x given to it. The machine may be a fairly complex one which behaves differently on different parts of the domain of definition, as in the case of Example 2.

EXERCISES

1. For $f(x) = x^2 - x - 4$, find $f(1), f(3)$, $f(-1)$, $f(-2)$, and $f(0)$.
2. For $f(x) = x^3 + 2x^2 - x + 1$, find $f(0), f(1), f(2), f(3)$, $f(-1)$, and $f(-2)$

3. For $f(x) = 3x^2 - 2x - 4$, find $f(-3)$, $f(\frac{1}{2}), f(3)$, and $f(10)$.

4. What is the domain of definition of the function $f(x) = \sqrt{x+1}$? Explain. What is $f(-2)$, $f(-1), f(3)$, and $f(2)$?
5. For $f(x) = -1$, find $f(0), f(1)$, $f(-1), f(10)$, and $f(-2)$.
6. What is the domain of definition of the function $f(x) = 1/(x-2)$? Explain.
7. For

$$f(x) = \begin{cases} 3x - 2, & \text{for } x \geq 1 \\ 4x, & \text{for } x < 1 \end{cases}$$

   find $f(1), f(2), f(3)$, and $f(-1)$.

8. For

$$f(x) = \begin{cases} 1, & \text{for } x \geq 2 \\ -1, & \text{for } x < 2 \end{cases}$$

   find $f(1), f(2)$, $f(\frac{1}{2}), f(0)$, and $f(-1)$.

3.2   GRAPHS OF FUNCTIONS

The **graph of a function** f is the graph of the equation $y = f(x)$, where x takes on only those values in its domain of definition.[*] A crude, but often satisfactory, approach to sketching the graph of a function f is to let the independent variable x take on a number of values, calculate the corresponding values of $y = f(x)$, plot the points $(x, f(x))$, and join these points with a smooth curve. It is an approach that must be used with caution since the graph of a function, or equation, may exhibit gaps and breaks (see, for example, section 2.2, Example 4, p. 48).

Particular attention must be paid to the domain of definition of the function, and care must be taken not to inadvertently allow the independent variable to take on "illegal" values.

EXAMPLE 1

Sketch the graph of the constant function $f(x) = 3$.

The domain of definition of $f(x) = 3$, or $y = 3$, is the collection of all real numbers. For $x = -2$, $y = 3$; for $x = 0$, $y = 3$; for $x = 10$, $y = 3$; and so on. The graph of $f(x) = 3$ is the horizontal line three units above the x-axis and parallel to the x-axis (see Figure 3.1).

More generally, the graph of the constant function $f(x) = b$ is a horizontal line b units above the x-axis if b is positive, b units below the x-axis if b is negative, and the x-axis itself if b is zero.

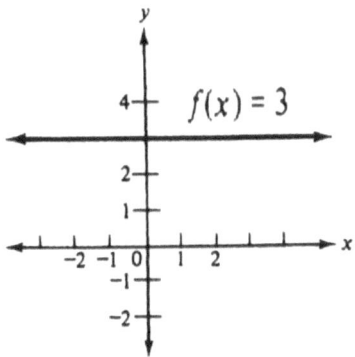

Figure 3.1

---

[*] Please review section 2.2, chapter 2, at this point.

The function f with the general form

$$f(x) = mx + b$$

where m and b are constants, is said to be a **linear function,** or **linear function of one variable.** If the domain of definition is the set of all real-numbers, then the graph of f is a line. If the domain of definition of f is some subcollection of real numbers, then the graph of f is part of a line.

EXAMPLE 2

Sketch the graph of $f(x) = 3x - 2$.

Since the domain of definition consists of all real-numbers and the function is linear it suffices to plot two points to determine its graph. For $x = 1$, $f(1) = 1$; for $x = 2$, $f(2) = 4$. Thus (1, 1) and (2, 4) on the graph, which is shown in Figure 3.2.

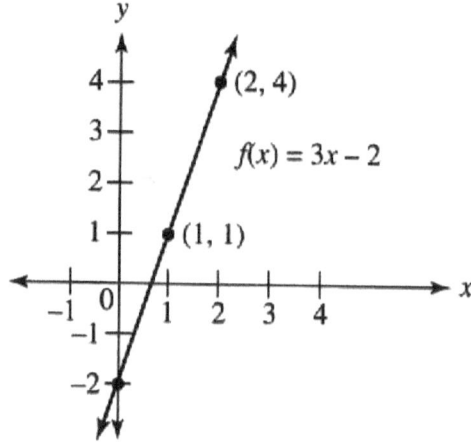

Figure 3.2

EXAMPLE 3

Sketch the graph of $f(x) = 3x - 2$, where $1 \le x \le 2$.

The domain of definition of this linear function consists of all real-numbers in the interval with end points 1 and 2, including 1 and 2. Thus the graph of $f(x) = 3x - 2$, where $1 \le x \le 2$, consists of all points on the line segment with end points $(1, f(1)) = (1,1)$ and $(2, f(2)) = (2,4)$ including these end points (see Figure 3.3).

To emphasize that these end points are part of the graph, we display them as solid dots.

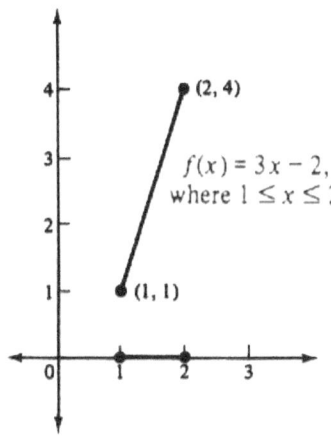

Figure 3.3

EXAMPLE 4

Sketch the graph of $f(x) = 3x - 2$, where $1 < x < 2$.

The domain of definition of this linear function consists of all real-numbers in the interval with end points 1 and 2, excluding the end points 1 and 2. The graph of $f(x) = 3x - 2$, where $1 < x < 2$, consists of all points on the line segment with end points (1, 1) and (2, 4), excluding these end points (see Figure 3.4).

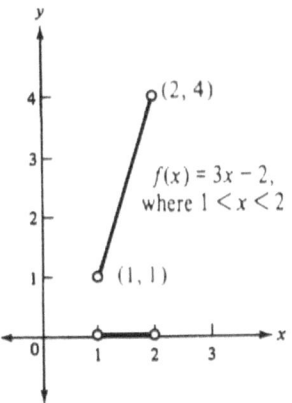

Figure 3.4

To properly sketch the graph, these end points must be shown, but at the same time it must be made clear that they are not on the graph. We do this by using open circles, as opposed to solid dots, to display them.

In connection with number intervals, a comment on notation is in order. The collection of all real-numbers between *a* and *b*, including the end points *a* and *b*, is often denoted by [*a*, *b*]. The collection of all real-numbers between *a* and *b*, excluding *a* and *b*, is denoted by (*a*, *b*), which should not be confused with the coordinates of a point (*a*, *b*). The context of the discussion makes clear which meaning is intended. Thus, to include an end point of an interval use a square bracket; to exclude an end point use a parenthesis. [*a*, *b*), therefore, denotes the collection of all real-numbers between *a* and *b*, including *a* but excluding *b*.

EXAMPLE 5

Sketch the graph of $f(x) = 3x - 2$, where the domain of definition consists of the values 1, 2, and 3.

Since the domain of definition consists of three values, the graph of $f(x)$ consists of three points on the line $y = 3x - 2$, that is, (1, 1), (2, 4), and (3, 7) (see Figure 3.5).

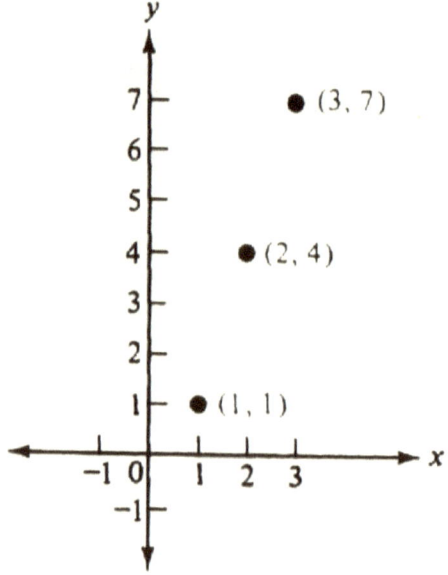

Figure 3.5

EXAMPLE 6

Sketch the graph of

$$g(x) = \begin{cases} 3x-2, & \text{for } x \neq 2 \\ 5, & \text{for } x = 2. \end{cases}$$

For $x = 2$ we obtain the point $(2, 5)$; for $x \neq 2$ we obtain points on the line $y = 3x - 2$. The point $(2, 4)$ on this line is not on the graph, and we therefore indicate $(2, 4)$ by an open circle. Since $(2, 5)$ is on the graph, we indicate it by a solid dot.

The graph, shown in Figure 3.6, consists of the point $(2, 5)$ together with all points on the line $y = 3x - 2$, except $(2, 4)$.

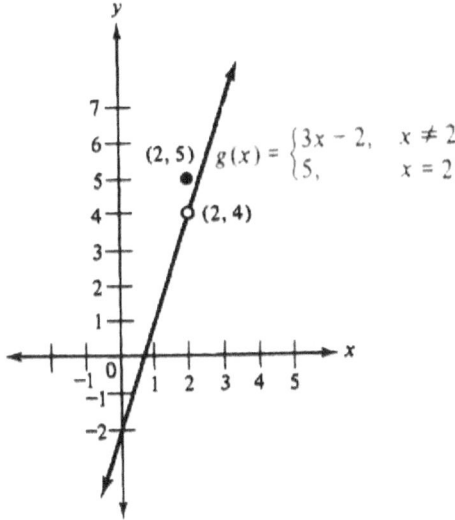

Figure 3.6

EXAMPLE 7

Sketch the graph of

$$f(x) = \begin{cases} 1, & \text{for } x \geq 2 \\ -1, & \text{for } x < 2. \end{cases}$$

For $x \geq 2$, $f(x) = 1$, so that we obtain all points on the ray one unit above the x-axis with end point (2, 1). The end point (2, 1) is on the graph and is thus indicated by a solid dot. For $x < 2$, $f(x) = -1$, so that we obtain all points on the ray one unit below the x-axis with end point $(2, -1)$. The end point $(2, -1)$ is not on the graph and is thus indicated by an open circle. The graph is shown in Figure 3.7.

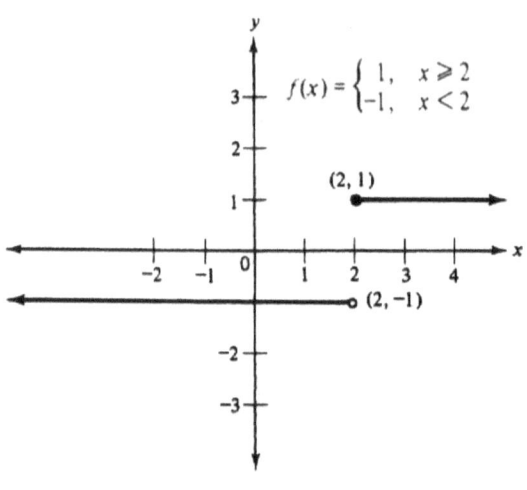

Figure 3.7

EXAMPLE 8

Sketch the graph of $f(x) = \dfrac{1}{(x-1)^2}$.

Let us first note that $f(1)$ is not defined, since division by zero, which is not defined, arises if substitution of 1 for x is attempted. To sketch the graph, we examine the behavior of $f(x)$ for values greater than 1 and for values less than 1. For $x = 2, 5, 11$, and 101, we have $f(2) = 1$, $f(5) = \dfrac{1}{16}$, $f(11) = \dfrac{1}{100}$, and $f(101) = \dfrac{1}{10,000}$. In general, as x takes on larger and larger values, $f(x)$ remains positive but gets closer and closer to 0, as shown in Figure 3.8(a).

If we give x values that are getting closer and closer to 1, but that are greater than 1, such as $1 + \dfrac{1}{10}$, $1 + \dfrac{1}{50}$, and $1 + \dfrac{1}{100}$, we obtain $f(1 + \dfrac{1}{10}) = 100$,

$f(1+\dfrac{1}{50}) = 2500$, and $f(1+\dfrac{1}{100}) = 10,000$, values that are getting larger and larger, as shown in Figure 3.8(b).

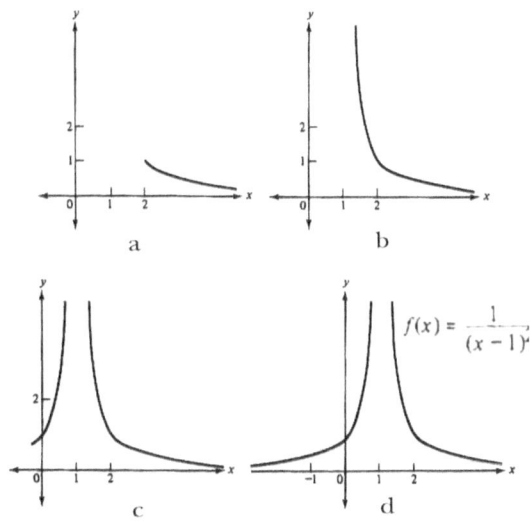

Figure 3.8

The behavior of $f(x)$ to the right of 1 is reproduced on the left of 1. If we give x values that are getting closer and closer to 1, but that are less than 1, such as $1-\dfrac{1}{10}$, $1-\dfrac{1}{50}$, and $1-\dfrac{1}{100}$, we obtain $f(1-\dfrac{1}{10}) = 100$, $f(1-\dfrac{1}{50}) = 2500$, and $f(1-\dfrac{1}{100}) = 10,000$, values that are getting larger and larger, as shown in Figure 3.8(c).

For $x = 0$, -9, and -99, we obtain $f(0) = 1$, $f(-9) = \dfrac{1}{100}$, and $f(-99) = \dfrac{1}{10,000}$, values that remain positive but get closer and closer to 0, as shown in Figure 3.8(d).

EXERCISES

Sketch the graphs of the following functions.

1. $f(x) = 2$
2. $f(x) = 3x + 2$
3. $f(x) = x^2$
4. $f(x) = -x^2 + 2$
5. $f(x) = \dfrac{1}{x^2}$
6. $f(x) = 2x^2 - 1$

7. $f(x) = 2x+1$, for $1 \le x \le 3$ 8. $f(x) = 2x+1$, for $1 < x < 3$

9. $f(x) = 2x+1$, for $x = 1,2,3$ 10. $f(x) = x^2$, for $-1 \le x \le 1$

11. $f(x) = \dfrac{1}{x^2}$, for $-1 < x < 1$ 12. $h(x) = \dfrac{x^2 - x}{x}$

13. $h(x) = \begin{cases} 3, & \text{for } x \ge 1 \\ -1, & \text{for } x < 1 \end{cases}$ 14. $f(x) = \begin{cases} x+1, & \text{for } x \ge 0 \\ -x+1, & \text{for } x < 0 \end{cases}$

15. $f(x) = \begin{cases} 2x+1, & \text{for } x \ge 1 \\ -x+3, & \text{for } x < 1 \end{cases}$ 16. $f(x) = \begin{cases} 1+x, & \text{for } x \ge 1 \\ 1-x, & \text{for } x < 1 \end{cases}$

17. $f(x) = \begin{cases} 3x+2, & \text{for } x \ge 1 \\ 5x, & \text{for } x < 1 \end{cases}$ 18. $f(x) = \begin{cases} 2x-3, & \text{for } x \ge 2 \\ 3x-1, & \text{for } x < 2 \end{cases}$

19. $f(x) = \dfrac{1}{x-1}$ 20. $f(x) = \dfrac{1}{(x-2)^2}$

21. $f(x) = \dfrac{1}{x-3}$ 22. $f(x) = \dfrac{1}{(x-3)^2}$

23. $f(x) = \begin{cases} x^2, & \text{for } x \ne 2 \\ 1, & \text{for } x = 2 \end{cases}$ 24. $f(x) = \begin{cases} 2x-1, & \text{for } x \ne 1 \\ 5, & \text{for } x = 1 \end{cases}$

25. $f(x) = \begin{cases} 2x-1, & \text{for } x < 2 \\ 5, & \text{for } x = 2 \\ -x+5, & \text{for } x > 2 \end{cases}$ 26. $f(x) = \begin{cases} 3, & \text{for } x < 2 \\ 2, & \text{for } x = 2 \\ 3, & \text{for } x > 2 \end{cases}$

3.3   FUNCTIONS FROM APPLICATIONS

Functions that arise from applied settings are more complex than those with no applied background in the sense that their domains of definition are more complex. We must now be sensitive to values that not only make mathematical sense, but sense in terms of the application.

EXAMPLE 1   COST FUNCTIONS

Consider a firm that is producing a single, uniform commodity. The function

$$c = c(x)$$

which describes the total cost c for the production of x units of output, is called the **cost function of the firm.** Thus, for example, suppose that

$$c(x) = \frac{1}{2}x^2 + 20x + 900$$

is the cost function of a coffee producer, where x is output in tons per day and $c(x)$ is the cost in dollars per ton. Then the cost of producing 10 tons per day is

$$c(10) = \frac{1}{2}(10)^2 + 20(10) + 900 = \$1150.$$

Values of $c(x)$ for selected values of x are shown in Table 3.1, and the resulting points are plotted in Figure 3.9(a). Different scales are required for the x and y axes because of the difference in magnitude between the output values x and cost values $c(x)$. While necessary in this case, these different scales have a distortion effect that we should be aware of and sensitive to. The points plotted in Figure 3.9(a) almost seem to lie on a line; in fact, they lie on a parabolic arc.

Table 3.1

x	0	4	8	12	16	20
$c(x)$	900	988	1092	1212	1348	1500

The smallest value of $c(x)$ is 900. Much space would be wasted if we were to indicate the full range of values between 0 and 900 on the y-axis. It is common practice to omit such a range and to indicate this omission by employing the symbol shown on the y-axis in Figure 3.9(a).

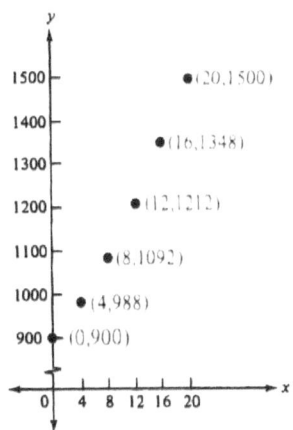

Figure 3.9(a)

A Fundamental Question

What is the domain of definition of $c(x)$? First, let us note that negative values for x make no sense in this situation. Moreover, output, being expressed in tons of coffee produced per day, is divisible up to a point. Output levels of 1/2 of a ton, 1/5 of a ton, 1/10 of a ton may be feasible from a technological point of view, but what about 1/100 of a ton, 1/1000 of a ton, 1/10,000 of a ton, and the like? At some point, sooner or later, the technological level of the production machinery will not be able to accomodate to such levels. What about irrational numbers such as $\sqrt{3}$, $\sqrt{105}$, $\sqrt{1001}$ and the like? It makes no sense to talk about production at output levels that correspond to such values. And then we should keep in mind that we can only increase output up to a certain level, a level which cannot be identified in a sharp, unequivocal way.

What it comes down to is that the domain of definition of a function that arises from a concrete applied setting is much more complex than one that arises from a mathematical setting in which the restriction is that the domain of definition must make mathematical sense (avoid values that lead to division by 0, for example).

What this example illustrates is the basic fact that reality is a more demanding taskmaster than mathematical considerations by themselves. It is a fact that we should not allow ourselves to forget, for we will have occasion to give attention to a number of reality situations.

An Idealized Mathematical Representation of $c(x)$.

To simplify the problem of dealing with the complexities of reality by making them mathematically manageable we take

$$c(x) = \frac{1}{2}x^2 + 20x + 900$$

defined for $x \geq 0$,

as our mathematical representation, called a **mathematical model**, for the real-world $c(x)$ with its more complex domain of definition. The graph of this mathematical model $c(x)$ is presented in Figure 3.9(b).

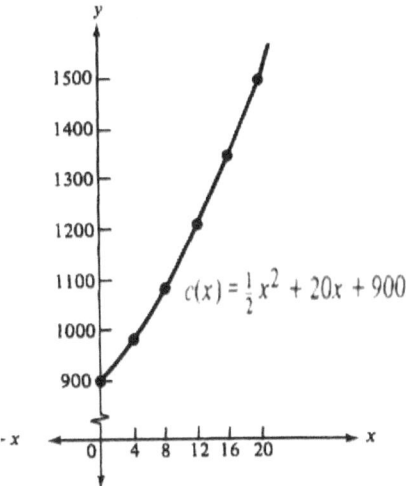

Figure 3.9 (b)

Real World Actualities vs. Mathematical Representations

This interplay between real-world actualities and simplified mathematical representations for them, called **mathematical models,** is characteristic of the applications of mathematics. We shall see this interplay at work in a number of situations throughout this book.

EXAMPLE 2   COST, REVENUE, AND PROFIT FUNCTIONS

The function

$$R = R(x)$$

which describes total revenue as a function of output, is called the **revenue function of the firm.** If

$$R(x) = 306x - 5x^2$$

is the revenue function of the coffee producer, where x is output in tons per day and $R(x)$ is revenue in dollars per ton, the revenue obtained for an output of 10 tons per day is

$$R(10) = 306(10) - 5(10)^2 = 2560 \text{ or } \$2560.$$

The domains of definition of these functions consist of all $x \geq 0$ for which $c(x)$ and $R(x)$ make economic sense.

The profit function $P(x)$ of the firm is the difference between the revenue and cost functions:

$$P(x) = R(x) - c(x)$$

If $c(x) = \frac{1}{2}x^2 + 20x + 900$ and $R(x) = 306x - 5x^2$ are math models of the coffee producer's cost and revenue functions, the math model of coffee producer's profit function is:

$$P(x) = (306x - 5x^2) - (\frac{1}{2}x^2 + 20x + 900)$$

$$= -\frac{1}{2}x^2 + 286x - 900, \quad \text{where} \quad x > 0$$

EXAMPLE 3   THE THOMAS COMPANY

The Thomas fast-food chain has 60 restaurants in New England, each doing an average of $20,000 worth of business per day. Studies conducted on the impact of opening new restaurants in the region indicate that the average amount of business done by each restaurant will drop by $200 per day for each new restaurant opened. These conditions are our starting point.

With respect to these conditions, express the total daily average income of all restaurants in the chain as a function of the number of new restaurants opened.

Let x denote the number of new restaurants to be opened. Then $60 + x$ expresses the total number of restaurants that will be in operation, and $20,000 - 200x$ expresses the average daily income of each restaurant. The total daily average income $I(x)$ is the product of the average income of one restaurant, $20,000 - 200x$, and the number of restaurants, $60 + x$. Thus

$$I(x) = (20,000 - 200x)(60 + x).$$

What is the domain of definition of $I(x)$? First of all, $x \geq 0$ since we will either add a certain number of restaurants to the region or not. Negative values for x make no sense because we are not considering selling off existing restaurants.

Fraction values, such as $\frac{1}{2}$ and $\frac{3}{4}$, clearly make no sense in this setting. And then there is the question of how large x can realistically be. It is clear that at the very least $I(x)$ must exceed zero if $I(x)$ is to be realistic. This leads us to consider the inequality

$$20,000 - 200x > 0,$$

from which we obtain $x < 100$. Thus, at the very least, x must be less than 100.

In summary, we have the non-negative values $0, 1, 2, \ldots$, perhaps extending as far as 99, as comprising the domain of definition of $I(x)$.

We define the math model function for $I(x)$ defined for $0, 1, \ldots, 99$ by $I(x)$ defined for $0 \leq x \leq 99$.

The management of the Thomas Company would like to determine the number of additional restaurants that should be added to the chain to maximize $I(x)$. This problem is addressed in section 8.2 of ch. 8, Example 5.

EXAMPLE 4

The monthly charge for natural gas in a certain region is $2 for any amount up to and including 5 units (1 unit $= 100$ cubic feet), $0.40 per unit for any amount exceeding 5 units up to and including 1000 units, and $0.50 per unit for any amount in excess of 1000 units.

Express cost as a function of the amount used.

Let x denote the number of units used and $c(x)$ the cost of x units. For x between 0 and 5 inclusive $(0 \leq x \leq 5)$, $c(x) = 2.00$; for x between 5 and 1000, including 1000 $(5 < x \leq 1000)$, $c(x) = (0.40)x$; for x greater than 1000 $(x > 1000)$, $c(x) = 400$ (for the first 1000 units) plus $0.50(x - 1000)$ (for the amount by which x exceeds 1000 units). By multiplying and collecting terms, we obtain:

$$400 + 0.50(x - 1000) = (0.50)x - 100$$

Thus for the math model function for cost, we have:

$$c(x) = \begin{cases} 2.00, & \text{for } 0 \leq x \leq 5 \\ (0.40)x, & \text{for } 5 < x \leq 1000 \\ (0.50)x - 100, & \text{for } x > 1000 \end{cases}$$

Why math model function? For what real-number values does cost not make sense?

This example illustrates a function whose rule cannot be stated in terms of one algebraic expression. Three algebraic expressions are needed, with the expression used depending on where in the domain of definition the number in question falls. For example, $c(0) = 2.00$; if no gas is used, you are still charged $2. $c(500) = (0.40)500 = 200$; if 500 units are used, you are charged $200.

It is a common error to equate function with algebraic expression. Although three algebraic expressions appear, we have one cost function, not three cost functions, that expresses cost as a function of the amount of gas used.

EXAMPLE 5

Each bacterium in a culture consisting of three *E. coli* organisms divides at the end of each second. Determine the function that describes the number of bacteria in the culture at the end of t seconds. Find the number of bacteria in the culture at the end of 5 seconds.

At time $t = 0$, the starting point of our study, the culture contains 3 bacteria. At time $t = 1$, each bacterium gives rise to 1 additional bacterium, so that the number of bacteria at time $t = 1$ is

$$3 + 3 = 3(2).$$

At time $t = 2$ each bacterium in the culture gives rise to 1 additional member, so that the number of bacteria in the culture at time $t = 2$ is the number present at time $t = 1$, 3(2), plus the size of the increase, 3(2). Thus at time $t = 2$ there are

$$3(2) + 3(2) = 3(2)[2] = 3(2)^2$$

organisms. At time $t = 3$ each bacterium in the culture gives rise to 1 additional member, so that the number of bacteria in the culture at time $t = 3$ is the number present at time $t = 2$, $3(2)^2$, plus the size of the increase, $3(2)^2$. Thus at time $t = 3$ there are

$$3(2)^2 + 3(2)^2 = 3(2)^2[2] = 3(2)^3$$

bacteria.

More generally, this analysis leads to the function

$$y = 3(2)^t$$

where y is the number of bacteria in the culture at the end of t seconds. The domain of definition consists of values ($t = 0$, 1, 2, etc.) up to a certain point.

> As t becomes larger and larger the accuracy of this growth function dimishes, since it does not reflect such factors as the limited ability of the environment to support life and loss through death.

The predicted number of bacteria in the culture at the end of 5 seconds is

$$y = 3(2)^5 = 96.$$

EXERCISES

In Exercises 1-8 state the domain of definition of the function determined for the stated problem and its corresponding mathematical model.

1. The monthly charge for telephone service in a certain region is $8 for any number of message units up to and including 50 and $0.09 for each additional message unit beyond 50. Express cost as a function of the number of message units accumulated.

2. $500 is invested at 8 percent per annum compounded semiannually. Express the compound amount on deposit as a function of time in years over the first 4 years of investment.

3. A taxi fleet uses 100 taxis to service Johnson City. Each taxi brings in an average of $200 per day in fares. If additional taxis are added to the fleet, it is estimated that the amount in fares brought in by each taxi will drop by $10 per day for each additional taxi added. Express the total daily average income of the taxi fleet as a function of the number of taxis added to the fleet. State the corresponding math model function.

4. An author signed a contract with a publishing company that stipulated a royalty rate of 15 percent of the list price of the book (estimated at $10 per copy) for the first 5000 copies sold, and 20 percent of list price for each copy over 5000 that is sold. Express royalty income as a function of the number of books sold.

5. A new government agency, formed to study the growth of bureaucracy in government, is beginning its operations with a staff of two persons. At the end of each month each person hires two assistants. By using an approach analogous to the one employed in Example 5, find a function that describes the number of staff in this agency at the end of t months. At this rate, what would the size of the staff be 1 year later?

6. The income tax paid in a certain region is $0.00 if taxable income is less than or equal to $5000, $20 plus 1 percent of the excess over $5000 if taxable income is over $5000 but not over $25,000, and $220 plus 3 percent of the excess over $25,000 if taxable income is over $25,000. Express tax to be paid as a function of taxable income.

7. A first-class letter whose weight does not exceed 3 ounces is to be sent from one point in the United States to another. Express postage cost as a function of weight.

8. Some property loses value over time. This decrease in value is called **depreciation.** Thus, if an item is purchased for $1000 and 1 year later its resale value is $600, its depreciation after 1 year is $400. One of several methods used for depreciating property is linear depreciation, which assumes that the loss in value over a specified time is a fixed percentage of the original value.
    Suppose that the cost of a machine is $2000 and that it depreciates linearly at the rate of 10 percent per year. Express the value of the machine as a function of time. Find the value of the machine after 3 years. More generally, suppose that an item costs C dollars and depreciates linearly at the rate r per annum. Express the value of the item as a function of time.

9. What is the domain of definition of the cost, revenue, and profit functions discussed in Example 2?

3.4 EXPONENTIAL FUNCTIONS

The functions

$$y = 2^x, \quad y = 10^x, \quad y = (\tfrac{1}{2})^x, \quad \text{and} \quad y = e^x$$

where $e \approx 2.71828$, illustrate functions with the general structure

$$y = b^x$$

where b is a positive constant other than 1. Functions of this type, with a fixed base and variable exponent, are called **exponential functions.** To obtain a geometric view of exponential functions, consider $y = 2^x$ and $y = (\frac{1}{2})^x$. Values of y for values of x for these functions are given in Tables 3.2 and 3.3, from which we obtain the graphs shown in Figures 3.10 and 3.11.

Table 3.2

x	-4	-3	-2	-1	0	1	2	3	4
2^x	$\frac{1}{16}$	$\frac{1}{8}$	$\frac{1}{4}$	$\frac{1}{2}$	1	2	4	8	16

Table 3.3

x	-4	-3	-2	-1	0	1	2	3	4
$(\frac{1}{2})^x$	16	8	4	2	1	$\frac{1}{2}$	$\frac{1}{4}$	$\frac{1}{8}$	$\frac{1}{16}$

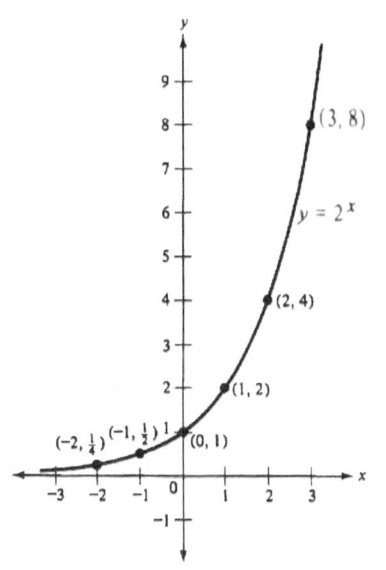

Figure 3.10   Figure 3.11

More generally, the graph of $y = b^x$, where $b > 1$, has the form shown in Figure 3.10, and the graph of $y = b^x$, where b is between 0 and 1 $(0 < b < 1)$, has the form shown in Figure 3.11.

Exponential functions and functions of exponential type (built up from exponential functions), appear in a number of situations in the worlds of biology, business, economics, finance, physics, and psychology. The psychologist C. L. Hull[*] was lead to the function

$$h(x) = 100(1 - e^{-ax})$$

where a is a positive constant, to describe habit strength in terms of repetitions. The graph of $h(x)$ is shown in Figure 3.12. In his now classic work on population dynamics, G. F. Gause[†] obtained the function

$$y = a(1 - e^{-kx})$$

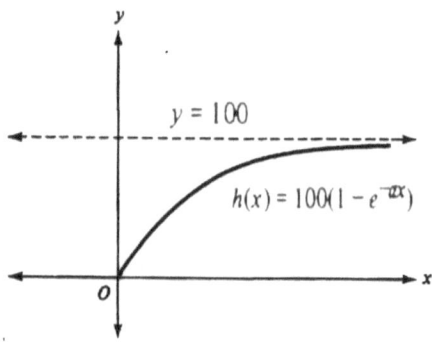

Figure 3.12

Figure 3.12

where a and k are positive constants, to describe the relative increase of predators as a function of the concentration of prey.

[*] C. L. Hull, *Principles of Behavior* (New York: Appleton-Century-Crofts, 1945).

[†] G. F. Gause, *The Struggle for Existence* (Baltimore, Md.: Williams & Wilkins Co., 1934; reprinted by Dover Publications, Inc., New York, 1971), p. 55.

EXERCISES

1. Sketch the graph of $f(x) = 3^x$.

2. Sketch the graph of $f(x) = (\frac{1}{2})^x$.

3.5   LOGARITHMIC FUNCTIONS

Logarithms were originally developed in the early seventeenth century in response to a need for simplifying extremely tedious computations that arose in astronomy. Although sophisticated calculators and computers have now made logarithms obsolete as a calculation tool, logarithmic functions arise in higher mathematics and its applications—calculus, in particular—and familiarity with the logarithm concept is essential.

> If b is a positive number, $b \neq 1$, and $b^y = x$, then the exponent y is called the **logarithm of x to the base b**. In symbols we write
>
> $$y = \log_b x$$
>
> Thus the logarithm of x to the base b is the exponent y to which b must be raised to obtain x.

For example:

$$\log_{10} 100 = 2, \quad \text{since } 10^2 = 100$$
$$\log_3 81 = 4, \quad \text{since } 3^4 = 81$$
$$\log_{10} 1 = 0, \quad \text{since } 10^0 = 1$$
$$\log_4 4 = 1, \quad \text{since } 4^1 = 4$$

EXERCISES

Express the following exponential statements in logarithmic notation.

1. $6^2 = 36$

2. $10^3 = 1000$

3. $25^{1/2} = 5$

4. $2^3 = 8$

5. $27^{1/3} = 3$

6. $10^{-2} = \dfrac{1}{100}$

7. $5^{-3} = \dfrac{1}{125}$
8. $32^{-1/5} = \dfrac{1}{2}$
9. $9^{3/2} = 27$

10. $5^0 = 1$
11. $81^{-3/4} = \dfrac{1}{27}$
12. $10^{-3} = \dfrac{1}{1000}$

Express the following logarithmic statements in exponential form.

13. $\log_2 16 = 4$
14. $\log_3 1 = 0$
15. $\log_5 125 = 3$

16. $\log_{12} 144 = 2$
17. $\log_{10} 0.1 = -1$
18. $\log_8 4 = \dfrac{2}{3}$

19. $\log_7 343 = 3$
20. $\log_{10} 10,000 = 4$
21. $\log_{10} 0.01 = -2$

Find the value of each of the following.

22. $\log_2 8$
23. $\log_{10} 1000$
24. $\log_{10} 0.001$

25. $\log_3 3$
26. $\log_9 27$
27. $\log_8 \dfrac{1}{2}$

28. $\log_4 64$
29. $\log_b b^2$
30. $\log_b b^{-1}$

Logarithms with respect to base e, whose value correct to five decimal places is 2.71828, are called **natural logarithms.**[*] They are of particular importance in mathematics and its applications. $\log_e x$ is denoted by **ln x.**

Properties of Logarithms

Logarithms have some interesting and important properties that are noteworthy. To introduce these properties let M and N denote positive numbers, and b a positive number that will serve as a base. Then M and N can be expressed as follows for suitable x and y.

$$M = b^x, \qquad\qquad N = b^y$$

Thus:

$$MN = b^{x+y}$$

---

[*] To define e we need the concept of limit of a function, which is taken up in the study of calculus. e is defined by $\lim\limits_{n \to \infty} (1 + \dfrac{1}{n})^n$. e is the value $(1 + \dfrac{1}{n})^n$ approaches as n gets larger and larger without bound.

From these relations and the definition of logarithm, we have:

$$log_b M = x$$
$$log_b N = y$$
$$log_b MN = x + y$$

Since $x + y = log_b M + log_b N$, we have the following fundamental property of logarithms:

1. $$log_b MN = log_b M + log_b N$$

That is, the logarithm of a product is the sum of the logarithms of the component factors. This principle can be extended to include the product of any number of factors.

Since

$$\frac{M}{N} = \frac{b^x}{b^y} = b^{x-y},$$

in logarithmic terms we have

$$log_b \left(\frac{M}{N} \right) = x - y$$
$$= log_b M - log_b N$$

Thus we have the following property:

2. $$log_b \left(\frac{M}{N} \right) = log_b M - log_b N$$

That is, the logarithm of a quotient is the logarithm of the numerator minus the logarithm of the denominator.

Starting with $M = b^x$ and raising both terms to the power r yields:

$$M^r = b^{xr}$$

In logarithmic terms this says

$$\log_b M^r = xr$$
$$= rx$$
$$= r(\log_b M)$$

since $x = \log_b M$. Thus we have the following property:

3. $$\log_b M^r = r(\log_b M)$$

That is, the logarithm of a number raised to a power equals the power times the logarithm of the number.

Another interesting property of logarithms is the following:

4. $$\log_b M = \frac{1}{\log_M b}$$

To illustrate these properties of logarithms, let us take as given that $\log_{10} 3 = .4771$ and $\log_{10} 4 = .6021$, approximately. By property 1,

$$\log_{10} 12 = \log_{10}(4 \cdot 3) = \log_{10} 4 + \log_{10} 3$$
$$= .6021 + .4771$$
$$= 1.0792, \text{ approximately}$$

By property 2:

$$\log_{10}(\frac{12}{3}) = \log_{10} 12 - \log_{10} 3$$
$$= 1.0792 - .4771$$
$$= .6021, \text{ approximately}$$

By property 3:

$$\log_{10} \sqrt{4} = \log_{10} 4^{1/2} = \frac{1}{2}(\log_{10} 4)$$
$$= \frac{1}{2}(.6021)$$
$$= .3011, \text{ approximately}$$

By property 4:

$$\log_4 10 = \frac{1}{\log_{10} 4} = \frac{1}{.6021} = 1.6609, \text{ approximately}$$

EXERCISES

Given that $\log_{10} 2 = .3010$, $\log_{10} 3 = .4771$, $\log_{10} 5 = .6990$, *and* $\log_{10} 7 = .8451$, *use properties 1 through 4 to find the following values.*

31. $\log_{10} 15$ 32. $\log_{10} 35$ 33. $\log_{10}(\frac{7}{5})$

34. $\log_{10} 30$ 35. $\log_{10}(\frac{3}{5})$ 36. $\log_{10} 0.3$

37. $\log_{10} 28$ 38. $\log_3 10$ 39. $\log_{10} \sqrt{5}$

40. $\log_{10} 7^4$ 41. $\log_{10} \sqrt[3]{5}$ 42. $\log_{10} 0.7$

43. $\log_5 10$ 44. $\log_{10} \sqrt[3]{35}$ 45. $\log_{10}\left(\dfrac{\sqrt{5}}{2}\right)$

46. $\log_{10} \sqrt[3]{35}$ 47. $\log_{10}\left(\dfrac{10}{\sqrt{3}}\right)$ 48. $\log_{10} \sqrt[4]{15}$

49. $\log_{10} 125\sqrt{2}$ 50. $\log_{10}\left(\dfrac{7}{5}\right)$ 51. $\dfrac{\log_{10} 7}{3}$

Logarithmic Functions

A function of the form

$$y = \log_b x$$

where base b is a positive constant and $b \neq 1$, is called a **logarithmic function.** Logarithmic functions are defined only for positive real numbers. To obtain a geometric view of logarithmic functions, consider $y = \log_2 x$ and $y = \log_{1/2} x$. Values of y for selected values of x for these functions are given in Tables 3.4 and 3.5, from which we obtain the graphs shown in Figures 3.13 and 3.14.

Table 3.4

x	$2^{-2} = \dfrac{1}{4}$	$2^{-1} = \dfrac{1}{2}$	$2^0 = 1$	$2^1 = 2$	$2^2 = 4$	$2^3 = 8$
$\log_2 x$	-2	-1	0	1	2	3

Table 3.5

x	$\left(\dfrac{1}{2}\right)^3 = \dfrac{1}{8}$	$\left(\dfrac{1}{2}\right)^2 = \dfrac{1}{4}$	$\left(\dfrac{1}{2}\right)^1 = \dfrac{1}{2}$	$\left(\dfrac{1}{2}\right)^0 = \dfrac{1}{1}$	$\left(\dfrac{1}{2}\right)^{-1} = 2$	$\left(\dfrac{1}{2}\right)^{-2} = 4$	$\left(\dfrac{1}{2}\right)^{-3} = 8$
$\log_{1/2} x$	3	2	1	0	-1	-2	-3

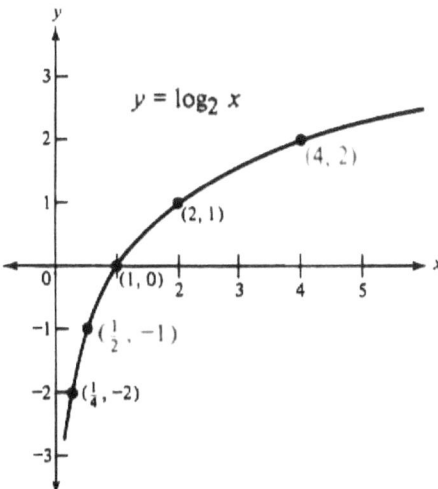

Figure 3.13

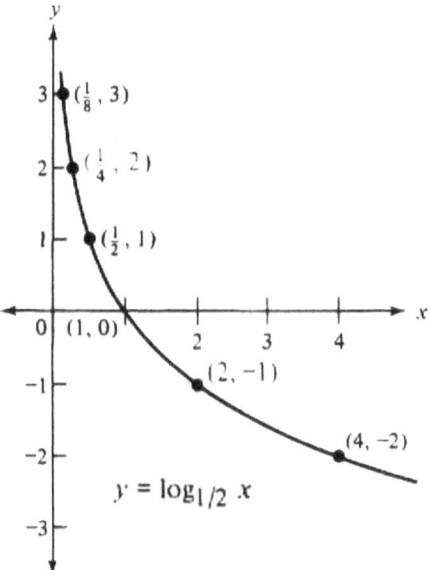

Figure 3.14

More generally, the graph of $y = \log_b x$, where $b > 1$ (shown in Figure 3.15) has the same form as $y = \log_2 x$, and the graph of $y = \log_b x$, where $0 < b < 1$ (shown in Figure 3.16), has the same form as $y = \log_{1/2} x$.

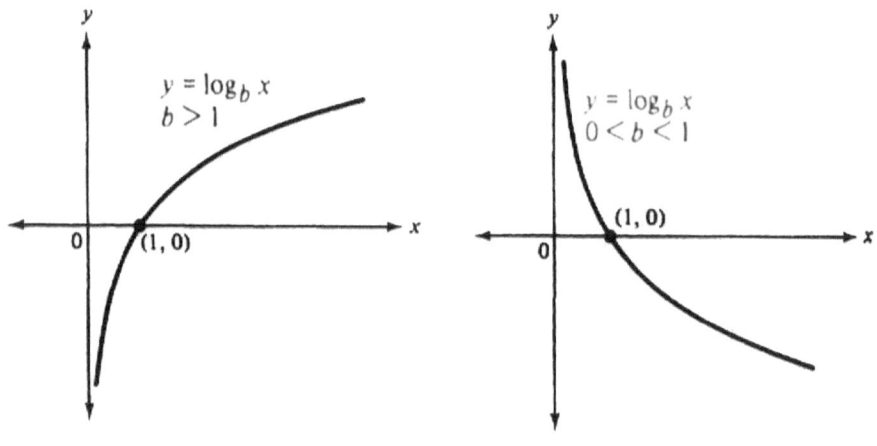

Figure 3.15   **Figure 3.16**

An interesting situation that gives rise to a logarithmic-type function involves potassium-argon dating. The potassium-argon dating technique is based on the fact that potassium-40, in the course of undergoing radioactive decay, is transformed into argon-40. This technique can be used to date potassium-bearing rocks and minerals that were as recently formed as 10,000 years ago and as old as 4 billion years. The potassium-argon age function is

$$t = (1.885 \cdot 10^9) \ln \ (9.068x + 1)$$

where t is age in years and x is the ratio of the amount of argon-40 to the amount of potassium-40 in the rock today.

A major archeological event occurred in 1959 when Dr. and Mrs. Louis Leakey discovered a fossil hominid skull, which they termed *Zinjanthropus boisei*, at Olduvai Gorge in Tanzania. Potassium-argon dating of the volcanic material associated with the hominid remains of Olduvai Gorge played a key role in establishing that the *Zinjanthropus* remains were about 1.75 million years old.

EXERCISES

52. Sketch the graph of $y = \log_3 x$.
53. Sketch the graph of $y = \log_{1/3} x$.

3.6   A MARKET EQUILIBRIUM QUESTION

Of interest in the world of economics is the function that expresses the demand D for a commodity (that is, the amount of the commodity absorbed by the market per unit of time) as a function of the unit price, x, of the commodity. The function

$$D = D(x)$$

which expresses demand D as a function of price x, is called a **demand function.**

The domain of definition of $D = D(x)$ consists of all values of x for which the function makes economic sense. These values generally occur in an interval, and, for **mathematical convenience**, the set of all values in some suitable interval is adopted as the domain.

In many situations, $D = D(x)$ is a linear function. To illustrate, let us suppose that

$$D = -2x + 150$$

expresses the demand D for coal in thousands of tons per month in a certain region as a function of x, the price of coal in dollars per ton. What is the domain of this function? Negative values of x do not make sense, and x such that D is zero or negative can be discarded as well. To determine x for which D is zero, solve $D = -2x + 150 = 0$ for x.

$$-2x + 150 = 0$$
$$-2x = -150$$
$$x = 75$$

Thus we take the interval $(0, 75)$ as the domain of definition of the **math model function** corresponding to our demand function. The graph of $D = -2x + 150$ is shown in Figure 3.17.

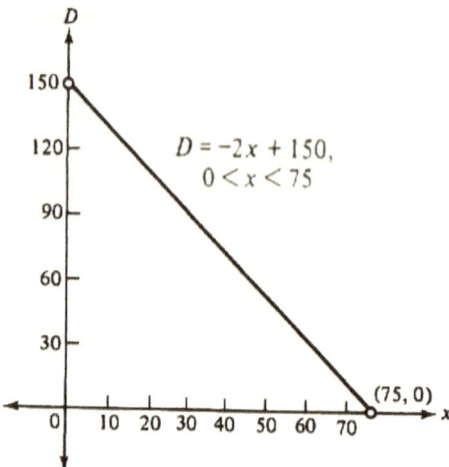

Figure 3.17

Another relationship of interest in economics is the function that expresses the supply S of a commodity (the amount of the commodity that producers make available to the market per unit time) as a function of the unit price, x, of the commodity. The function

$$S = S(x)$$

which expresses supply S as a function of price x, is called a **supply function.** In many situations, $S = S(x)$ is a linear function. For example, let us assume that

$$S = 3x$$

expresses the supply of coal in thousands of tons per month as a function of x, the price of coal in dollars per ton. We take all nonnegative values in a suitable interval as the domain of definition of the **math model function** corresponding to our supply function. The graph of $S = 3x$ is shown in Figure 3.18.

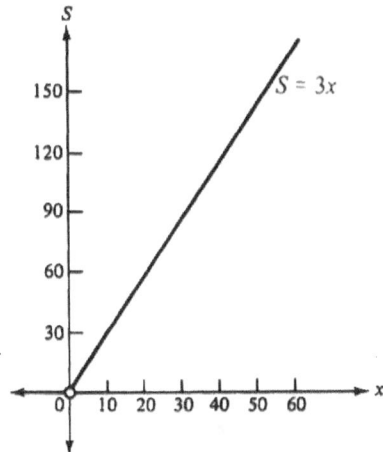

Figure 3.18

If both $D = D(x)$ and $S = S(x)$ pertain to the same market, and x, D, and S are expressed in the same units, then **market equilibrium** corresponds to that point at which the demand and supply curves intersect. The corresponding price and quantity are called the **equilibrium price** and **equilibrium quantity**, respectively. Under certain conditions that define what economists call **pure competition**, there is a tendency for the price to adjust itself until market equilibrium is attained. The equilibrium price is determined by setting $D(x)$ equal to $S(x)$ and solving for x. From our demand and supply functions for coal, we obtain:

$$3x = -2x + 150$$
$$5x = 150$$
$$x = 30$$

For $x = 30$, $D(30) = S(3) = 90$. Thus the equilibrium price is \$30 per ton, the equilibrium quantity is 90 thousand tons of coal per month, and the market equilibrium point is $E(30, 90)$.

The situation can be seen in geometric terms by graphing the demand and supply functions on the same coordinate system (see Figure 3.19).

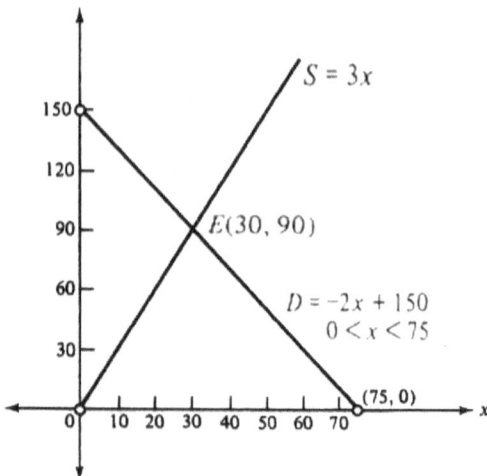

Figure 3.19

Math vs. Reality

This is an issue that comes up whenever an application of mathematics arises. We have the market equilibrium point $E(30, 90)$, mathematically speaking, but its real world accuracy, we should keep in mind, hinges on the real world accuracy of the demand and supply functions $D = -2x + 150$ and $S = 3x$.

EXERCISES

1. The demand and supply functions for beef in Astin City are $D = -5x + 50$ and $S = 15x$, where x is the price in dollars per pound, D is demand in thousands of pounds per week, and S is supply in thousands of pounds per week.

   (a) Sketch the graphs of the math model functions of the demand and supply functions on the same coordinate system.

   (b) Determine the market equilibrium point.

   (c) Does this analysis establish the real world accuracy of the market equilibrium point? Explain

2. The demand and supply functions for orange juice in the summer in Bell City are $D = -4x + 400$ and $S = 4x$, where x is the price in cents per gallon, D is demand in thousands of gallons per day, and S is supply in thousands of gallons per day.

   (a) Sketch the graphs of the math model functions of the demand and supply functions on the same coordinate system.

   (b) Determine the market equilibrium point.

   (c) Does this analysis establish the real world accuracy of the market equilibrium point? Explain.

3. Let us suppose that the cost and revenue functions of a chocolate producer are given by $C = 1000 + 100x$ and $R = 150x$, respectively, where x is output in tons per week, and C and R are in dollars per ton. **Break-even analysis** is concerned with determining the output for which the firm breaks even, that is, the output for which cost equals revenue.

   (a) Sketch the graphs of the math model functions of the cost and revenue functions on the same coordinate system.

   (b) Determine the break-even point.

   (c) Does this analysis establish the real world accuracy of the break-even point? Explain.

3.7   MULTIVARIABLE FUNCTIONS

The idea of multivariable functions is a natural extension of the concept of function of one independent variable. For example, u is said to be a function of **two independent variables,** x and y, if for each ordered pair of numbers (x, y) there is determined, by means of some rule, exactly one value of u. If the rule of the function is denoted by f, then the value assigned to (x, y) is denoted by $f(x, y)$. The **domain of definition,** or **domain,** of the function consists of all ordered pairs of numbers (x, y) to which a value is assigned by the rule. Here, too, when the domain is not explicitly stated, it is understood to be the collection of all ordered pairs of numbers for which the rule makes sense.

The function defined by

$$f(x,y) = 2xy - x^2 \text{ or } u = 2xy - x^2$$

for example, is defined for all ordered pairs of numbers. For $(-1, 2)$, we have

$$f(-1, 2) = 2(-1)2 - (-1)^2 = -5$$

while to $(2, -1)$ there is assigned the value

$$f(2, -1) = 2(2)(-1) - (2)^2 = -8.$$

Taking this a step further, u is said to be a function of **three independent variables,** x, y, and z, let us say, if for each ordered triple of numbers (x, y, z) there is determined, by means of some rule, exactly one value of u. The **domain** of such a function is the collection of all ordered triples of numbers (x, y, z) to which a value is assigned by the rule.

If, for example,

$$f(x, y, z) = 3x + yz^2 \text{ or } u = 3x + yz^2$$

then:

$$f(1, 2, -1) = 3(1) + 2(-1)^2 = 5$$
$$f(-2, 4, -1) = 3(-2) + 4(-1)^2 = -2$$

The concept of **function of n independent variables** is defined in an analogous way. Functions of more than one independent variable are called **multivariable functions.**

EXAMPLE 1

Consider the function defined by $f(x, y) = x^2 y - y^2$, where $x^2 + y^2 < 25$.

The inequality $x^2 + y^2 < 25$ describes the domain of definition. It tells us that the rule of the function can only be applied to ordered pairs of numbers (x, y) that satisfy $x^2 + y^2 < 25$.

Geometrically, this domain of definition consists of all points interior to the circle defined by $x^2 + y^2 = 25$ (see Figure 3.20).

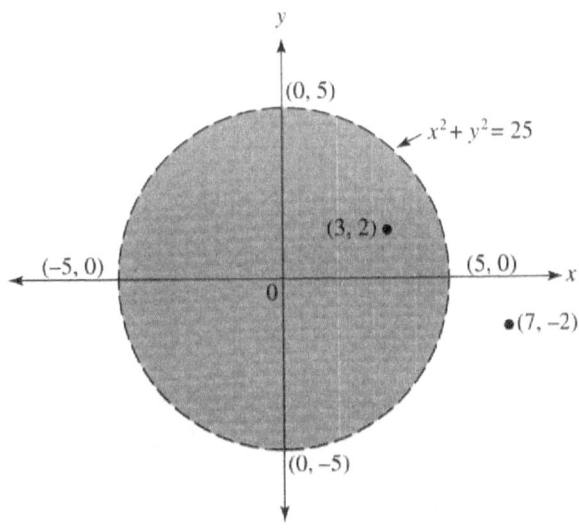

Figure 3.20

Thus, for example, $f(7,-2)$ is undefined, since $(7,-2)$ is not in the domain of $f(x,y)$. $(7^2 + (-2)^2 = 54 > 25)$. The point $(3, 2)$, on the other hand, is in the domain of definition of $f(x,y)$. $(3^2 + 2^2 = 13 < 25)$, and $f(3,2) = 3^2(2) - (2)^2 = 14$.

EXAMPLE 2

Consider the function defined by $f(x,y) = 10x + \dfrac{1}{2}y$, where (1) $x \geq 0$, (2) $y \geq 0$, and (3) $x + 2y \leq 10$.

The stated inequalities describe the domain of definition. The function can only be applied to ordered pairs of numbers that satisfy all three conditions. The point $(2, 5)$, for example, satisfies conditions 1 and 2, but fails to satisfy condition 3 $[2 + 2(5) = 12 > 10]$. Thus $f(2, 5)$ is not defined. Since $(2, 4)$ satisfies conditions 1 through 3, $f(2, 4)$ is defined and is given by $f(2,4) = 10(2) + \dfrac{1}{2}(4) = 22$.

Any function with the structure

$$f(x,y) = Ax + By + C$$

where A, B, and C are constants, is said to be a **linear function in two variables.**
Thus $f(x, y) = 10x + \dfrac{1}{2} y$ is a linear function, with $A = 10$, $B = \dfrac{1}{2}$, and $C = 0$.

Linear functions in any number of variables are defined analogously. Linear functions are of particular importance in mathematics and its applications, and in Chapter 6 they will be seen to provide the structural foundation for linear programming.

EXERCISES

1.   For $f(x, y) = 3x^2 - y^2x$, determine $f(7, 3)$, $f(-1, 5)$, and $f(2, 6)$.
2.   For $f(x, y) = (2x + y)/ y^2$, where $x^2 + y^2 \le 16$, determine $f(3,2)$, $f(-3,1)$, $f(4, 0)$, and $f(-4,1)$.
3.   For $f(x, y, z) = 3xy - z^2$, where $x^2 + y^2 + z^2 \le 36$, determine $f(1, -1, 3)$, $f(4, 2, 1)$, $f(-5, 4, 2)$, and $f(-2, 4, 3)$.

SELF-TESTS FOR CHAPTERS 1-3

Allow 75 or so minutes for each self-test. Go over each one before proceeding to the next.

Self-Test 1

1.   Find the sum: $(-2)3 + 4(-2) + -(-3)$.

2.   Simplify by combining terms: $5xyz - 2xyz + 3xyz$.

3.   Remove parentheses and combine terms: $-(x + 3xy + 2) - x + 4xy + 5$.

4.   Factor $3xz + 4xy + x$.

5.   Factor $4xy^2 - 2y^2 + 6zy^2$.

6.   Does -1 have a multiplicative inverse? Explain.

7.   Remove parentheses and combine terms: $-2(4a - 2b + c) - (4a - b - 2c)$.

8.   Simplify, if possible, by canceling common factors.

   (a) $\dfrac{4x + 2}{x + 2}$ (b) $\dfrac{3 - x}{x - 3}$ (c) $\dfrac{2xy - x}{2y - 1}$

9. Express the product in simplest terms.

   (a) $\dfrac{x-2}{4} \cdot \dfrac{2x}{x-2}$

   (b) $\dfrac{1}{x-2} \cdot \dfrac{2-x}{x}$

10. Is $0 \div 4$ defined? Explain.

11. Determined: (a) $(\sqrt[4]{3})^4$ (b) $25^{3/2}$

12. The cost and revenue functions of a tea producer are $c(x) = x^2 + 10x + 100$ and $R(x) = 500x - 4x^2$, where x is output in tons per day and cost and revenue are in dollars per day.

   (a) Determine the cost and revenue for outputs of 20 and 49 tons per day.

   (b) Determine the profit function of the tea producer.

   (c) What are the domains of definition of the afore functions?

13. What will $3000 grow to in 8 years if invested at 10 percent per annum compounded semiannually?

14. Solve $2(3-x) \le 12 + x$. Sketch a graph of its solutions.

15. Sketch the graph of $5x - 2y \le 10$.

Self-Test 2

1. Factor completely: $3rst + 6rs + 12rsk$.

2. Does 0 have a multiplicative inverse? Answer YES or NO and explain.

3. Barbara Allen borrows $500 for one year at a simple discount rate of 10% per annum. What interest rate is actually involved?

4. For

$$f(x) = \begin{cases} 4x - 1, & \text{for } x \le -2 \\ 3x + 2, & \text{for } x > -2 \end{cases}$$

find $f(-3)$, $f(-1)$, $f(-2), f(1)$, and $f(3)$.

5. Are the functions $f(x)-(x^2-x)/(x-1)$ and $g(x)=x$ equal? Explain.

6. What is the domain of definition of $f(x)=\sqrt{x+1}$? Explain.

7. A limousine service charges \$1 for any distance up to and including the first mile and \$0.20 for each additional quarter-mile. Express cost as a function of distance traveled not exceeding 2 miles. What are the domains of definition of the function and its corresponding math model function?

8. Divide and express the quotient in simplest terms.

    (a) $\dfrac{y}{x-y}\div\dfrac{1}{xy-y^2}$ (b) $\dfrac{2}{a+3}\div\dfrac{a}{ab+3b}$

9. Add or subtract and simplify.

    (a) $\dfrac{2}{x}+\dfrac{3}{x+1}$ (b) $\dfrac{3x-1}{x}-\dfrac{1-4x}{4x}$

10. Evaluate each of the following.

    (a) $2(3^{-2})$ (b) $(-2)^{-4}$ (c) $\dfrac{(-2)^{-2}}{2(3^{-3})}$

11. Simplify by using properties of exponents.

    (a) $(x^{-1})^{-3}$ (b) $\dfrac{2x^2y}{x^{-2}y^2}$ (c) $\dfrac{(x^2y^3)^2}{xy}$

12. Determine the value of each of the following.

    (a) $(\sqrt{81})^{-2}$ (b) $\sqrt[3]{-216}$ (c) $\dfrac{1}{(\sqrt[3]{-8})^2}$

    (d) $(125)^{2/3}$ (e) $64^{-1/3}$ (f) $2^{1/2}\cdot2^{-5/2}$

13. Solve $5(x-2)=2x-1$.

14. \$2000 is borrowed for 4 months at simple interest of 9 percent per annum. Find the amount.

Self-Test 3

1. Simplify $\sqrt{4x^2 + 16y^2}$ to whatever extent possible.

2. Evaluate: $27^{-1/3}$

3. Define 1.

4. Subtract and simplify: $\dfrac{1}{x-3} - \dfrac{x}{x+2}$

5. Are there any values of x for which the fraction stated in question 4 is not defined? Explain.

6. Is $-1/\pi$ the negative of ϖ? Answer YES or NO and explain.

7. What does it mean to say that the effective rate corresponding to 12% per annum compounded 6 times a year is 12.6%?

8. The Marsden Company will need $5000 in 4 years to replace its computer. How much should be initially invested at 15% per annum compounded 3 times a year to meet the expense?

9. A chain of dairy-product outlets has 30 stores in a certain region, each doing an average of $10,000 worth of business per day. Studies indicate that if new stores are opened in the region, the average amount of business done by each store will drop by $200 per day for each new store that is opened.

    Express the total daily amount of business done by the dairy outlets per day as a function of the number of new stores opened. What is the domain of definition? What is the domain of definition of this function's corresponding math model function?

10. Sketch the graph of $f(x) = x^2 + 2$, where $-2 < x < 1$.

11. Sketch the graph of:

$$f(x) = \begin{cases} -x+3, & \text{for } x < 1 \\ 3, & \text{for } x = 1 \\ 2x, & \text{for } x > 1 \end{cases}$$

12. Sketch the graph of $f(x) = 1/(x+1)$.

13. For $f(x, y, z) = (x^2 + y)/z$, where $x^2 + y^2 \leq 16$, find $f(1, 2, -5)$, $f(-4, 1, 3), f(3, 1, 6)$, and $f(3, 2, -3)$.

Self-Test 4

1. Determine: (a) $\dfrac{-3}{\sqrt[3]{125}}$ (b) $\dfrac{(-3)^{-2}}{4^{-3}}$

2. (a) Would it be correct to say that $f(x) = \dfrac{x-3}{(x-2)(2x+1)}$ is not defined for $x = 3$? Explain.

   (b) Are there any values of x for which $f(x)$ is not defined? Explain.

3. Solve for z: $2m^2 z + xm - 3 = 4$.

4. Arnold Forman borrows $2000 for 1 year at a simple discount rate of 15% per annum. How much does he receive?

5. Sketch the graph of $2x - y \leq 4$.

6. (a) Determine the effective rate corresponding to 15% per annum compounded 3 times a year.

   (b) What is the meaning of the value determined in part (a)?

7. Esther Thomas has a car loan of $5000 which is to be paid in semi-annual installments of $2500 plus simple interest at the rate of 10% per annum on the principal outstanding during each payment period. What is her total interest.

8. Simplify $\sqrt{9m^2 + 81n^2}$ as completely as possible.

9. To determine a student's course grade Professor William Frost counts the average of three class tests T_1, T_2 and T_3 60% and the final exam grade T_4 40%. State the course grade y as a function of T_1, T_2, T_3 and T_4.

10. $f(x) = \dfrac{x-2}{2-x}$ and $g(x) = -1$. Does $f(x) = g(x)$ for all values of x? Explain.

11. Sketch the graph of $f(x) = 2x - 1$, where $1 \le x < 4$.

12. Evaluate: (a) $(4^{2/3})^6$ (b) $\dfrac{5(2^{-2})}{4(-3)^{-3}}$

13. Express x in terms of p and q: $\dfrac{3}{x} + 2pq = 12$.

14. The income tax to be paid in Bell City is $0.00 if taxable income is less than or equal to $15,000 and 8% of the excess over $15,000 if taxable income is over $15,000. Express tax to be paid as a function of taxable income.

15. The demand and supply functions for sugar in Bell City are $D = -5x + 200$ and $S = 5x$, respectively, where x is price in cents per pound, D is demand in thousands of pounds per week, and S is supply in thousands of pounds per week.

   (a) Sketch the graphs of the math model functions of the demand and supply functions on the same coordinate system.

   (b) Determine the market equilibrium point.

   (c) Does this analysis establish the real world accuracy of the market equilibrium point? Explain.

CHAPTER 4

Linear Structures

4.1 LINEAR EQUATIONS AND LINES

In section 2.2 we noted that the graph of every linear equation in two variables is a line, and that every line is the graph of a linear equation in two variables. The problem we turn to now is that of determining the equation of a given line. Let L denote a **nonvertical line** (see Figure 4.1).

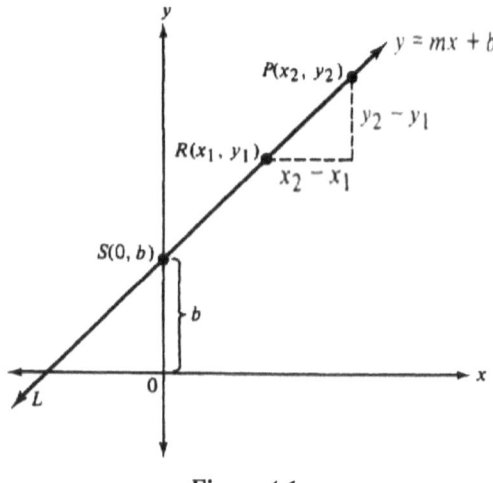

Figure 4.1

L is described by some linear equation that, upon solving for y in terms of x, can be written as

$$y = mx + b$$

where m and b are constants. Let $R(x_1, y_1)$ and $P(x_2, y_2)$ denote two points on L. Then we have

$$y_2 = mx_2 + b$$

and

$$y_1 = mx_1 + b.$$

Subtracting and simplifying yields:

$$y_2 - y_1 = mx_2 + b - (mx_1 + b)$$
$$y_2 - y_1 = mx_2 + b - mx_1 - b$$
$$y_2 - y_1 = mx_2 - mx_1$$
$$y_2 - y_1 = m(x_2 - x_1)$$

By dividing both sides by $x_2 - x_1$, we obtain:

$$\frac{y_2 - y_1}{x_2 - x_1} = m$$

The constant m is called the **slope** of line L. Geometrically, m expresses the ratio of the vertical climb to the horizontal run in going from any one point $R(x_1, y_1)$ on L to any other point $P(x_2, y_2)$ on L. The slope m is a measure of the inclination of line L to the x-axis and is a fundamental characteristic of the line. To determine the equation of L, it suffices to know the slope of L and a point on L, or two points on L.

Let us note that if we substitute 0 for x in the equation $y = mx + b$, $y = b$ is obtained. The constant b is called the **y-intercept** of line L since it is the y-value of the point of intersection of L with the y-axis (see Figure 4.1).

EXAMPLE 1

Find an equation of the line L that passes through $R(1, 3)$ and has slope 2.

If $P(x, y)$ is any point on L, then:

$$y = 2x + b$$

Since $R(1, 3)$ is on L, we have:

$$3 = 2(1) + b$$
$$1 = b$$

Thus

$$y = 2x + 1$$

is an equation of line L (see Figure 4.2).

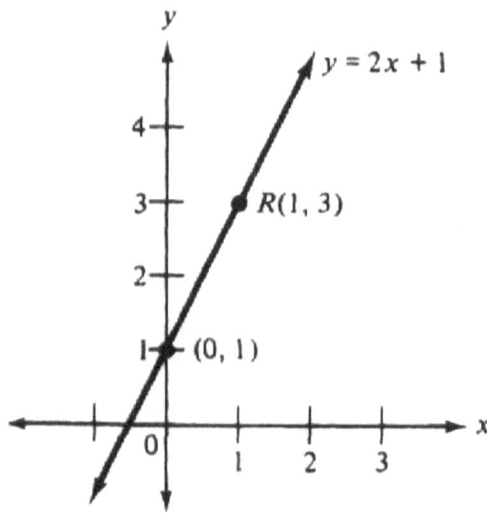

Figure 4.2

EXAMPLE 2

Find an equation of the line L that passes through $R(1, 1)$ and $S(3, 2)$.

We first determine the slope of L from the points $R(1, 1)$ and $S(3, 2)$.

$$m = \frac{2-1}{3-1} = \frac{1}{2}$$

Thus:

$$y = \frac{1}{2}x + b$$

Since $R(1, 1)$ is on L, we have:

$$1 = \frac{1}{2}(1) + b$$

$$\frac{1}{2} = b$$

Thus

$$y = \frac{1}{2}x + \frac{1}{2}$$

is an equation for L (see Figure 4.3).

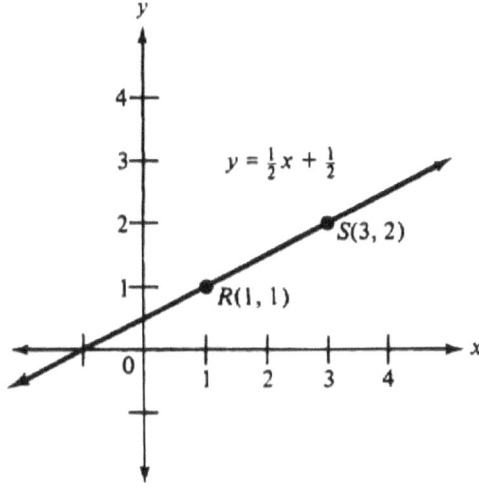

Figure 4.3

Horizontal and Vertical Lines

If L is a horizontal line that passes through the point $S(0, b)$ (see Figure 4.4),

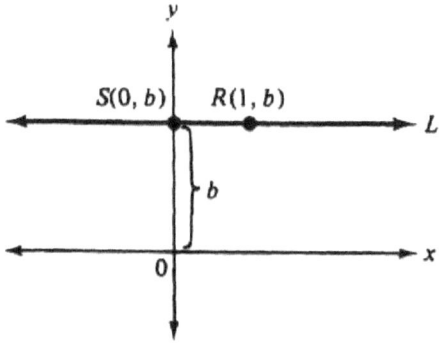

Figure 4.4

then all points on L have y-value b. Thus $R(1, b)$ is another point on L, and the slope of L is:

$$m = \frac{b-b}{1-0} = \frac{0}{1} = 0$$

As an equation of L, we obtain:

$$y = mx + b = 0 + b = b$$

Therefore, $y = 2$ is an equation of the horizontal line that passes through $(0, 2)$, $y = -3$ is an equation of the horizontal line that passes through $(0, -3)$, and $y = 0$ is an equation of the horizontal line that passes through the origin $(0, 0)$, the x-axis (see Figure 4.5).

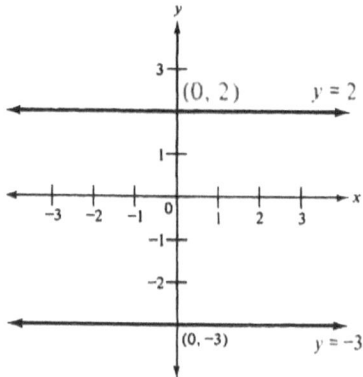

Figure 4.5

The slope of a vertical line, on the other hand, is undefined. To see why this is the case, consider the vertical line L that passes through the point $(1, 0)$ (see Figure 4.6).

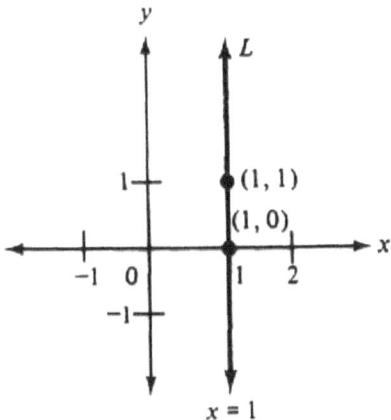

Figure 4.6

Since all points on L have the same x-value, 1, this line is described by the equation $x = 1$. The slope of L is undefined, since the difference between the x-values for any two points is $1 - 1 = 0$, and division by 0 is undefined.

More generally, the vertical line L that passes through $(b, 0)$ (see Figure 4.7) is described by the equation $x = b$.

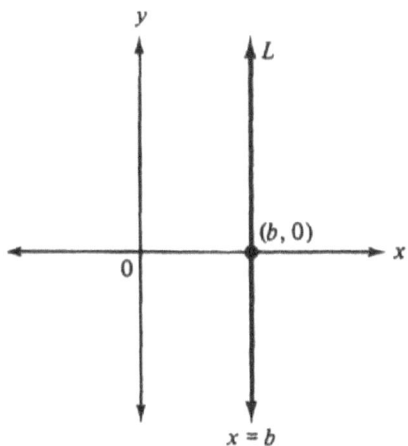

Figure 4.7

EXERCISES

Write an equation of the line satisfying the following conditions.

1. Passing through (1, 4) with slope 2.

2. Passing through $(-1, 3)$ with slope 3.

3. Passing through (2, 1) with slope -1.

4. Passing through $(4, -1)$ with slope $-\dfrac{1}{2}$.

5. Passing through $(1, -3)$ with slope -2.

6. Passing through $(-1, 2)$ and (3, 4).

7. Passing through (1, 1) and (4, 2).

8. Passing through $(-2, 1)$ and (4, 1).

9. Passing through (1, 4) and (1, 5).

10. Passing through $(-1, -1)$ and (2, 5).

4.2   SYSTEMS OF LINEAR EQUATIONS: A FIRST LOOK

In many problems of interest a number of linear equations arise, and it is required to find values of the variables that satisfy all the equations. For example, consider the equations:

$$2x + 3y = 14 \qquad (4.1)$$
$$x - 2y = -7 \qquad (4.2)$$

These equations, considered together as a unit, are said to form a **system of two linear equations**. A **solution of such a system** is any ordered pair of numbers, one for x and the other for y, that satisfies both equations in the system. Thus $x = 1$, $y = 4$, also denoted by (1, 4), is a solution of this system since substitution of 1 for x and 3 for y in Equations (4.1) and (4.2) yields

$$2(1)+3(4)=14$$
$$1-2(4)=-7$$

both of which are correct statements. Although $(7, 0)$ satisfies Equation (4.1), it does not satisfy Equation (4.2), and thus is not a solution of the system consisting of (4.1) and (4.2).

To find solutions of systems of linear equations, we shall employ, in a systematic way, techniques that leave the solutions of a system unchanged. Consider the following principles.

> 1. If both sides of an equation are multiplied by a nonzero constant, then the resulting equation has the same solutions as the original equation.

For example, the equations $x-2y=-7$ and $-2(x-2y=-7)$, that is, $-2x+4y=14$, have the same solutions.

> 2. If two equations have a common solution, then this common solution is also a solution of the sum of the two equations.

For example, the equations $2x+3y=14$ and $x-2y=-7$ have (1, 4) as a solution in common. Their sum, $3x+y=7$, also has (1, 4) as a solution. Principles 1 and 2 taken together yield the following:

> 3. If two equations have a common solution, then this common solution is also a solution of the sum of one of the equations and a nonzero constant multiple of the other.

For example, the equations $2x+3y=14$ and $x-2y=-7$ have (1, 4) as a solution in common. The sum of $2x+3y=14$ and $-2(x-2y=-7)$ is $0x+7y=28$, which also has (1, 4) as a solution.

These principles lead to the following rules of operation, which leave the solutions of a system unchanged.

Rule 1.   An equation in a system of linear equations may be multiplied by a **nonzero constant.**

Rule 2.   Any equation in a system of linear equations may be added to any other equation in the system.

Rule 3.   A **nonzero constant multiple** of any equation in a system of linear equations may be added to any other equation in the system.

To illustrate one way of employing these rules to determine solutions, consider again the system

$$2x + 3y = 14 \qquad (4.1)$$
$$x - 2y = -7 \qquad (4.2)$$

We begin with the **basic assumption** that this system has a solution.

The difficulty that faces us is that there are two variables to contend with. Thus we shall direct our efforts toward taking one of the variables out of the scene and reducing the problem to solving one linear equation in one unknown. The easiest way to proceed is to take x out of the picture by multiplying Equation (4.2) by -2, yielding

$$-2x + 4y = 14 \qquad (4.3)$$

and adding (4.3) to (4.1), thereby obtaining:

$$7y = 28$$
$$y = 4$$

To determine x, we substitute 4 for y in one of the equations, $x - 2y = -7$, for example, and solve for x. This yields:

$$x - 2(4) = -7$$
$$x = 1$$

Thus, we may conclude, **if our system has a solution**, then (1, 4) is the solution. We saw earlier that (1, 4) is indeed a solution since it satisfies both (4.1) and (4.2).

EXAMPLE 1

Solve the system:

$$x + 3y = 90 \qquad (4.4)$$
$$2x + y = 80 \qquad (4.5)$$

Again we begin with the **basic assumption** that this system has a solution. Technically speaking, whether x or y is eliminated from the scene is up to us. For the sake of argument, let's eliminate y. To do so we multiply (4.5) by -3, thus yielding:

$$x + 3y = 90 \qquad (4.6)$$
$$-6x - 3y = -240 \qquad (4.7)$$

By adding (4.6) and (4.7) we obtain:

$$-5x = -150$$
$$x = 30$$

To determine y we substitute 30 for x in one of our equations, $2x + y = 80$, let us say, and solve for y. We have:

$$2(30) + y = 80$$
$$y = 20$$

Thus, **if our system has a solution**, (30, 20) is the solution. To check if the solution candidate (30, 20) is a solution, we substitute 30 for x and 20 for y in (4.4) and (4.5) and see whether they are satisfied. This yields:

$$30 + 3(20) = 30 + 60 = 90$$
$$2(30) + \quad 20 = 60 + 20 = 80$$

Thus (30, 20) is the solution of our system.

EXAMPLE 2

Solve the system:

$$3x - 4y = 10 \qquad (4.8)$$
$$2x - 5y = 9 \qquad (4.9)$$

Our **basic assumption** is that this system has a solution.

Let us begin by eliminating x. To do so we must multiply both equations by constants so that x drops out when the resulting equations are added. The simplest way to achieve this is to multiply (4.8) by -2 and multiply (4.9) by 3. We obtain:

$$-6x + 8y = -20 \qquad (4.10)$$
$$6x - 15y = 27 \qquad (4.11)$$

Adding (4.10) and (4.11) yields:

$$-7y = 7$$
$$y = -1$$

Substituting $y = -1$ into (4.8) yields:

$$3x - 4(-1) = 10$$
$$x = 2$$

By substituting 2 for x and -1 for y in (4.8) and (4.9), it is easily verified that the solution candidate $(2, -1)$ is the solution of this system.

EXAMPLE 3

Solve the system:

$$x + y = 2 \qquad (4.12)$$
$$x + y = 1 \qquad (4.13)$$

Once again, our **basic assumption** is that this system has a solution.

It is clear, however, by inspection, that this system has no solution since the sum of two numbers cannot be equal to 2 [required by (4.12)] and at the same time be equal to 1 [required by (4.13)].

But let's see what happens when we eliminate a variable, x, for example. Multiplying (4.13) by -1 and adding the result to (4.12) yields:

$$0 = 1 \qquad (4.14)$$

This contradiction to the structure of the real-number system (recall the assumed condition that 0 and 1 are **not** the same.) establishes that our **basic assumption** that this system has a solution is untenable.

EXAMPLE 4

Solve the system:

$$2x + 4y = 12 \qquad (4.15)$$
$$-x - 2y = -6 \qquad (4.16)$$

As before, our **basic assumption** is that this system has a solution.

To eliminate x we multiply (4.16) by 2 and add the result to (4.15). This yields

$$0x + 0y = 0 \qquad (4.17)$$

which has all ordered pairs of numbers as solutions. (For all values of x and y, $0x + 0y = 0$ is satisfied.) Let us observe that (4.15) is a multiple of (4.16); by multiplying (4.16) by -2 we obtain (4.15).

Thus our system has **infinitely many solutions**—all ordered pairs of numbers that satisfy $2x + 4y = 12$. By infinitely many solutions we mean that there are more solutions than can be described by any positive integer.

> The solution structures that emerged in these examples illustrate the possibilities in general. **A system of two linear equations in two variables has either one solution, no solution, or infinitely many solutions.**

From a geometric point of view, the one-solution system corresponds to a pair of lines that intersect in one point (whose coordinates describe the solution), the no-solution system corresponds to a pair of parallel lines (no points of intersection), and the infinitely many solution system corresponds to two lines that coincide. In connection with Examples 2 through 4, these possibilities are shown in Figures 4.8 through 4.10.

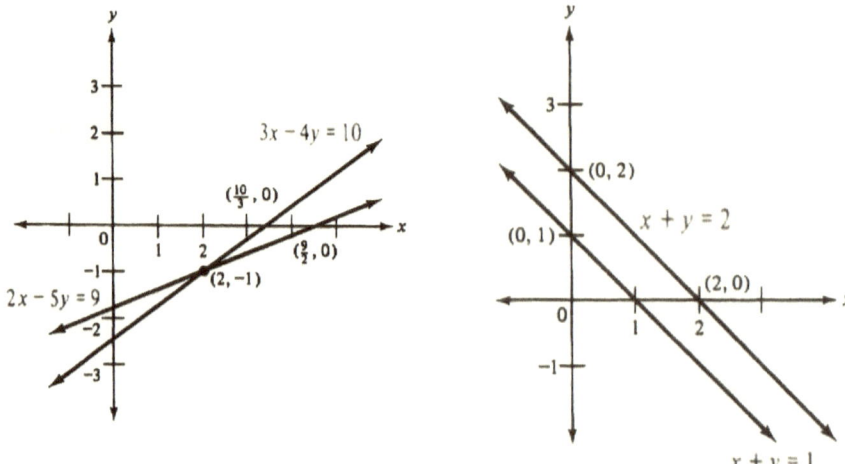

Figure 4.8   Figure 4.9

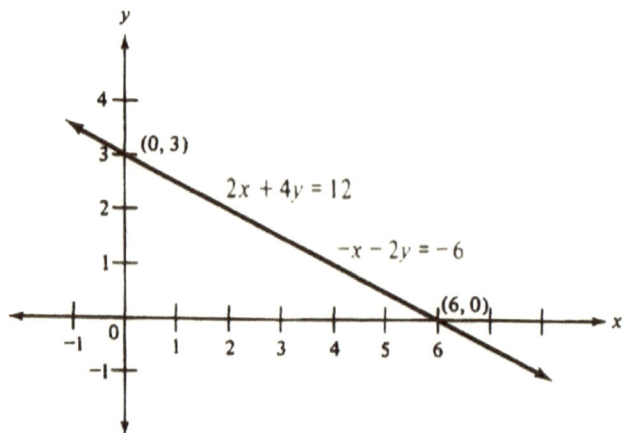

Figure 4.10

EXERCISES

Determine the solutions of the following systems of linear equations.

1. $x + y = 9$
   $x - y = 1$

2. $x + y = 25$
   $-x + 4y = 0$

3. $x + 3y = 13$
   $2x - y = -2$

4. $3x - y = 1$
   $2x - 2y = 6$

5. $2x + 3y = 13$
   $5x - 2y = 4$

6. $3x + 6y = 12$
   $2x - 4y = -8$

7. $2x + y = 23$
   $4x + 5y = 75$

8. $s + 2t = 10$
   $3s + t = 15$

9. $4x + 3y = 320$
   $5x + 2y = 330$

10. $2x + 3y = 120$
    $3x + 2y = 150$

11. $2x + 3y = 120$
    $x + y = 55$

12. $x + 1.2y = 12$
    $x + 2y = 16$

13. $5x + 3y = 95$
    $3x + 5y = 88$

14. $3.25x + 2y = 225$
    $4x + 3y = 320$

4.3 AN APPLICATION TO DETERMINING A REGRESSION LINE*

The marketing department of the Rasa Company wants to predict **average sales volume** on the basis of advertising expenditures. To establish a quantitative relationship between these variables, data (shown in Table 4.1) were obtained from a **sample** of reports for eight monthly periods **chosen at random**** from the company's files.

Letting x denote advertising expenditure, y

---

* Interesting, but may be omitted without loss of continuity.

** A sample of values is part of a larger collection of values called a population. To say that a **sample is chosen at random** means that the sample is chosen without bias that favors certain samples being chosen over others.

Table 4.1

Advertising Expenditure (x) (thousands of dollars)	Sales Volume (y) (thousands of dollars)
2	120
4	150
7	180
9	195
12	210
17	240
22	275
27	300

denoting sales volume, and plotting the eight pairs of values obtained yields the scatter diagram shown in Figure 4.11.

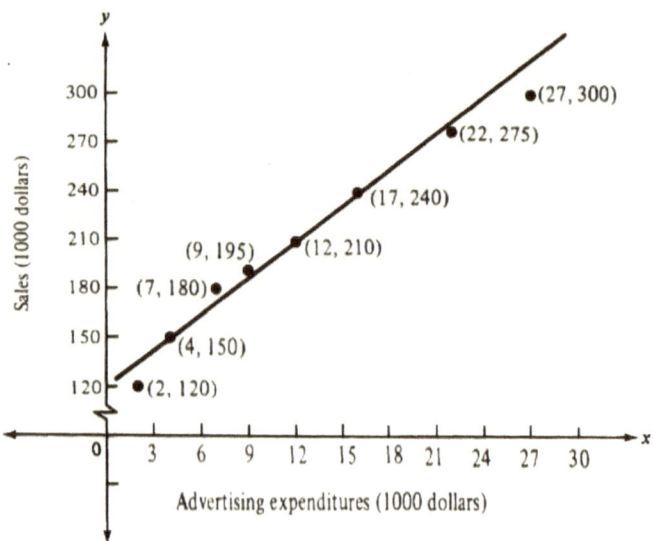

Figure 4.11

Although not all points lie on a line, it does seem that a linear function offers a reasonably good description of the relationship between sales volume and advertising expenditure. Assuming this to be the case, the problem of determining the line of best fit to the eight sample points arises. The most widely employed criterion for best fit is provided by the principle of least squares.

To describe the idea that underlies the least-squares principle, consider the three points (X_1, Y_1), (X_2, Y_2), (X_3, Y_3), and line $y = mx + b$ shown in Figure 4.12.

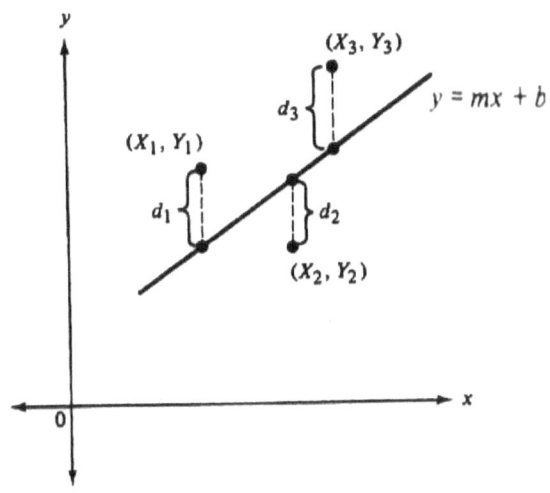

Figure 4.12

The criterion defining line of best fit should, in some sense, minimize the deviation of the points from the line. The least-squares principle does this by defining the **least-squares line of best fit** as that line for which the sum of the squares of the vertical distances from the data points to the line, that is,

$$d_1^2 + d_2^2 + d_3^2$$

is smallest.

It can be shown that for the least-squares line of best fit, denoted by

$$y = mx + b,$$

m and b satisfy the system of equations

$$3b + Am = B$$
$$Ab + Cm = D$$

where 3 is the number of data points,

$$A = X_1 + X_2 + X_3$$

is the sum of the x-values of the data points,

$$B = Y_1 + Y_2 + Y_3$$

is the sum of the y-values of the data points,

$$C = X_1^2 + X_2^2 + X_3^2$$

is the sum of the squared x-values of the data points, and

$$D = X_1 Y_1 + X_2 Y_2 + X_3 Y_3$$

is the sum of the mixed products of corresponding x and y data values.

More generally, if there is a sample of n data points, the same system of equations, with 3 replaced by n, is obtained. That is, we have

$$nb + Am = B$$
$$Ab + Cm = D$$

where A, B, C, and D are the same above defined sums, only with n, instead of 3, terms. This system looks more unfriendly than it actually is; the constants A, B, C, and D for a given situation are easily obtained when the data are organized in column form.

To illustrate, we present the data for the Rasa Company's problem in the first two columns of Table 4.2.

Table 4.2

X_i	Y_i	X_i^2	$X_i Y_i$
2	120	4	240
4	150	16	600
7	180	49	1260

9	195	81	1755
12	210	144	2520
17	240	289	4080
22	275	484	6050
27	300	729	8100
Sums: $A = 100$	$B = 1670$	$C = 1796$	$D = 24,605$

Column 3, containing X_1^2 values, is obtained by squaring each of the X_i values in column 1; column 4, containing X_iY_i values, is obtained by multiplying corresponding X_i and Y_i entries in columns 1 and 2. The sums A, B, C, and D are the sums of these four respective columns. Since there are 8 data points, $n = 8$, and we obtain the following system of equations to be solved for m and b:

$$8b + 100m = 1670 \qquad (4.18)$$
$$100b + 1796m = 24,605 \qquad (4.19)$$

To simplify matters we multiply Equation (4.18) by $\dfrac{1}{2}$, thus obtaining:

$$4b + 50m = 835 \qquad (4.20)$$
$$100b + 1796m = 24,605 \qquad (4.19)$$

Since the easiest variable to eliminate is b, we multiply (4.20) by -25, thus obtaining:

$$-100b - 1250m = -20,875 \qquad (4.21)$$
$$100b + 1796m = 24,605 \qquad (4.19)$$

Adding (4.21) and (4.19) gives us:

$$546m = 3730$$

$$m = \frac{3730}{546}$$

$$= 6.8$$

Substituting $m = 6.8$ into (4.20) and solving for b yields:

$$4b + 50(6.8) = 835$$
$$4b + 340 = 835$$
$$4b = 495$$
$$b = 123.8$$

If you have not dropped your calculator recently and are on otherwise friendly terms with it, it is easily verified that $m = 6.8$ and $b = 123.8$ satisfy the system of Equations (4.18) and (4.19). Thus **the (least squares) line of best fit to our sample of data is:**

$$y = 6.8x + 123.8 \qquad\qquad (4.22)$$

y is interpreted as denoting the **average of y in terms of x** in this application.

To minimize the risk of confusion we shall use a different symbol, $\hat{y}$, read y-hat, rather than y, when our intent is to interpret y as denoting the **average of y in terms of x.** For the Rasa Company's situation we have:

$$\hat{y} = 6.8x + 123.8 \qquad\qquad (4.22a)$$

Since the eight pairs of data values are a sample of the totality, or population, of such pairs of values, Equation (4.22a) is termed **a sample regression line.** It serves as an estimate for the **population regression line**

$$\hat{y} = Mx + E .$$

Other randomly chosen samples yield other estimates for the population regression line.

To illustrate the use of Equation (4.22a) as a tool for prediction, let us suppose that the Rasa Company is interested in predicting the sales volume corresponding to advertising expenditures of $15 thousand per month. Substituting 15 for x in Equation (4.22a) yields:

$$\hat{y} = 6.8(15) + 123.8$$
$$= 225.8$$

For advertising expenditures of $15 thousand per month, we predict an **average monthly sales volume of $225.8 thousand**. We are not, we should keep in mind, predicting $225.8 thousand in sales for each month that $15 thousand is spent on advertising.

Since the system of equations

$$nb + Am = B \tag{4.23}$$
$$Ab + Cm = D, \tag{4.24}$$

often referred to as the system of **normal equations,** occurs in connection with all linear regression problems, it would be advantageous to obtain a general solution for m and b in terms of $n, A, B, C,$ and D that could be applied to any problem.

We begin by eliminating b; multiplying (4.23) by $-A$ and (4.24) by n yields:

$$-nAb - A^2m = -AB \tag{4.25}$$
$$nAb + nCm = nD \tag{4.26}$$

Adding (4.25) and (4.26) gives us:

$$nCm - A^2m = nD - AB$$

By factoring m we obtain:

$$(nC - A^2)m = nD - AB$$

Dividing both sides by $nC - A^2$ yields:

$$m = \frac{nD - AB}{nC - A^2}$$

By solving Equation (4.23) for b, we obtain:

$$nb + Am = B$$
$$nb = B - Am$$
$$b = \frac{B - Am}{n}$$

$\vdots$

In summary, then, we have:

$$m = \frac{nD - AB}{nC - A^2} \qquad (4.27)$$

$$b = \frac{B - Am}{n} \qquad (4.28)$$

That is:

$$m = \frac{n(\text{sum of } X_i Y_i) - (\text{sum of } X_i)(\text{sum of } Y_i)}{n(\text{sum of } X_i^2) - (\text{sum of } X_i)^2} \qquad (4.27)$$

$$b = \frac{(\text{sum of } Y_i) - (\text{sum of } X_i)m}{n} \qquad (4.28)$$

To use these expressions for the computation of m and b, m must be determined first from (4.27) and then substituted into (4.28) for the computation of b.

EXAMPLE 1

The institutional research department of a large university wishes to predict the average income of graduates six years after graduation on the basis of the graduate's grade-point average (GPA). For this purpose the department chose a **random sample** of 10 alumni who had graduated six years earlier. The results of this sample are shown in Table 4.3.

Table 4.3

GPA (x)	2.0	2.3	2.5	2.5	3.0	3.2	3.3	3.5	3.5	3.8
Income (y) (1000 dollars)	19	25	25	30	32	35	37	42	45	40

Corresponding to these data, find the estimated linear regression equation that would enable us to predict income, on average, on the basis of grade-point average. Find the predicted **average income** for a grade—point average of 3.4 and interpret the value obtained.

We begin by arranging the data in column form, determining the X_i^2 and X_iY_i columns, and summing these columns to obtain A (the sum of the X_i values), B (the sum of the Y_i values), C (the sum of X_i^2 values), and D (the sum of X_iY_i values). From the data obtained in Table 4.4, we obtain m and b from Equations (4.27) and (4.28).

Table 4.4

GPA (X_i)	Income (Y_i)	X_i^2	X_iY_i
2.0	19	4.0	38.0
2.3	25	5.3	57.5
2.5	25	6.3	62.5
2.5	30	6.3	75.0
3.0	32	9.0	96.0
3.2	35	10.2	112.0
3.3	37	10.9	122.1
3.5	42	12.3	147.0
3.5	45	12.3	157.5
3.8	40	14.4	152.0
Sums: $A = 29.6$	$B = 330$	$C = 91$	$D = 1019.6$

$$m = \frac{n(\text{sum of } X_iY_i) - (\text{sum of } X_i)(\text{sum of } Y_i)}{n(\text{sum of } X_i^2) - (\text{sum of } X_i)^2}$$

$$= \frac{10(1019.6) - (29.6)(330)}{10(91) - (29.6)^2}$$

$$= \frac{10,196 - 9768}{910 - 876.2}$$

$$= \frac{428}{33.8}$$

$$= 12.7$$

$$b = \frac{(\text{sum of } Y_i) - (\text{sum of } X_i)m}{n}$$

$$= \frac{330 - 375.9}{10}$$

$$= -4.6$$

Thus the sample linear regression equation, based on the sample given, is

$$\hat{y} = 12.7x - 4.6.$$

Substitution of 3.4 for x in this regression equation yields $\hat{y} = 38.6$.

For students with a 3.4 grade-point average, the predicted **average** income six years after graduation is $38.6 thousand.

Here, too, we should be careful to note that we are not predicting the income of a specific student six years after graduation, but rather the average income of all students six years after graduating from the university. Specific incomes making up that average may differ considerably from each other.

Limitations

The nature of the problems considered makes it clear that discretion must be exercised in choosing values of x to be substituted into the regression equation. In general, the further removed x is from the data values, the less reliable we can expect the predicted average value of y to be. It is well known, for example, that wheat yields improve as rainfall increases, and one could determine a regression line for given data.

But it is also well known that wheat yields improve up to a certain point, and that too much rain has a destructive rather than beneficial effect. If the regression line were used to predict average wheat yields for suitably large rainfall values, the predicted average yield would differ considerably from the average yield actually obtained.

Cause and Effect

Let us also note that in establishing a linear regression equation relating x and y for prediction purposes, we are **not claiming that the behavior of one variable causes the behavior of the other variable.** Cause and effect is an entirely separate issue altogether.

In connection with Example 1, it is clear that, although grade-point average might be used to predict average income, income level is not caused by grade-point average.[*]

---

[*] For further discussion of regression analysis see, for example, W. J. Adams, I. Kabus, M. P. Preiss, *Statistics: Basic Principles and Applications*, 2nd ed. (Dubuque: Kendall/Hunt Pub. Co., 2000), ch. 12 or 2nd ed., revised (Philadelphia: Xlibris, 2008). This book is available on the web at webpage.pace.edu/wadams.

EXERCISES

1. The following data show the annual salary increment (y) in hundreds of dollars at a certain company and the number of times (x) the president of the company was complimented per week for his intellectual keenness.

   (a) Find the sample linear regression equation specifying the **average** value of y in terms of x.

x	1	2	3
y	3	7	12

   (b) Find the predicted **average** value of y for $x = 5$ and interpret the result obtained.

   (c) Plot the original data as well as the sample regression line on one diagram.

2. The following table shows equipment maintenance expenditures (x) and profit before taxes (y) for a **random sample** of 10 firms in the same industry. All figures are in thousands of dollars.

x	5	10	13	15	18	21	27	31	37	41
y	24	36	28	36	36	54	52	76	76	100

   (a) Find the sample linear regression equation specifying the **average** value of y in terms of x.

   (b) Find the predicted **average** value of y corresponding to an equipment maintenance expenditure of $50 thousand and interpret the result obtained.

   (c) Plot the original data as well as the sample regression line on one diagram.

3. The following table shows scores (x) on an aptitude test for salesmen employed by the Veronika Company and first-year sales (y) in thousands of dollars for a **random sample** of eight dossiers chosen from the company's files.

x	24	16	20	17	15	25	13	25
y	310	160	280	200	200	290	140	320

(a) Find the sample linear regression equation specifying the **average** value of y in terms of x.

(b) Find the predicted **average** value of y corresponding to a test score of 30 and interpret the result obtained.

(c) Plot the original data as well as the sample regression line on one diagram.

4. The following table shows scores (x) on a mathematics proficiency exam given to all entering freshmen at Ecap University and the final exam grade (y) in a calculus course for a **random sample** of 10 students.

x	20	25	30	32	40	50	60	60	65	70
y	50	50	60	64	70	75	80	85	90	90

(a) Find the sample linear regression equation specifying the **average** value of y in terms of x.

(b) Find the predicted **average** value of y for $x = 55$ and interpret the result obtained.

(c) Plot the original data as well as the sample regression line on one diagram.

5. The following table shows the grade point average (y) in the last two years of university studies and the student's age at the time of graduation (x) for a **random sample** of five students in a large university.

x	20	24	26	30	40
y	2.2	2.4	2.4	3.0	3.2

(a) Find the sample regression equation specifying the average value of y in terms of x.

(b) Find the predicted **average** general point average for age 35 and interpret the result obtained.

(c) Plot the original data as well as the sample regression line on one diagram.

4.4 INTRODUCTION TO THE TABLEAU METHOD

Although the elimination-of-a-variable technique works well for systems of two linear equations in two unknowns, its extension to larger systems of linear equations is awkward. We now turn our attention to developing a general solution method that is effective for a wide spectrum of systems of linear equations and whose procedures do not depend on the number of equations or variables in the system. First, some preliminary definitions.

Two or more linear equations involving the same variables, considered as a unit, are said to form a **system of linear equations**. If there are m equations and n unknowns, the system is called an m **by** n **system**. A **solution of an** m **by** n **system** is an ordered collection of n numbers that satisfies each of the m equations in the system. For example,

$$x - 2y + 2z = -1 \qquad (4.29)$$
$$3x + 2y + 2z = 9 \qquad (4.30)$$
$$2x - 3y - 3z = 6 \qquad (4.31)$$

is a 3 by 3 system. The ordered triple $(3, 1, -1)$ is a solution of this system since it satisfies all three equations of the system. The 2 by 3 system

$$x + 2y + z = 2 \qquad (4.32)$$
$$-x - y + 2z = 4 \qquad (4.33)$$

has infinitely many solutions, two of which are $(0, 0, 2)$ and $(-10, 6, 0)$.

As is the case with 2 by 2 systems, a general m by n system has either one solution, no solution, or infinitely many solutions.

The **tableau method** for solving systems of linear equations, or **Gauss-Jordan elimination procedure**, as it is sometimes called, is based on the three earlier cited rules of operation, which, while changing the form of a system, leave its solutions undisturbed. We restate these rules here for convenience.

Rule 1.   Any equation in a system of linear equations may be multiplied by a **nonzero constant**.

Rule 2. Any equation in a system of linear equations may be added to any other equation in the system.

Rule 3. A **nonzero constant** multiple of any equation in a system of linear equations may be added to any other equation in the system.

In short, the application of rules 1 through 3 reduce a given system of linear equations to an **equivalent system,** that is, one with the same solutions. To illustrate the tableau method, consider the following system:

$$4x+3y=10 \qquad (4.34)$$
$$2x+5y=12 \qquad (4.35)$$

If we could reduce this system to an equivalent system with the structure

$$0x+1y=2$$
$$1x+0y=1$$

then the solution candidate common to both systems could easily be read off as $x=1$, $y=2$.

The tableau method is a systematic procedure for applying rules 1 through 3 to replace the given system by an equivalent system, with a structure characterized by an arrangement of 0's and 1's such that the solution can be read off at a glance. There are two key features to this arrangement:

1. **Each column involving a variable has one 1 with all other values being 0.**

2. **No row contains more than one 1 as a multiplier of a variable.**

To obtain this structure, we begin by choosing a term in either (4.34) or (4.35), $2x$ in Equation (4.35), for example, and convert the coefficient 2 to 1. This is done by applying rule 1 and multiplying Equation (4.35) by $\frac{1}{2}$, the reciprocal of 2. The result is shown in Figure 4.13.

$$4x + 3y = 10 \tag{4.34}$$
$$②x + 5y = 12 \tag{4.35}$$
$$\tag{4.36}$$
$$1x + \tfrac{5}{2}y = 6 \tag{4.37}$$

Figure 4.13

The value to be converted to 1 is called the **pivot value**. It is useful to indicate the chosen pivot value by circling it, as shown in Figure 4.13. To **pivot** on the chosen pivot value is to convert it to 1 by multiplying the equation containing the pivot value by its reciprocal. It is desirable to maintain relative positions of equations, so that Equation (4.35) is replaced by Equation (4.37) as shown.

The next step is to convert the term $4x$ in the column of our pivot value to $0x$ (see Figure 4.13); in short, convert 4 in the column of 2 to 0. To do this we work with Equation (4.37) by applying rule 3; multiply (4.37) by -4, thus obtaining $-4x - 10y = -24$, and add this result to (4.34), $4x + 3y = 10$, thereby obtaining $0x - 7y = -14$, which is recorded as (4.36) in Figure 4.14.

$$4x + \quad 3y = 10 \tag{4.34}$$
$$②x + \quad 5y = 12 \tag{4.35}$$
$$0x + ⊖7y = -14 \tag{4.36}$$
$$1x + \quad \tfrac{5}{2}y = 6 \tag{4.37}$$

Figure 4.14

Our next step is to repeat the process. Choose a pivot value and convert it to 1, at the same time observing the guideline that no row should contain more than one 1 as a multiplier of a variable. Thus we must stay out of Equation (4.37), and the only candidate available for the office of pivot value is -7 of $-7y$ in Equation (4.36). We indicate -7 as our choice of pivot value by circling it as shown in Figure 4.14. To pivot on -7, multiply Equation (4.36) by $-\dfrac{1}{7}$, the reciprocal of -7, and record the result as Equation (4.38), as shown in Figure 4.15.

$$4x + \quad 3y = 10 \tag{4.34}$$
$$(2)x + \quad 5y = 12 \tag{4.35}$$

$$0x + (-7)y = -14 \tag{4.36}$$
$$1x + \tfrac{1}{2}y = 6 \tag{4.37}$$

$$0x + \quad 1y = 2 \tag{4.38}$$
$$\tag{4.39}$$

Figure 4.15

Next we convert the term $\dfrac{5}{2}y$ in the column of our pivot value of $0y$; in short, convert $\dfrac{5}{2}$ in the column of -7 to 0. To do this we work through Equation (4.38) by applying rule 3; multiply (4.38) by $-\dfrac{5}{2}$, thus obtaining $0x - \dfrac{5}{2}y = -5$, and add this result to (4.37), $1x + \dfrac{5}{2}y = 6$, thereby obtaining $1x + 0y = 1$, which is recorded as (4.39) in Figure 4.16.

$$0x + 1y = 2, \quad y = 2 \tag{4.38}$$
$$1x + 0y = 1, \quad x = 1 \tag{4.39}$$

Figure 4.16

From Figure 4.16 we obtain the solution candidate $x = 1$, $y = 2$. To confirm our basic assumption that the system has a solution and as a check against error, it is easily verified that $(1, 2)$ is a solution of the original system (4.34), (4.35).

Let us observe that in carrying out these procedures the arithmetic operations are performed on the numerical values that define the equations. The variables x and y are just carried along for the ride, so to speak, and play no part in these procedures.

Therefore, there is no need to continually write down the variables; it suffices to make note of which columns belong to which variables, identify the column of constants, and perform the indicated procedures on the resulting tableau of numbers.

Rewriting (4.34) and (4.35), leaving out variables x and y and the equality sign, yields the tableau of numbers shown in Figure 4.17.

	Rows	x	y	Constant column
(4.34) $4x + 3y = 10$	①	4	3	10
(4.35) $2x + 5y = 12$	②	2	5	12

Figure 4.17

The procedures described earlier in terms of equations would now be described in terms of the rows in the tableau that describe the equations.

1. **Choose a pivot value in the part of the tableau corresponding to the variables, ② in row 2, for example, and pivot on it.**

This is done by multiplying row ② by $\dfrac{1}{2}$, thus yielding row ④, $[1 \quad \dfrac{5}{2} \quad 6]$, shown in Figure 4.18.

	Rows	x	y	Constant column
T_1	①	4	3	10
	②	②	5	12
T_2	③			
	④	1	$\frac{5}{2}$	6

Figure 4.18

2. **Convert 4 in the column of pivot value 2 to 0.**

To do this we work through row ④ (which we shall term the **one-row** since it is the result of pivoting). Multiply row ④ by -4, thus obtaining $[-4 \quad -10 \quad -24]$, and add this result to row ①, $[4 \quad 3 \quad 10]$, thereby obtaining $[0 \quad -7 \quad -14]$, which is recorded as row ③ in Figure 4.19.

	Rows	x	y	Constant column
T_1	①	4	3	10
	②	②	5	12
T_2	③	0	−7	−14
	④	1	$\frac{5}{2}$	6

Figure 4.19

---

3. **Repeat the process on tableau T_2. Choose a pivot value in a row that does not contain a 1 from a previous pivot operation, and pivot on it.**

Row ④ is excluded by this guideline, and thus −7 in row ③ is the only available pivot value. We pivot on −7 by multiplying row ③ by $-\frac{1}{7}$, the reciprocal of −7, and record the result as row ⑤, [0 1 2], in Figure 4.20.

	Rows	x	y	Constant column
T_1	①	4	3	10
	②	②	5	12
T_2	③	0	−7	−14
	④	1	$\frac{5}{2}$	6
T_3	⑤	0	1	2
	⑥			

Figure 4.20

---

4. **Convert $\dfrac{5}{2}$ in the column of the pivot value −7 to 0.**

To do this we work through row ⑤, our one row. Multiply row ⑤ by $-\dfrac{5}{2}$, thus obtaining $[0 \quad -\dfrac{5}{2} \quad -5]$, and add to row ④, $[0 \quad \dfrac{5}{2} \quad 6]$, thus obtaining $[1 \quad 0 \quad 1]$, which is recorded as row ⑥ in Figure 4.21. From rows ⑤ and ⑥ in tableau T_3, we read off the solution, $x = 1$, $y = 2$.

	Rows	x	y	Constant column	
T_1	①	4	3	10	
	②	②	5	12	
T_2	③	0	⟨-7⟩	-14	
	④	1	$\frac{5}{2}$	6	
T_3	⑤	0	1	2	$y = 2$
	⑥	1	0	1	$x = 1$

Figure 4.21

In summary, the procedures of the tableau method consist of the following sequence of steps.

---

1. **Express the given system of equations in tableau form.**

2. **Choose any column associated with a variable of the system, select a nonzero number as pivot value, and pivot on that value (that is, convert it to 1).**

The resulting row, called the **one-row,** is recorded in the analogous position in the next tableau and serves as the tool for converting all other numbers in the column of the pivot value to 0.

---

3. **Convert all nonzero numbers in the column of the pivot value to 0.**

To convert a number c to 0, multiply the one-row by $-c$ and add the result to the row containing c.

---

4. **Repeat steps 2 and 3, observing the guideline that pivot values should not be chosen from rows that contain a 1 from a previous pivot operation.**

5. **Steps 2 and 3 are repeated until every column (other than the column of constants) contains exactly one 1 and all other values in the column are 0. Such a column is said to be in zero-one form. When all columns are in zero-one form, the solution candidate is read off.**

For some systems it is not possible to convert all columns to zero-one form. Such situations will be discussed later (sec. 4.7).

In brief, the basis of the tableau method is the conversion of all possible columns connected with variables of the system to **zero-one form**. To further illustrate, we turn to the following examples.

EXAMPLE 1

Solve the system:

$$x+3y = 90 \qquad\qquad (4.40)$$
$$2x+ y = 80 . \qquad\qquad (4.41)$$

The tableau corresponding to this system is shown in Figure 4.22.

			Rows	x	y	
$x + 3y = 90$			①	1	3	90
$2x + y = 80$	$\longleftrightarrow$		②	2	1	80

Figure 4.22

We begin by choosing as our pivot value the number 1 in row ①. In choosing a pivot value, it's a good idea to look around for 1's and -1's since they are easiest to work with. Row ① is carried over to tableau T_2 as row ③ (see Figure 4.23).

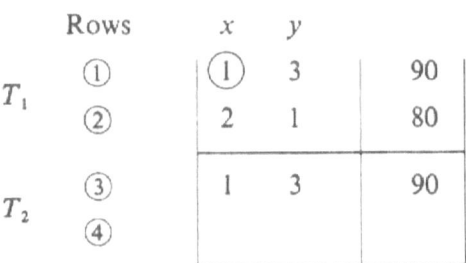

	Rows		x	y	
T_1	①		①	3	90
	②		2	1	80
T_2	③		1	3	90
	④				

Figure 4.23

Row ③, our one-row, serves as our tool for converting 2 in the column of our pivot value to 0. To convert 2 to 0 we multiply row ③ by -2, thus obtaining $[-2 \quad -6 \quad -180]$, and add to row ②, $[2 \quad 1 \quad 80]$, thereby obtaining row ④, $[0 \quad -5 \quad -100]$, shown in Figure 4.24.

	Rows		x	y		
T_1	①		①	3	90	
	②		2	1	80	
T_2	③		1	3	90	row ①
	④		0	-5	-100	(-2)row ③ + row ②

Figure 4.24

To help the reader keep track of the sequence of operations, the origin of each row is indicated on the right side of the tableaus. This, of course, is not necessary in the privacy of one's study.

Since not all appropriate columns in tableau T_2 are in zero-one form, we repeat the conversion steps on T_2. The value -5 in row ④ is the only eligible pivot candidate. (3 in row ③ is ineligible since row ③ contains a 1 from a previous pivot operation.) We pivot on -5 by multiplying row ④ by $-\frac{1}{5}$, the reciprocal of -5, thereby obtaining row ⑥ in Figure 4.25.

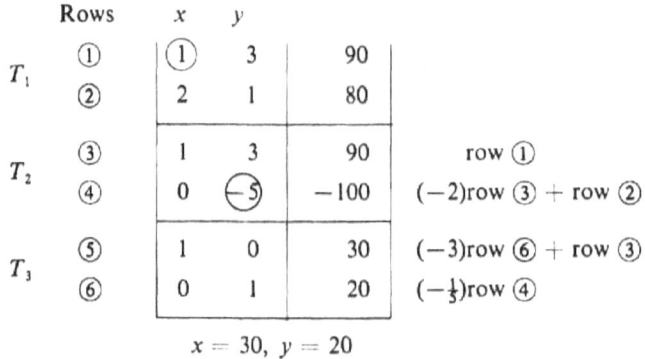

Figure 4.25

Row ⑥, our one-row, now serves as our tool for converting 3 in the column of our pivot value to 0. To convert 3 to 0, we multiply row ⑥ by -3, thus obtaining $[0 \quad -3 \quad -60]$, and add to row ③, $[1 \quad 3 \quad 90]$, thereby obtaining row ⑤, $[1 \quad 0 \quad 30]$, shown in Figure 4.26.

Rows		x	y		
T_1	①	①	3	90	
	②	2	1	80	
T_2	③	1	3	90	row ①
	④	0	⑤	-100	(-2)row ③ + row ②
T_3	⑤	1	0	30	(-3)row ⑥ + row ③
	⑥	0	1	20	$(-\frac{1}{3})$row ④

$x = 30,\ y = 20$

Figure 4.26

Since the x and y columns in tableau T_3 are in zero-one form, we can read off the solution candidate $x = 30$, $y = 20$. It is easily verified that solution candidate $(30, 20)$ satisfies the given system of Equations (4.40), (4.41).

Let us suppose that in applying the tableau method to a system of equations there arises, as illustrated by Figure 4.27,

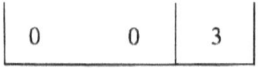

Figure 4.27

a row of zeros, except for a non-zero term in the constant column. In terms of the variables of this system, x and y, let us say, this row represents the equation $0x + 0y = 3$. Since this equation has no solution, the system to which it belongs has no solution. Thus:

> **The appearance of a row of zeros, except for a non-zero value in the constant column, implies that the given system of equations has no solution.**

On the other hand, suppose that a row arises consisting entirely of zeros, as illustrated by Figure 4.28.

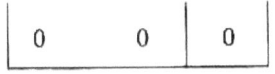

Figure 4.28

This row represents the equation $0x + 0y = 0$, which is satisfied by all values of x and y.

Such an equation, and its corresponding row may be omitted from the system, or tableau, since it has no further point of interest. Any common solution to the other equations in the system will satisfy this equation as well and there is no need to reproduce it.

> **Discard from a tableau any row consisting entirely of zeros and continue the solution generation process.**

Systems of two equations in two unknowns serve as a vehicle here for the introduction of the row and column procedures that underlie the tableau method. The real power of this technique is to be felt in dealing with larger systems, which we take up in the three sections that follow.

For a more complete perspective on the row and column operations introduced here, it is noteworthy that these procedures find application elsewhere. We develop one application to matrix inversion and matrix solutions to systems of linear equations in sections 11.3 and 11.4.

EXERCISES

Determine the solutions of the following systems of linear equations by the tableau method.

1. $x+y=9$
    $x-y=1$

2. $x+\ y=25$
    $-x+4y=\ 0$

3. $x+3y=13$
    $2x-\ y=-2$

4. $3x-\ y=1$
    $2x-2y=6$

5. $2x+\ y=23$
    $4x+5y=75$

6. $x-2y=\ \ 6$
    $-3x+6y=-18$

7. $3x+\ y=9$
    $x+2y=8$

8. $5x+3y=1400$
    $4x+\ y=\ \ 756$

9. $2x-\ y=-5$
    $-6x+3y=15$

10. $2x+3y=15$
    $x+\ y=\ 6$

11. $x+y=250,000$
    $2x+y=400,000$

4.5   THE TABLEAU METHOD FOR LARGER SYSTEMS

In applying the tableau method to larger systems we proceed in the same way. It's just that there is more to be done.

EXAMPLE 1

Solve the system

$$x-2y+2z=-1 \qquad (4.42)$$
$$3x+2y+2z=9 \qquad (4.43)$$
$$2x-3y-3z=6 \qquad (4.44)$$

The tableau corresponding to this system is shown in Figure 4.29.

	Rows	x	y	z	
$x-2y+2z=-1$	①	1	-2	2	-1
$3x+2y+2z=\ \ 9$ ⟷	②	3	2	2	9
$2x-3y-3z=\ \ 6$	③	2	-3	-3	6

Figure 4.29

We begin by choosing as our pivot value the number 1 in row ①. Row ① is carried over to tableau T_2 as row ④ (see Figure 4.30).

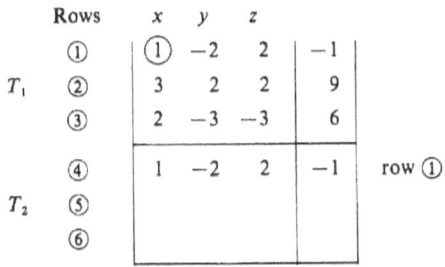

	Rows	x	y	z		
	①	①	-2	2	-1	
T_1	②	3	2	2	9	
	③	2	-3	-3	6	
	④	1	-2	2	-1	row ①
T_2	⑤					
	⑥					

Figure 4.30

Row ④, our one-row, serves as our tool for converting 3 and 2 in the first column of T_1 to 0. To convert 3 to 0, we multiply row ④ by -3, thus obtaining $[-3 \quad 6 \quad -6 \quad 3]$, and add to row ②, thereby obtaining row ⑤, $[0 \quad 8 \quad -4 \quad 12]$, shown in Figure 4.31. To convert 2 to 0, we multiply row ④ by -2 and add to

		x	y	z		
	①	①	-2	2	-1	
T_1	②	3	2	2	9	
	③	2	-3	-3	6	
	④	1	-2	2	-1	row ①
T_2	⑤	0	8	-4	12	(-3)row ④ + row ②
	⑥					

Figure 4.31

row ③, thereby obtaining row ⑥ in Figure 4.32.

		x	y	z		
	①	①	-2	2	-1	
T_1	②	3	2	2	9	
	③	2	-3	-3	6	
	④	1	-2	2	-1	row ①
T_2	⑤	0	8	-4	12	(-3)row ④ + row ②
	⑥	0	①	-7	8	(-2)row ④ + row ③

Figure 4.32

Since not all appropriate columns in tableau T_2 are in zero-one form, we repeat the process on T_2. Staying out of row ④, which has a 1 from a previous pivot operation, we choose a pivot value, that is 1, in row ⑥. (The other eligible pivot candidates are -7, -4, and 8, but 1 is easier to work with.) Row ⑥ is carried over to tableau T_3 as row ⑨ (see Figure 4.33).

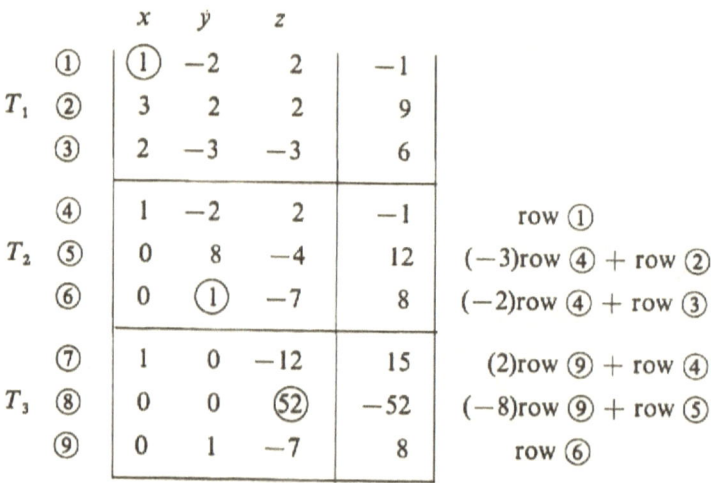

Figure 4.33

Row ⑨ serves as our tool for converting 8 and -2 in the column of our pivot value to 0.

To convert 8 to 0, we multiply row ⑨ by -8 and add to row ⑤, thus obtaining row ⑧ in tableau T_3. To convert -2 to 0, we multiply row ⑨ by 2 and add to row ④, thus obtaining row ⑩ in tableau T_3.

Since not all appropriate columns in T_3 are in zero-one form, we repeat the process on T_3. The number 52 in row ⑧ is the only eligible pivot candidate (why?); we pivot on 52 by multiplying row ⑧ by $\dfrac{1}{52}$, thereby obtaining row ⑩ in tableau T_4 (see Figure 4.34).

		x	y	z		
	①	①	-2	2	-1	
T_1	②	3	2	2	9	
	③	2	-3	-3	6	
	④	1	-2	2	-1	row ①
T_2	⑤	0	8	-4	12	(-3)row ④ + row ②
	⑥	0	①	-7	8	(-2)row ④ + row ③
	⑦	1	0	-12	15	(2)row ⑨ + row ④
T_3	⑧	0	0	$\boxed{52}$	-52	(-8)row ⑨ + row ⑤
	⑨	0	1	-7	8	row ⑥
	⑩	1	0	0	3	(12)row ⑪ + row ⑦
T_4	⑪	0	0	1	-1	$(\frac{1}{52})$row ⑧
	⑫	0	1	0	1	(7)row ⑪ + row ⑨

$$x = 3,\ y = 1,\ z = -1$$

Figure 4.34

Row ⑪ serves as our tool for converting -12 and -7 in the column of our pivot value to 0. To convert -12 to 0, we multiply row ⑪ by 12 and add to row ⑦, thus obtaining row ⑩ in tableau T_4. To convert -7 to 0, we multiply row ⑪ by 7 and add to row ⑨, thus obtaining row ⑫ in T_4.

Since the x, y, and z columns of T_4 are in zero-one form, we can read off the solution candidate $x = 3$, $y = 1$, $z = -1$. It is easily verified that $(3,1,-1)$ satisfies the given system of equations.

EXAMPLE 2

Solve the system

$$x + 2y - 4z = -4$$
$$-x - 4y - 2z = -2$$
$$4x + 2y - 3z = 4.$$

A sequence of tableaus leading to the solution is shown in Figure 4.35.

		x	y	z		
T_1	①	①	2	-4	-4	
	②	-1	-4	-2	-2	
	③	4	2	-3	4	
T_2	④	1	2	-4	-4	row ①
	⑤	0	-2	-6	-6	row ④ + row ②
	⑥	0	-6	13	20	(-4)row ④ + row ③
T_3	⑦	1	0	-10	-10	(-2)row ⑧ + row ④
	⑧	0	1	3	3	$(-\frac{1}{2})$row ⑤
	⑨	0	0	㉛	38	(6)row ⑧ + row ⑥
T_4	⑩	1	0	0	$\frac{70}{31}$	(10)row ⑫ + row ⑦
	⑪	0	1	0	$-\frac{21}{31}$	(-3)row ⑫ + row ⑧
	⑫	0	0	1	$\frac{38}{31}$	$(\frac{1}{31})$row ⑨

$$x = \tfrac{70}{31},\, y = -\tfrac{21}{31},\, z = \tfrac{38}{31}$$

Figure 4.35

Tableau T_2 is obtained by pivoting on 1 in row ① and converting -1 and 4 in the column of 1 to 0. Pivoting on -2 in tableau T_2 avoids the introduction of fractions. Thus -2 is the best choice among the available pivot candidates $(-2, -6, 13, -6)$ since messy arithmetic is avoided.

In general, whenever possible, it's a good idea to choose a pivot value that divides into all the numbers in its row so as to avoid the introduction of fractions.

Tableau T_3 is obtained by pivoting on -2 in T_2 and converting 2 and -6 in the column of -2 to 0. Since not all columns in tableau T_3 are in zero-one form, we must continue the conversion process. The value 31 in row ⑨ of T_3 is the only available pivot choice (why?), and thus we must pivot on 31.

Tableau T_4 is obtained by pivoting on 31 and converting 3 and -10 in the column of 31 to 0. The $x, y,$ and z columns in T_4 are in zero-one form, and thus the solution candidate $x = \dfrac{70}{31}$, $y = -\dfrac{21}{31}$, $z = \dfrac{38}{31}$ can be read off.

Our last stop is to confirm that this solution candidate is the solution.

EXAMPLE 3

Solve the system

$$x - 2y + z = 3$$
$$2x - 3y - z = 7$$
$$5x - 8y - z = 20.$$

A sequence of solution tableaus is shown in Figure 4.36. Since row ⑨ consists of 0's, except for the nonzero constant term, 3, our system has no solution.

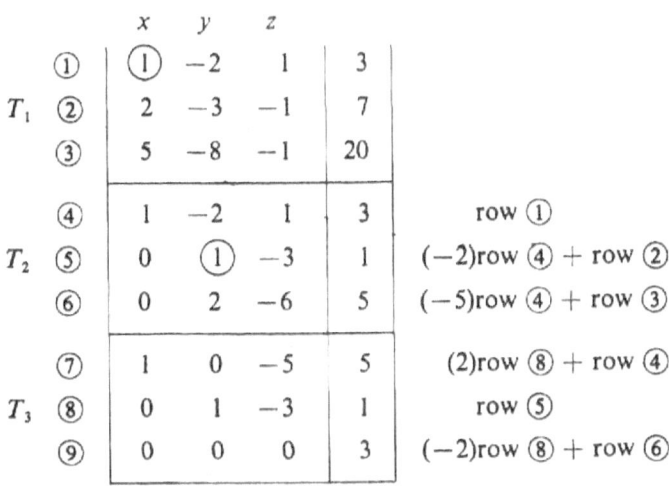

Figure 4.36

EXERCISES

Determine the solutions of the following systems by the tableau method.

1. $x + 3y + 2z = -1$
    $2x + 3y + 3z = -2$
    $-2x + 2y - 3z = 7$

2. $x + 2y - 2z = 1$
    $2x + 3y + 2z = -3$
    $2x - 3y - z = 6$

3. $2s + t - u = 2$
    $3s - 2t + u = 7$
    $s - 3t - 2u = -7$

4. $4u + v - 2w = -3$
    $2u + 3v - 2w = 7$
    $7u + 2v + 5w = -13$

5. $2x + 3y + z = 4$
    $x + y - 3z = 11$
    $4x + 5y - 5z = 9$

6. $\begin{aligned} 8s + t - u &= -7 \\ 4s + 3t + v &= -3 \\ 2s + t + 4u + 4v &= 14 \\ 4s + 2t - 3u - v &= -12 \end{aligned}$

7. $\begin{aligned} 2s + t + 4u &= 8 \\ 3s - 2t - 7u &= 12 \\ -4s + 5t + 10u &= -16 \end{aligned}$

We next examine two applications from the world of accounting which lead to large systems. Knowledge of accounting is not assumed for discussion of these applications.

4.6 NOTEWORTHY APPLICATIONS*

EXAMPLE 1 *SERVICE CHARGE ALLOCATION*

The Arkin Company, which makes television sets, has three production departments, which we shall denote by P_1, P_2, and P_3, and four service departments—accounting, maintenance, marketing, and purchasing. Each service department's cost must be distributed to the production departments and to other service departments based on their respective usages of the services provided. For each service department listed in the leftmost column of Table 4.5, the fraction of its total cost assigned to the service and production departments of the firm is given. Thus from row 1 we have that 2 percent of the total cost of accounting is assigned to maintenance, 10 percent of the total cost of accounting is assigned to marketing, and so on. Column 1 specifies the fraction of the service departments' costs that is assigned to accounting, column 2 specifies the

Table 4.5

SERVICE DEPARTMENT	SERVICE DEPARTMENTS				PRODUCTION DEPARTMENTS			JANUARY OVERHEAD
	Acc.	*Main.*	*Mar.*	*Pur.*	P_1	P_2	P_3	
Accounting	0	0.02	0.10	0.10	0.24	0.26	0.28	$20,000
Maintenance	0.10	0	0.20	0.10	0.20	0.20	0.20	$18,000
Marketing	0	0	0	0	0.30	0.30	0.40	$80,000
Purchasing	0.10	0.10	0.10	0	0.20	0.20	0.30	$10,000

---

*The applications considered here are examined from a matrix algebra point of view in section 11.5.

fraction of the service departments' costs that is assigned to maintenance, and so on. (Such determinations can be made in various ways; one way is based on the floor space occupied by each department.)

The problem is to determine the service departments's total costs (overhead plus charges for services provided by other departments) and to allocate these costs to the production departments.

Let x, y, z, and w denote the total cost of the accounting, maintenance, marketing, and purchasing departments, respectively, for January.

From column 1 we have that the costs of the maintenance, marketing, and purchasing departments that are assigned to accounting are $0.10y$, $0z$, and $0.10w$, respectively. The overhead of the accounting department (cost that is directly assigned to accounting such as salaries of employees in accounting, equipment, supplies, etc.) for January is $20,000. Thus x, the total cost of accounting, must satisfy the following condition:

$$x = 20,000 + 0x + 0.1y + 0z + 0.1w$$

x	$20,000$	$0x$	$0.1y$	$0z$	$0.1w$
total cost of acc.	Jan. overhead for acc.	cost of acc. assigned to acc.	cost of main. assigned to acc.	cost of mar. assigned to acc.	cost of pur. assigned to acc.

which reduces to:

$$x = 20,000 + 0.1y + 0.1w$$

A similar analysis for the maintenance, marketing, and purchasing departments yields the following conditions:

$$y = 18,000 + 0.02x + 0.1w$$
$$z = 80,000 + 0.1x + 0.2y + 0.1w$$
$$w = 10,000 + 0.1x + 0.1y$$

Rearranging terms gives us the following system:

$$x - 0.1y \qquad\qquad -0.1w = 20,000$$
$$-0.02x + \quad y \qquad\quad -0.1w = 18,000$$
$$-0.1x - 0.2y + z - 0.1w = 80,000$$
$$-0.1x - 0.1y + \qquad\quad w = 10,000$$

A sequence of tableaus leading to the solution of this system is shown in Figure 4.37. From tableau T_4 we obtain the total cost values for the service departments.

T_1	①	−0.1	0	−0.1	20,000
	−0.02	1	0	−0.1	18,000
	−0.1	−0.2	1	−0.1	80,000
	−0.1	−0.1	0	1	10,000
T_2	1	−0.1	0	−0.1	20,000
	0	(0.998)	0	−0.102	18,400
	0	−0.21	1	−0.11	82,000
	0	−0.11	0	0.99	12,000
T_3	1	0	0	−0.1102204	21,843.69
	0	1	0	−0.1022044	18,436.87
	0	0	1	−0.1314629	85,871.74
	0	0	0	(0.9787576)	14,028.06
T_4	1	0	0	0	23,423.43
	0	1	0	0	19,901.72
	0	0	1	0	87,755.93
	0	0	0	1	14,332.52

$$x = 23{,}423.43 \qquad z = 87{,}755.93$$
$$y = 19{,}901.72 \qquad w = 14{,}332.52$$

Figure 4.37

Accounting:	$23,423
Maintenance:	$19,902
Marketing:	$87,756
Purchasing:	$14,333

From these cost values and the percentages given in columns 5, 6, and 7 of Table 4.5, we obtain the following allocation of service departments' costs to the production departments:

Department P_1: $(0.24)(23,423) + (0.20)(19,902) + (0.30)(87,756)$
$+ (0.20)(14,333) = \$38,795$

Department P_2: $(0.26)(23,423) + (0.20)(19,902) + (0.30)(87,756)$
$+ (0.20)(14,333) = \$39,264$

Department P_3: $(0.28)(23,423)+(0.20)(19,902)+(0.40)(87,756)$
$+(0.30)(14,333) = \$49,941$

If we were to replace the January overhead values of $20,000, $18,000, $80,000, and $10,000 by values for another month, we would have made a small change in the overall nature of the problem. Mathematically, the change is not small because we would be faced by a different system of equations.

The question is, can we develop a way that would allow us to change the overhead values with only a small change needed in the mathematical analysis needed to deal with the modified system of equations? The answer is yes, and the means for doing this is developed in section 11.5.

EXAMPLE 2   *AN INCOME CONSOLIDATION PROBLEM*

The accompanying affiliation diagram shows the interdependency structure between the Russel, Ferrara, and Thomas Companies.

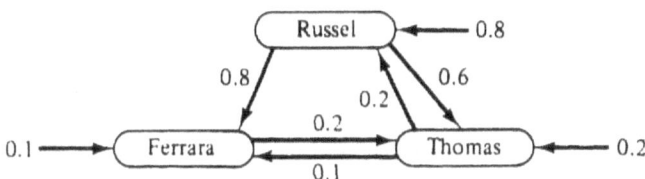

This diagram shows, for example, that the Russel Company owns 60 percent of the corporate stock of the Thomas Company, while the Thomas Company owns 20 percent of the corporate stock of the Russel Company. Eighty percent of the Russel Company's stock is not controlled by affiliates.

Let us suppose that the net incomes of the Russel, Ferrara, and Thomas Companies from their own operations are $100,000, $80,000, and $60,000, respectively. The consolidated basis net income of a company is its net income determined in terms of its interdependency with other affiliates.

The problem we consider here is that of determining the net incomes of the Russel, Ferrara, and Thomas Companies on a consolidated basis and the allocation of net incomes to these affiliate companies.

To address this problem we introduce the following notation:

1. Let x denote the net income of the Russel Company on a consolidated basis.

2. Let y denote the net income of the Ferrara Company on a consolidated basis.

3. Let z denote the net income of the Thomas Company on a consolidated basis.

Then we have the following relations:

$$x = 100,000 + 0.8y + 0.6z$$
$$y = 80,000 + 0.2z$$
$$z = 60,000 + 0.2x + 0.1y$$

Transposing and combining similar terms yields the following system:

$$x - 0.8y - 0.6z = 100,000$$
$$y - 0.2z = 80,000$$
$$-0.2x - 0.1y + z = 60,000$$

A sequence of tableaus leading to the solution of this system is shown in Figure 4.38.

	x	y	z	
	①	−0.8	−0.6	100,000
T_1	0	1	−0.2	80,000
	−0.2	−0.1	1	60,000
	1	−0.8	−0.6	100,000
T_2	0	①	−0.2	80,000
	0	−0.26	0.88	80,000
	1	0	−0.76	164,000
T_3	0	1	−0.2	80,000
	0	0	⓪.828	100,800
	1	0	0	256,521.74
T_4	0	1	0	104,347.82
	0	0	1	121,739.13

$x = 256,521.74, y = 104,347.82, z = 121,739.13$

Figure 4.38

From tableau T_4 we obtain the following net income statements on a consolidated basis:

Net income of the Russel Company on a consolidated basis = $256,521.74

Net income of the Ferrara Company on a consolidated basis = $104,347.82

Net income of the Thomas Company on a consolidated basis = $121,739.13

In determining the allocation of net incomes of affiliate companies, only the nonaffiliate shareholders in the parent company constitute the majority interest. For example, 20 percent of the Russel Company's stock is held by the Thomas Company; accordingly, the outside interest in the Russel Company of 80 percent is the equity multiplier in determining the consolidated net income of this company.

In summary, we have the following allocation of net incomes of affiliate companies:

Consolidated Net Income

Russel Company: 80 percent of the Russel Company's consolidated basis net income (80 percent of $256,521.74) = $205,217.39.

Ferrara Company: 10 percent of the Ferrara Company's consolidated basis net income (10 percent of $104,347.82) = $10,434.78.

Thomas Company: 20 percent of the Thomas Company's consolidated basis net income (20 percent of $121,739.13) = $24,347.83.

EXERCISES

1. The Sonin Company, which makes computers, has two production departments, P_1 and P_2, and three service departments, S_1, S_2, and S_3. Each service department's total cost must be distributed to the production departments and to the other service departments based on their respective usages of the services provided. For each service department listed in the leftmost column of Table 4.6, the fraction of its total cost assigned to the service and production departments of the firm is given.

Table 4.6

SERVICE DEPARTMENT	SERVICE DEPARTMENT			PRODUCTION DEPARTMENT		MARCH OVERHEAD
	S_1	S_2	S_3	P_1	P_2	
S_1	0	0.10	0.05	0.40	0.45	$40,000
S_2	0.10	0	0.10	0.40	0.40	$30,000
S_3	0.20	0.05	0	0.35	0.40	$20,000

(a) Set up the system of equations that describes the conditions to be satisfied by the total costs of the service departments for March.

(b) Solve this system to determine the total costs of the service departments.

(c) Determine the allocation of the total costs of the service departments to the production departments.

2. The accompanying affiliation diagram shows the interdependency structure between the Ramunė, Algis, and Charles companies.

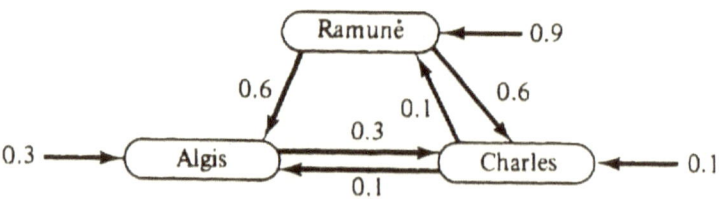

These companies have net incomes from their own operations of $80,000, $60,000 and $50,000, respectively.

(a) Set up and solve a system of linear equations to determine the net incomes of these affiliate companies on a consolidated basis.

(b) Determine the allocation of net incomes to these affiliate companies.

3. Letting x denote the state income tax owed, y denote the state capital stock tax owed, and z denote the federal income tax owed, a tax consultant developed the following relationships among these taxes for a client corporation.

$$x = 0.095(27,257 - y)$$
$$y = 0.01(152,970 - x - y - z)$$
$$z = 0.48(27,263 - x - y) - 833$$

Set up a system of linear equations from these relations and solve it to determine the taxes owed.

4. The federal income tax of the Sommers Company is computed after deducting payroll bonuses and the contribution to the profit-sharing plan. Payroll bonuses are based on income after deducting the profit-sharing contribution and federal income tax. The profit-sharing contribution is based on income after deducting bonuses and taxes. The tax rate is 50 percent, the bonus rate is 10 percent, the profit-sharing contribution is 5 percent, and profit, before consideration of taxes, bonuses, and profit sharing, is $2,000,000.

   (a) Set up a system of linear equations to describe the interrelationships among taxes, bonuses, and the profit-sharing contribution.

   (b) Solve this system to determine these amounts.

4.7   SYSTEMS WITH INFINITELY MANY SOLUTIONS[*]

EXAMPLE 1

Solve the system

$$\begin{aligned} x + 2y + z &= 2 & (4.45) \\ -x - y + 2z &= 4 & (4.46) \\ 3x + 4y - 3z &= -6. & (4.47) \end{aligned}$$

A sequence of three solution tableaus for this system is shown in Figure 4.39.

---

[*] Interesting, but may be omitted without loss of continuity.

		x	y	z		
	①	①	2	1	2	
T_1	②	−1	−1	2	4	
	③	3	4	−3	−6	
	④	1	2	1	2	row ①
T_2	⑤	0	①	3	6	row ④ + row ②
	⑥	0	−2	−6	−12	(−3)row ④ + row ③
	⑦	1	0	−5	−10	(−2)row ⑧ + row ④
T_3	⑧	0	1	3	6	row ⑤
	⑨	0	0	0	0	(2)row ⑧ + row ⑥

Figure 4.39

Observe that row ⑨ in tableau T_3 consists entirely of 0's and may thus be omitted. At the same time, another curious feature emerges; the third column of tableau T_3 cannot be converted to zero-one form without violating our standing rule that a pivot value is not to be chosen from a row that has a 1 from a previous pivot operation.

This signals us to stop and write the equations described by the tableau. Doing this yields:

$$x \quad\quad -5z = -10$$
$$y + 3z = \quad 6$$

By transposing the z term in both equations, we obtain:

$$x \quad\quad = -10 + 5z$$
$$y = \quad 6 - 3z$$

If we give z the value 1, we obtain $x = -5$ and $y = 3$; if we give z the value 0, we obtain $x = -10$ and $y = 6$.

Our system has infinitely many solution candidates, and a specific solution candidate can be obtained by giving z any specific value of our choice and determining x and y from $x = -10 + 5z$ and $y = 6 - 3z$. We describe these solution candidates by writing:

$$x = -10 + 5z$$
$$y = 6 - 3z$$
$$z \text{ is arbitrary}$$

We establish that these solution candidates are the solutions of our system by substituting $-10+5z$ for x and $6-3z$ for y in (4.45) through (4.47), and verifying that 2, 4, and -6, respectively, are obtained.

$$(-10+5z)+2(6-3z)+z=-10+5z+12-6z+z= \quad 2 \quad (4.45)$$

$$-(-10+5z)-(6-3z)+2z=10-5z-6+3z+2z= \quad 4 \quad (4.46)$$

$$3(-10+5z)+4(6-3z)-3z=-30+15z+24-12z-3z=-6 \quad (4.47)$$

When a tableau is obtained in which not all columns connected with variables are in zero-one form, and it is not possible to convert the remaining columns to zero-one form, write the equations that correspond to the tableau.

By transposing those terms that arise from the columns not in zero-one form, we obtain a description of the infinitely many solution candidates of the system.

EXAMPLE 2

Solve the system

$$x-3y+4z+ w=8 \tag{4.48}$$
$$-x+4y- z+2w=4. \tag{4.49}$$

A sequence of solution tableaus for this system is shown in Figure 4.40.

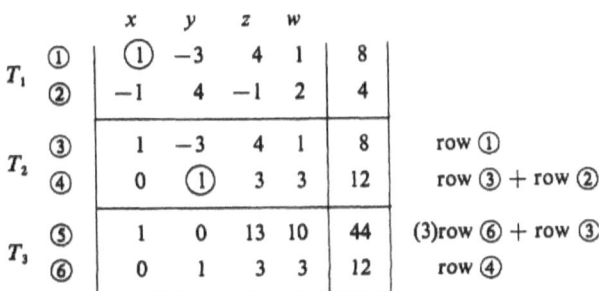

Figure 4.40

Writing the equations that correspond to tableau T_3 yields:

$$x +13z+10w=44$$
$$y+ 3z+ 3w=12$$

By transposing the z and w terms in both equations, we obtain the following description of the solution candidates of our system:

$$x = 44 - 13z - 10w$$
$$y = 12 - 3z - 3w$$
$$z \text{ is arbitrary}$$
$$w \text{ is arbitrary}$$

Our system has infinitely many solution candidates, and a specific solution candidate can be obtained by giving z and w specific values of our choice, and determining x and y from $x = 44 - 13z - 10w$ and $y = 12 - 3z - 3w$.

If we give z and w the value 0, we obtain $x = 44$ and $y = 12$; for $z = 1$ and $w = 2$, we obtain $x = 11$ and $y = 3$.

To establish in general that the cited solution candidates describe the solutions of our system, substitute $44 - 13z - 10w$ for x and $12 - 3z - 3w$ for y in (4.48) and (4.49), and verify that 8 and 4, respectively, are obtained.

$$44 - 13z - 10w - 3(12 - 3z - 3w) + 4z + w$$
$$= 44 - 13z - 10w - 36 + 9z + 9w + 4z + w \qquad (4.48)$$
$$= 8$$

$$-(44 - 13z - 10w) + 4(12 - 3z - 3w) - z + 2w$$
$$= -44 + 13z + 10w + 48 - 12z - 12w - z + 2w \qquad (4.49)$$
$$= 4$$

EXAMPLE 3   A TRAFFIC NETWORK PROBLEM

Part of a traffic network being designed to service a component of the Johnson City Airport is shown in Figure 4.41.

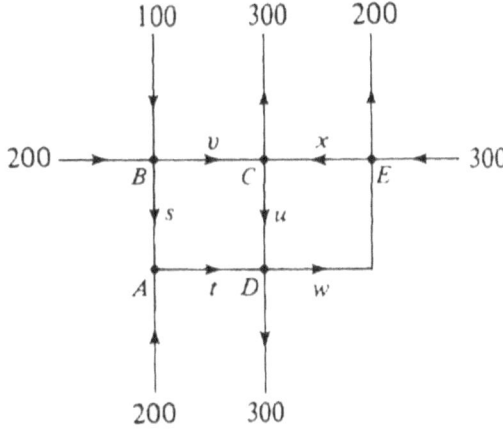

Figure 4.41

The roads are one way, as shown by the arrows, and the given values express the expected number of cars entering and leaving the network per hour during a heavy load period. The total number of cars entering the network, 800, equals the total number leaving the network, so that a basic traffic-flow equilibrium condition is satisfied. Another basic equilibrium condition is that the total number of cars entering each intersection point equal the total number of cars leaving the intersection point.

Assuming this to be the case, let us explore conclusions that can be drawn about traffic flow in the interior branches of the network.

To do so, we introduce variables s, t, u, v, w, and x as shown in Figure 4.41. Variable x denotes the number of cars passing between intersection points E and C per hour, t denotes the number of cars passing between intersection points A and D per hour, and so on. Since $200+s$ cars enter A while t cars leave A, for equilibrium we have:

$$A: \quad s+200 = t$$

Similarly, for B, C, D, and E we obtain

$$B: \quad 200+100 = s+v$$
$$C: \qquad v+x = u+300$$
$$D: \qquad t+u = w+300$$
$$E: \quad w+300 = x+200$$

Rewriting these equations so that the variables appear on one side of an equation and the constant appears on the other side yields the system:

$$
\begin{aligned}
s - t &&&&&= -200 & (4.50)\\
s &&+v &&&= 300 & (4.51)\\
&&-u+v &&+x &= 300 & (4.52)\\
&t+u &&-w &&= 300 & (4.53)\\
&&&&-w+x &= 100 & (4.54)
\end{aligned}
$$

A sequence of solution tableaus is shown in Figure 4.42.

		s	t	u	v	w	x		
T_1	①	(1)	−1	0	0	0	0	−200	
	②	1	0	0	1	0	0	300	
	③	0	0	−1	1	0	1	300	
	④	0	1	1	0	−1	0	300	
	⑤	0	0	0	0	−1	1	100	
T_2	⑥	1	−1	0	0	0	0	−200	row ①
	⑦	0	(1)	0	1	0	0	500	(−1)row ⑥ + row ②
	⑧	0	0	−1	1	0	1	300	row ③
	⑨	0	1	1	0	−1	0	300	row ④
	⑩	0	0	0	0	−1	1	100	row ⑤
T_3	⑪	1	0	0	1	0	0	300	row ⑫ + row ⑥
	⑫	0	1	0	1	0	0	500	row ⑦
	⑬	0	0	−1	1	0	1	300	row ⑧
	⑭	0	0	(1)	−1	−1	0	−200	(−1)row ⑫ + row ⑨
	⑮	0	0	0	0	−1	1	100	row ⑩
T_4	⑯	1	0	0	1	0	0	300	row ⑪
	⑰	0	1	0	1	0	0	500	row ⑫
	⑱	0	0	0	0	−1	1	100	row ⑲ + row ⑬
	⑲	0	0	1	−1	−1	0	−200	row ⑭
	⑳	0	0	0	0	(−1)	1	100	row ⑮
T_5	㉑	1	0	0	1	0	0	300	row ⑯
	㉒	0	1	0	1	0	0	500	row ⑰
	㉓	0	0	0	0	0	0	0	row ㉕ + row ⑱
	㉔	0	0	1	−1	0	−1	−300	row ㉕ + row ⑲
	㉕	0	0	0	0	1	−1	−100	(−1)row ⑳

Figure 4.42

Writing the equations that correspond to tableau T_5 yields

$$
\begin{aligned}
s \quad +v \quad\quad &= \ 300 \\
t \quad +v \quad\quad &= \ 500 \\
u-v \quad -x &= -300 \\
w-x &= -100
\end{aligned}
$$

By transposing the terms arising from columns in T_5 that are not in zero-one form (v and x columns), we obtain the following description of the solution candidates of our system.

$$
\begin{aligned}
s &= \ 300-v \\
t &= \ 500-v \\
u &= -300+v+x \\
w &= -100+x \\
&v \text{ is arbitrary} \\
&x \text{ is arbitrary}
\end{aligned}
$$

We leave it as an exercise to verify that these relations describe the solutions of (4.50) through (4.54).

Although this system has infinitely many solutions, only a small number of them make sense in terms of the network. Since the variables express the number of cars passing per hour between branches connecting intersection points of the network, the values given to these variables must, at the very least, be restricted to nonnegative integers (0, 1, 2, etc.). The requirement $s \geq 0$ yields:

$$
s = 300-v \geq 0, \qquad v \leq 300
$$

From $w \geq 0$ we have:

$$
w = -100+x \geq 0, \qquad x \geq 100
$$

From $u \geq 0$ we obtain:

$$
u = -300+v+x \geq 0, \qquad v+x \geq 300
$$

Thus traffic equilibrium cannot be maintained if more than 300 cars per hour pass between B and C, or fewer than 100 cars per hour pass between E and C, or the sum of the number of cars passing per hour between B and C and E and C is less than 300. This tells us something about the practical feasibility of the network.

EXERCISES

Solve the systems of equations stated in 1-6. Set up and solve the traffic network problem described in 7.

1.
$$x+3y-2z = 5$$
$$2x- y+3z = 10$$
$$-3x+5y-8z =-15$$

2.
$$3s+ t- 2u = 8$$
$$2s-3t+ 5u = 15$$
$$-s+7t-12u =-22$$

3.
$$2x+2y-2z = 8$$
$$x- y+3z =16$$

4.
$$x+2y-3z = 6$$
$$-2x+ y- z = 3$$
$$3x+ y+4z = 3$$
$$4x+3y+7z = 9$$

5.
$$x-2y+3z- w=2$$
$$-2x+5y- z+2w=4$$
$$-x+ y-2z+ w=2$$

6.
$$2s+ t+2u+ w=3$$
$$5s+3t- u+2w=2$$

7. Part of a traffic network being designed to service an envisioned shopping center is shown in Figure 4.43.

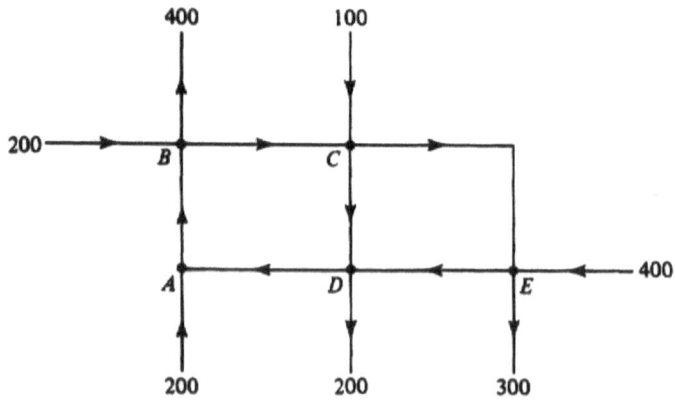

Figure 4.43

The total number of cars entering the network per hour during a peak period, 900, equals the total number leaving the network per hour, so that a basic equilibrium condition is satisfied.

On the basis of the equilibrium condition that the total number of cars entering each intersection point equals the total number leaving the intersection point, set up a system of linear equations to describe traffic flow in the interior of the network. Solve this system of equations and interpret the results obtained in terms of the network.

4.8   SYSTEMS OF LINEAR INEQUALITIES

Two or more linear inequalities involving the same variables, considered as a unit, are said to form a **system of linear inequalities**. For example,

$$2x-3y \leq 6 \tag{4.55}$$
$$5x+2y \geq 10 \tag{4.56}$$

and

$$x \geq 0 \tag{4.57}$$
$$y \geq 0 \tag{4.58}$$
$$x+3y \leq 90 \tag{4.59}$$
$$2x+ y \leq 80 \tag{4.60}$$

are systems of 2-variable linear inequalities while

$$x+2y+ z \leq 8 \tag{4.61}$$

$$2x-y-\frac{1}{2}z \leq 6 \tag{4.62}$$

and

$$x \geq 0 \tag{4.63}$$
$$y \geq 0 \tag{4.64}$$
$$z \geq 0 \tag{4.65}$$
$$2x+3y+ z \leq 11 \tag{4.66}$$
$$x+ y+3z \leq 10 \tag{4.67}$$
$$2x+2y+ z \leq 10 \tag{4.68}$$

are systems of three-variable linear inequalities.

The notion of solution of a system of inequalities is analogous to the notion of solution of a system of equations. For example, any ordered pair of numbers that satisfies all inequalities in the system

$$2x-3y \leq 6 \qquad (4.55)$$
$$5x+2y \geq 10 \qquad (4.56)$$

is said to be a solution of this two-variable system. Thus $(2, 1)$ is a solution, since the substitution of 2 for x and 1 for y in (4.55) and (4.56) yields

$$1 \leq 6$$
$$12 \geq 10$$

both of which are correct statements. The ordered pair $(1, 0)$ is not a solution since the substitution of 1 for x and 0 for y in (4.55) and (4.56) yields the statements

$$2 \leq 6$$
$$5 \geq 10$$

not both of which are correct.

Any ordered triple of numbers that satisfies all inequalities in the system

$$x \geq 0 \qquad (4.63)$$
$$y \geq 0 \qquad (4.64)$$
$$z \geq 0 \qquad (4.65)$$
$$2x+3y+ z \leq 11 \qquad (4.66)$$
$$x+ y+3z \leq 10 \qquad (4.67)$$
$$2x+2y+ z \leq 10 \qquad (4.68)$$

is said to be a solution of this three-variable system. Thus $(1, 2, 1)$ is a solution, since it satisfies inequalities (4.63) through (4.68), whereas $(2, 1, 3)$ is not a solution since it does not satisfy (4.67) (substitution yields $12 \leq 10$).

More generally, a **solution** of an n-variable system of linear inequalities is any ordered collection of n numbers that satisfies all inequalities in the system.

Graphs of Systems of Two-Variable Linear Inequalities[*]

To graph a system of two-variable inequalities we graph the individual inequalities on the same coordinate system and pick out the region which is common to them—that is, the overlap.

EXAMPLE 1

Sketch the graph of the system $\quad x+2y\le8$
$$3x+\ y\le9.$$

We first graph $x+2y\le8$. The boundary line $x+2y=8$ is shown in Figure 4.44. Since the test point $(0, 0)$ satisfies $x+2y\le8$, the graph of $x+2y\le8$ consists of all points on and below $x+2y=8$ as shown in Figure 4.45.

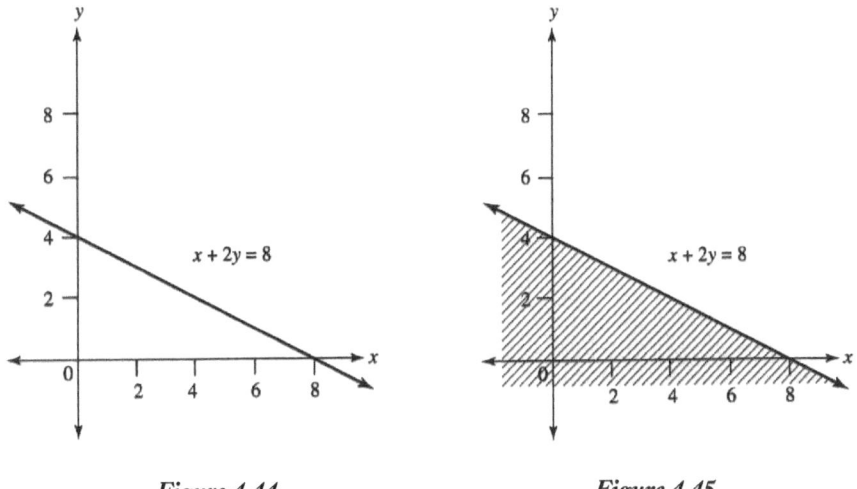

Figure 4.44   Figure 4.45

Sketching the graph of $3x+y\le9$ on the same coordinate system yields Figure 4.46. The overlap consists of all points on and below both boundary lines, shown in Figure 4.47.

---

[*] Please review section 2.5 at this point.

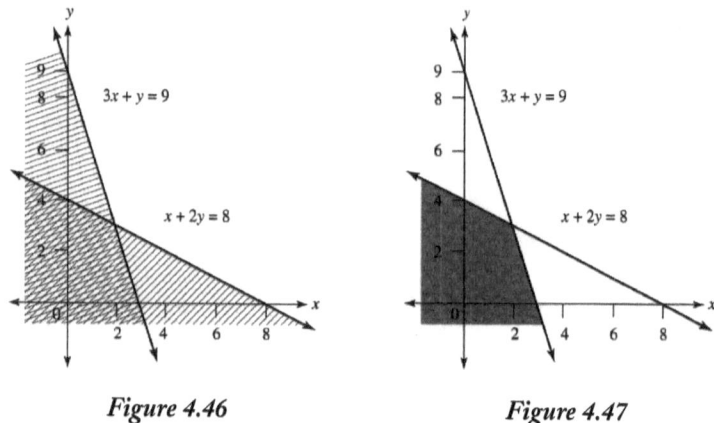

Figure 4.46   Figure 4.47

The nonnegativity conditions $x \geq 0$, $y \geq 0$ restrict us to the first quadrant. It is useful to make mental note of this because many problems of interest require such conditions. We encounter such problems in Chapter 6.

EXAMPLE 2

Sketch the graph of the system

$$x \geq 0$$
$$y \geq 0$$
$$x + 2y \leq 8$$
$$3x + y \leq 9.$$

All we have to do is restrict the graph shown in Figure 4.47 to the first quadrant. The result is shown in Figure 4.48.

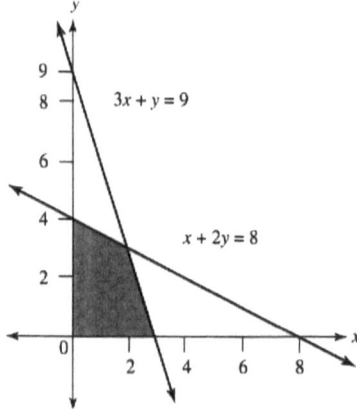

Figure 4.48

EXAMPLE 3

Sketch the graph of the system

$$x \geq 0$$
$$y \geq 0$$
$$2x + y \geq 23$$
$$3x + 5y \geq 75.$$

Since $x \geq 0$ and $y \geq 0$ restrict us to the first quadrant, it suffices to graph $2x + y \geq 23$ and $4x + 5y \geq 75$ restricted to the first quadrant. The graph of $2x + y \geq 23$ restricted to the first quadrant is shown in Figure 4.49(a), and the graphs of $2x + y \geq 23$ and $4x + 5y \geq 75$, restricted to the first quadrant, are shown in Figure 4.49(b).

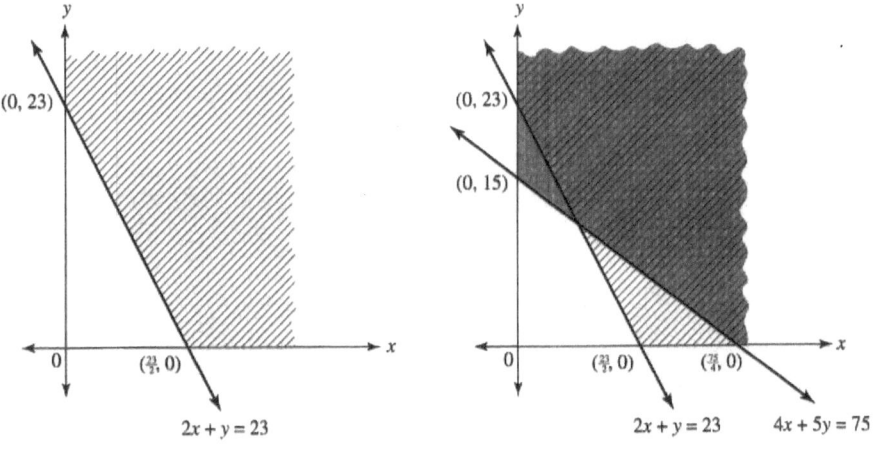

Figure 4.49(a)   Figure 4.49(b)

The graph of our system is the region in the first quadrant that is both shaded and lined in (the region in the first quadrant above and including both boundary lines) and is shown in Figure 4.50.

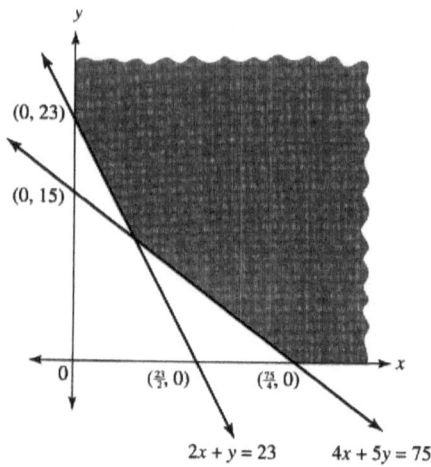

Figure 4.50

EXERCISES

1. Which of $(2, 3)$, $(4, 1)$, $(-1, 3)$, $(\frac{1}{2}, 2)$, $(6, -1)$, and $(3, 1)$ are solutions of the following system? Explain.

$$2x + 4y \le 15 \qquad\qquad (4.69)$$
$$3x + \ y \le 10 \qquad\qquad (4.70)$$

2. Which of $(8, 2)$, $(8, 17)$, $(20, 5)$, $(9, 12)$, $(5, 4)$, $(12, 2)$, and $(14, 7)$ are solutions of the following system? Explain.

$$x \ge \ 0 \qquad\qquad (4.71)$$
$$y \ge \ 0 \qquad\qquad (4.72)$$
$$x + \ y \le 25 \qquad\qquad (4.73)$$
$$-x + 4y \ge \ 0 \qquad\qquad (4.74)$$
$$x \ge \ 8 \qquad\qquad (4.75)$$

3. Which of $(2, 1, 3)$, $(1, 4, 1)$, $(4, 3, 2)$, $(3, 1, 1)$, $(4, 2, 2)$, $(1, 5, 1)$, $(1, 2, 1)$, and $(0, 4, 3)$ are solutions of the following system? Explain.

$$x + 2y + 3z \le 12 \qquad\qquad (4.76)$$
$$5x - \ y + \ z \le \ 9 \qquad\qquad (4.77)$$

4. Which of $(1, 1, 2)$, $(0, 1, 3)$, $(2, 1, 1)$, $(3, -1, 2)$, $(4, 1, 0)$, $(2, 1, 2)$, $(1, 3, 1)$, and $(0, 2, 3)$ are solutions of the following system? Explain.

$$x \geq 0, y \geq 0, z \geq 0$$
$$2x + 3y + z \leq 11$$
$$x + y + 3z \leq 10$$
$$2x + 2y + z \leq 10$$

Sketch the graphs of the following systems of inequalities.

5. $x \geq 0, y \geq 0$
    $x + y \geq 3$

6. $x \geq 0, y \geq 0$
    $2x + y \geq 6$
    $x \geq 3$

7. $x + y \leq 5$
    $2x + 3y \leq 12$

8. $x \geq 0, y \geq 0$
    $x + y \leq 5$
    $2x + 3y \leq 12$

9. $2x + y \geq 10$
    $x + 2y \leq 12$

10. $2x + y \leq 10$
    $x + 2y \geq 12$

11. $x + y \geq 5$
    $2x + 3y \geq 12$

12. $x + y \leq 5$
    $2x + 3y \geq 12$

13. $x + 4y \leq 10$
    $3x + y \leq 8$

14. $x \geq 0, y \geq 0$
    $x + 2y \geq 10$
    $3x + y \geq 15$
    $x + 2y \geq 4$

15. $x \geq 0, y \geq 0$
    $x + 2y \geq 4$
    $x - 2y \geq 2$

16. $x \geq 0, y \geq 0s$
    $2x + y \geq 4$
    $x + y \geq 3$

17. $x \geq 0, y \geq 0$
    $x + y \leq 25$
    $-x + 4y \geq 0$
    $x \geq 8$

18. $x \geq 0, y \geq 0$
    $5x + 4y \leq 120$
    $x \geq 8, x \leq 16,$
    $y \geq 6$

19. $x \geq 0, y \geq 0$
    $4x + 3y \leq 320$
    $5x + 2y \leq 330$

4.9   PREFACE TO LINEAR PROGRAMMING

A **linear function in two variables**, x and y, say, is a function with the structure

$$F(x, y) = Ax + By + C$$

where A, B, and C are constants.

Linear functions in n variables are defined in an analogous manner. Thus a **linear function in three variables,** x, y, and z, is one with the structure:

$$F(x, y, z) = Ax + By + Cz + D$$

Of particular interest in linear programming (chapter 6) are linear functions with domains of definition described by systems of linear inequalities.

EXAMPLE 1

Consider $F(x, y) = 5x + 10y$, where x and y must satisfy the following conditions:

$$x \geq 0 \tag{4.78}$$
$$y \geq 0 \tag{4.79}$$
$$2x + 3y \leq 24 \tag{4.80}$$
$$x + 4y \leq 16 \tag{4.81}$$

$F(x, y) = 5x + 10y$ is a linear function with $A = 5$, $B = 10$, and $C = 0$. Inequalities (4.78) through (4.81) describe the domain of definition of this function, and Figure 4.51 gives us a geometric view of this domain of definition.

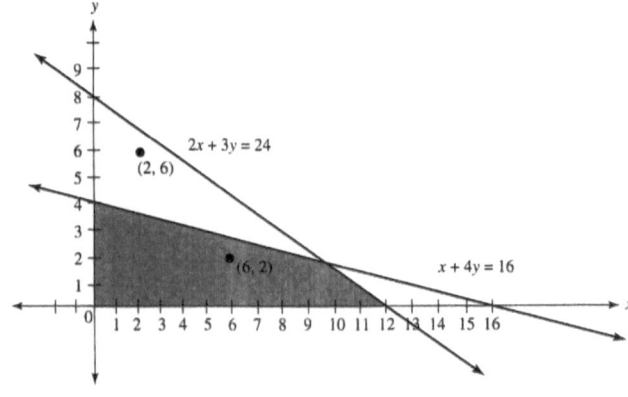

Figure 4.51

$F(x, y) = 5x + 10y$ is defined only for points that satisfy (4.78) through (4.81), that is, for points in the region shown in Figure 4.51. For example, $(6, 2)$ satisfies

(4.78) through (4.81); thus $F(6,2)$ is defined and equals $5(6)+10(2)=50$. The point $(2, 6)$, on the other hand, is not in the domain of definition, (4.81) is not satisfied, and thus $F(2, 6)$ is not defined.

Linear programming problems are concerned with finding the largest or smallest value of a linear function where the variables are required to satisfy linear conditions. In terms of Example 1 a linear programming problem emerges when we ask: What values of x and y satisfy (4.78) through (4.81), that is, are contained in the region shown in Figure 4.51, and yield the largest value for $F(x,y)=5x+10y$?

We apply the algebra developed in this chapter to develop a method for solving such problems in Chapter 6, and also examine a spectrum of situations which lead to such problems.

As a prelude to these developments we look at the nature of mathematical modeling in the next chapter.

EXERCISES

Which of the following are linear functions? Explain.

1. $F(x,y)=3x+\dfrac{1}{y}$ 2. $F(x,y,z)=3x+4y+2xz$

3. $F(x,y)=\dfrac{1}{2}x-3y-10$ 4. $F(x,y,z)=3x+2y+z^2$

5. For $C(x,y)=5x+4y$, where x and y satisfy $x\geq 0$, $y\geq 0$, $2x+y\geq 23$,

    $4x+5y\geq 75$, determine $C(0,15)$, $C(0,23)$, $C(\dfrac{23}{2},0)$, $C(\dfrac{20}{2},\dfrac{29}{3})$, and $C(\dfrac{75}{4},0)$.

6. For $C(x,y)=2x+3y+150,000$, where x and y satisfy $x\geq 0$, $y\geq 0$,

    $x+y\leq 15,000$, $x\leq 10,000$, $y\leq 12,000$, $x+y\geq 2000$, determine $C(0,$

    $2000)$, $C(0, 12,000)$, $C(3,000, 12,000)$, $C(15,000, 0)$, and $C(10,000, 5000)$.

7. For $I(x,y)=0.09x+0.06y$, where x and y satisfy $x\geq 0$, $y\geq 0$, $x+y\leq 25$,

    $-x+4y\geq 0$, $x\geq 8$, determine $I(8, 2)$, $I(25, 0)$, $I(8, 17)$, and $I(20, 5)$.

CHAPTER 5

More on Mathematical Modeling

5.1   WHAT IS THE COST OF SMOKING?

For many years an artist friend of mine declined to act on the health warnings of his doctor about the consequences of smoking. But recently the doctor inquired as to whether he had any idea about how much it was costing him, and proceeded to note a few figures. My friend gave up smoking immediately.

Jules Warner's Math Model for the Cost of Smoking

To be sure, not everyone is prepared to follow in my friend's footsteps, but I believe it would be informative to give attention to the cost of smoking from a math modeling perspective. Motivated by a desire to be "cool" and fit in with the crowd, young people sometimes take up smoking while in their teens.

Jules Warner considered the case of a teenager who takes up smoking at fifteen and persists with the practice for 50 years. The price of a pack of cigarettes will vary over 50 years, but considering what the current price range of a pack is, suppose $4.00 is taken as an estimate of the average price of a pack over the next 50 years, Jules reflected. A new smoker may begin modestly with a few cigarettes a day, and then work up to a pack, 2 packs, and perhaps eventually 3 packs or more a day. Take a pack and a half as an estimate of the average amount smoked per day, over 50 years, Jules thought it reasonable to assume based on the data he had obtained. This yields a cost figure of 1.5 packs per day, times $4.00 per pack, times 365 days, equals $2,190 per year, on average. Multiplying this figure by 50 yields $109,500 over a period of 50 years.

If our smoker gets married and has children he would probably give serious thought to obtaining life insurance, Jules considered. Insurance rates differ for smokers and non-smokers, and what might cost a smoker $1000 a year might only be on the order of $700 a year for a non-smoker, for a $300 difference, Jules' data suggested to him. If the insurance is held over 30 years, say, the additional cost for a smoker comes to $9000.

And then there are cleaning bills. If our smoker is not happy about his clothes, furniture, carpets, and draperies reeking of the smell of smoke and decides to do something about it, the extra cleaning bills might come to the order of $400 per year. Assume 30 years for this too, Jules thought, and the cost comes to $12,000.

And then there is the matter of clean teeth. If our smoker is sensitive about yellow teeth, he might require at least one extra cleaning a year at, Jules thought it reasonable to assume, $90 a cleaning. If 30 years is taken for this too, we have $2,700 as the total cost for teeth cleaning.

Jules' assumptions serve as a starting point for his analysis. Such are called **postulates**, or **axioms**. If we take Jules' postulates as a starting point, then we must accept his conclusions as following as inescapable consequences of them. They are called **valid conclusions**, or **theorems**. They are correct in a deductive logical sense. Mathematical methods, whether based on familiar addition, subtraction, multiplication and division or are the ultimate in technical mathematical sophistication, yield valid conclusions/theorems. Whether or not these theorems are true/realistic descriptions of real-world doings is another matter entirely. Mathematical validity and real-world truth/realism issues arise in every situation involving the application of mathematics to real-world doings.

In summary, Jules' math model portrait for the cost of smoking consists of the following postulates and theorems:

Jules Warner's Postulates and Theorems

Postulates

P1. A person takes up smoking at fifteen and continues the practice for 50 years.

P2. The estimated average cost of a pack of cigarettes over the 50 year period is $4.00.

P3. The estimated average amount smoked over 50 years is a pack and a half a day.

P4. The additional cost of life insurance for a smoker is $300 a year. This applies over a period of 30 years.

P5. The additional cost of cleaning clothes, furniture coverings, drapes, carpets, etc. for a smoker is, on average, $400 a year over 30 years.

P6. The additional cost of teeth cleaning for a smoker is, on average, $90 a year over 30 years.

Theorems

T1. The estimated cost of cigarettes over 50 years is $109,500.

T2. The estimated additional cost of life insurance over 30 years is $9,000.

T3. The estimated additional cost of cleaning clothes, furniture coverings, drapes, carpets, etc. over 30 years is $12,000.

T4. The estimated additional cost of teeth cleaning over 30 years is $2,700.

T5. The estimated total cost of smoking over 50 years is, rounded off, $133,000.

It all looks so precise, and in a deductive-logical sense, which defines the meaning of mathematical sense, it is.

> If you grant Jules' postulates P1 through P6 as a starting point, you must also grant his theorems T1 through T5 as being valid consequences of them.

T1 follows from P1, P2, and P3; T2 follows from P4; T3 follows from P5; T4 follows from P6. The mathematical tool in each of these cases is our good old friend multiplication. T5 follows from T1 through T4 by another good old friend, addition. We should be careful not to allow familiarity to breed a take-for-granted attitude. As humble as these good old friends might strike us, they are as successful in getting the job done in the sense of validity as the most sophisticated cutting edge mathematical tools around today.

Additional Strings Attached

There are additional strings attached to this story which we should not overlook, and these are concerned with the features Jules left out. The $133,000 figure is in terms of today's dollars. Jules did not attempt to adjust for inflation over 50 years. He also did not consider other factors which may play a role in the cost of smoking, and perhaps this should be explicitly noted as postulate P.

> **P.   Other factors that may play a role in the cost of smoking are not being considered.**

It is not current practice to explicitly list P with the postulates of a math model, but we would be wise not to fall into the out-of-sight, out-of-mind frame of mind.

What other factors are not being considered in this model? In an article which prompted the refinement developed here, Hubert Herring[*] notes some of the following:

---

[*] Hubert Herring, "Where There's Smoke, There's Outlay," *The New York Times*, April 27, 1997.

F1. In a study conducted by the National Bureau of Economic Research it was found that smokers earn 4 to 8 percent less than non-smokers. As herring points out, "this is a tricky statistic," and for this reason this dimension to the cost of smoking is not pursued here.

F2. Some 200,000 fires a year are started by smoking materials, which makes smokers more vulnerable to catastrophic loss.

F3. Expensive puffs. Some smokers have been known to take connecting flights so they could have a cigarette break at an airport, rather than taking a smoke-free cross-country flight.

F4. Smokers generally eat less than non-smokers, which translates to a saving on food.

F5. Smokers are generally more vulnerable to a number of health related problems, which translate to a cost. One dimension of this is seen in the difference in life insurance costs for smokers and non-smokers.

F6. Smokers' family members are generally more vulnerable to health related problems caused by secondary smoke, which translates to a cost.

Disagreement

On examining the postulates of Jules' model. Janet Wright, based on data she had obtained, was prompted to take issue with the realism of P2, P3, P5, and P6. Janet's data and study led her to introduce another math model for the cost of smoking by replacing P2, P3, P5, and P6 by P2a, P3a, P5a, and P6a, and introducing postulate P7 on health cost.

Postulates

P1. A person takes up smoking at fifteen and continues the practice for 50 years.

P2a. The estimated average cost of a pack of cigarettes over the 50 year period is $6.00.

P3a. The estimated average amount smoked over 50 years is two packs a day.

Janet Wright's Math Model

P4. The additional cost of life insurance for a smoker is $300 a year. This applies over a period of 30 years.

P5a. The additional cost of cleaning of clothes, furniture coverings, drapes, carpets, etc. for a smoker is, on average, $600 a year over 30 years.

P6a. The additional cost of teeth cleaning for a smoker is, on average, $150 a year over 30 years.

P7. The additional cost of health maintenance for a smoker is, on average, $600 a year over 50 years.

Here too, implicit in the formulation of Janet's postulates is the **assumption** that other factors which may play a role in the cost of smoking are not being considered (Postulate P).

Theorems

T1. The estimated cost of cigarettes over 50 years is $219,000.

T2. The estimated additional cost of life insurance over 30 years is $9,000.

T3. The estimated additional cost of cleaning clothes, furniture coverings, drapes, carpets, etc. over 30 years is $18,000.

T4. The estimated additional cost of teeth cleaning over 30 years is $4,500.

T5. The estimated additional cost of health maintenance over 50 years is $30,000.

T6. The estimated total cost of smoking over 50 years is, rounded off, $281,000.

Who's "Right"?

Jules: The total cost of smoking over 50 years is $133,000. Janet: the total cost of smoking over 50 years is $281,000. Who's "right"? Could both be "wrong"? What does "right" mean? What does "wrong" mean?

EXERCISES

1. Address the afore questions.

2. Arnold Jacob set up a math model for the cost of smoking based on the following postulates:

Jacob's postulates

P1. A person takes up smoking at fifteen and continues the practice for fifty years.

P2. The estimated average cost of a pack of cigarettes over the 50 year period is $5.00.

P3. The estimated average amount smoked over 50 years is 2.5 packs per day.

P4. The additional cost of life insurance for a smoker is $400 a year. This applies over a period of 30 years.

P5. The additional cost of health maintenance for a smoker is, on average, $1000 a year over 50 years.

   (a) What theorems can be deduced from Jacob's postulates concerning the cost of cigarettes, the additional cost of life insurance, the cost of health maintenance, and the total cost of smoking?

   (b) Jacob's theorem on the total cost of smoking disagrees with the theorems obtained by Warner and Wright in their models. Does this establish that Warner and Wright's theorems are not valid? Explain.

   (c) Jacob's postulates do not take into account the additional cost for smokers of cleaning of clothes, furniture coverings, etc., and teeth, whereas Warner and Wright's models do. Does this mean that Warner and Wright's conclusions on the total cost of smoking are valid, whereas Jacob's conclusion is not valid? Explain:

   (d) Does the fact that Warner and Wright's models take into account the additional cost for smokers of cleaning of clothes, furniture coverings, etc., and teeth make their models more realistic for the cost of smoking than Jacob's model? Explain.

   (e) "Cigarettes Up to $7 a Pack With New Tax," stated the headline of a *New York Times* article (July 1, 2002; B1). Does this result have significance for (i) Jacob's model? (ii) Warner's model? (iii) Wright's model? Explain.

5.2   MATH MODELS FOR A VACATION TRIP

Andy Plans a Vacation Trip

Recently, Andy was engaged in planning a car trip from home in Brooklyn, New York, to the popular vacation town of Kennebunkport, Maine, in mid-August. His problem was to set up a math model for the trip that would enable him to predict the total time required for the journey.

The setting of any such problem presents numerous features and characteristics, many of which are irrelevant or unessential to the focus of the problem. In developing his math model, Andy had to sort this out and decide on which features were fundamental and which were negligible. This required discretion and judgment, the most controversial aspect of the math model development process; one person's essential might be another's irrelevancy.

Andy examined a map and laid out a route. Based on data provided by friends who had recently made the trip, he made assumptions about departure time, weather conditions, the traffic flow to be expected along various points, speeds that would be possible, and the number of rest stops to be made and their duration. Such considerations led him to a math model consisting of a line segment 330 miles long joining points representing Brooklyn and Kennebunkport, and the problem of determining how long it would take an object moving at an average speed of 55 miles per hour to cover this distance.

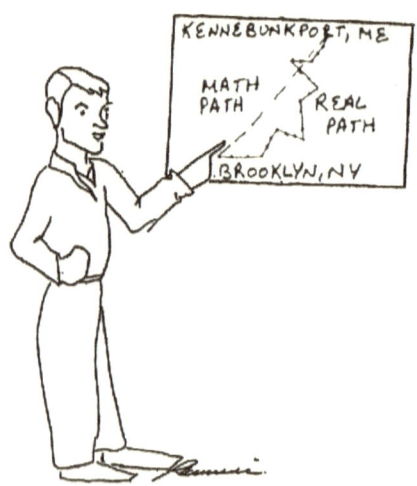

Andy's math model is an idealized, abstract rendering of the real situation involving a trip from Brooklyn to Kennebunkport. It is intended to capture the main features involved in taking such a trip and reflects these features as he sees them and the assumptions that he was led to make. As in the case of Jules vs. Janet concerning the cost of smoking, it is possible that someone else planning such a trip would see things in another light and compose a very different math model.

By employing division, we obtain the valid conclusion that an object moving along the idealized path of Andy's model at an average speed of 55 miles per hour would take 330 / 55 = 6, which translates to 6 hours to make the journey.

In summary, Andy's math model consists of the following:

Andy's Math Model

Andy's Postulates

P1: A line segment of length 330 miles joining points representing Brooklyn and Kennebunkport is taken as an idealized representation for the actual path of the trip.

P2: A point moving at an average speed of 55 miles per hour from the point representing Brooklyn to the point representing Kennebunkport along the line segment is taken as an idealized representation of the car trip itself.

Implicit in the formulation of Andy's postulates is the **assumption** that other factors that may play a role in the time it would take to make the trip are not being considered (Postulate P).

Andy's Theorem

T1: The time required to make the trip is 6 hours.

Andy's math model is a math portrait of his envisioned journey. It obviously is not a photographic likeness, but as with people portraits, which may vary considerably depending on the artist, we are prompted to inquire: Is it accurate?

One grip on this question is obtained through Andy's theorem. The acid test involves undertaking the journey, noting the time required, and comparing it with the projected time of 6 hours from Andy's theorem. If there is a "small" discrepancy between the actual and projected times, this would establish that Andy's theorem is realistic in this case and, by reflection, be evidence in support of the realism of his math model portrait of the journey. If there is a "large" discrepancy between the actual and projected times, this would establish that Andy's theorem is not realistic in this case and, reflecting back, lead us to conclude that his model is not a realistic portrait of the journey at hand.

> The status of Andy's theorem as a valid conclusion derived from his postulates is not at stake here. Division, yielding 330 / 55 = 6, establishes the validity of Andy's conclusion and confers on it the status of theorem.

To make a judgment on whether his valid conclusion is realistic or not, we must turn to the real world. This reduces to taking the trip and observing the outcome.

Henry's Math Model

Andy's friend Henry formulated an alternative math model for Andy's trip based on a route that would take Andy north on the Belt Parkway through Manhattan (see Fig. 5.1)

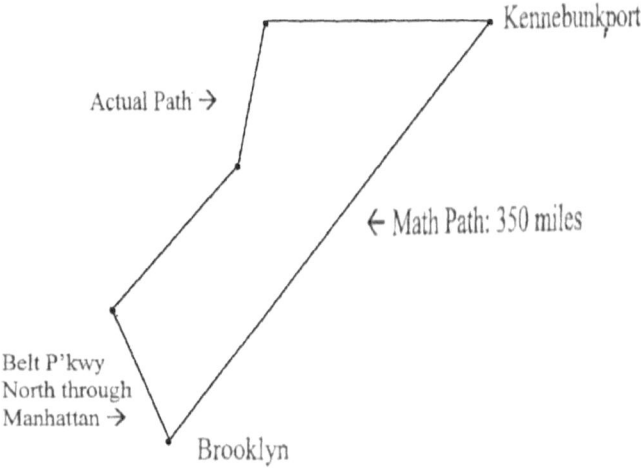

Figure 5.1

In summary, Henry's Model consists of the following:

Henry's Math Model

Henry's Postulates

P1a: A line segment of length 350 miles joining points representing Brooklyn and Kennebunkport is taken as an idealized representation for the actual path of the trip.

P2a: A point moving at an average speed of 50 miles per hour from the point representing Brooklyn to the point representing Kennebunkport is taken as an idealized representation of the car trip itself.

P.   Factors other than those considered by Andy and me that may play a role in the time it takes to make the trip are negligible.

Henry's Theorem

T1a: The time required to make the trip is 7 hours.

A number of questions arise.

EXERCISES

1. Prove Henry's theorem T1a. Does your "proof" of Henry's theorem establish that if Andy follows Henry's route the trip time will be approximately 7 hours? Explain.

2. How could Andy determine which, if either, of the models before him is realistic? Explain.

3. Is it possible that neither model is realistic? Explain.

4. Andy reflected on his two models and asked his cousin Algis for his advice. "Which one should I implement to take my vacation trip?" "It's a no brainer", Algis replied. "You want to make the trip in the shortest time, so implement the model you formulated." Do you agree with Algis? Explain.

5. Andy implemented the model he had formulated and the trip took him 5 hours and 55 minutes. "That proves it," said Algis to Andy. "Proves what?" asked Andy. "It proves that Henry's conclusion T1a on the travel time is not valid."

    (a) Do you agree with Algis? Explain.
    (b) If you do not agree with Algis, what does the 5 hour and 55 minutes travel time prove, if anything? Explain.

Rasa's Vacation Trip

Andy's sister Rasa was planning to take a brief vacation trip to Kennebunkport during Labor Day weekend. Andy's math model worked well for him and she decided to follow the route it prescribed, expecting the journey to take around 6 hours.

Rasa took the trip Labor Day weekend as planned, but it took her 7 hours. This actual trip differs considerably from the projected 6 hour trip time of Andy's theorem so that something clearly went wrong; but what? "Your theorem stinks," Rasa shouted at her brother in a somewhat agitated manner. "It's wrong, it's not valid," she continued.

> **Rasa's experience proved Andy's theorem wrong in terms of reality, but not validity.**

The distinction, as we have observed, is a fundamental one. In the confrontation between what actually happened and what the theorem says should happen, what actually happened—reality—wins. Andy's theorem is a false statement as a description of the travel time to Kennebunkport on a Labor Day weekend, but it remains a theorem. It is still an inescapable consequence of Andy's model—more specifically, his postulates ($330 \div 55$ is still 6) and this is what makes a theorem a theorem.

Rasa's experience sends us a signal that Andy's postulates are not realistic for travel to Kennebunkport on a Labor Day weekend. In reexamining Andy's postulates, we find that they do not realistically take into account unusually heavy traffic delays, characteristic of holiday weekends, around the tollgates of the Whitestone Bridge. Further examination of Rasa's actual trip shows that this is where she had the difficulty.

Rasa undertook to modify Andy's model to make it more realistic for travel to Kennebunkport on a Labor Day weekend. She reviewed Andy's assumptions and took into account data on traffic delays around the tollgates of the Whitestone Bridge on holiday weekends. This led her to the following refinement of Andy's model.

Rasa's Refinement of Andy's Math Model

Rasa's Postulates

P1: A line segment of length 330 miles joining points representing Brooklyn and Kennebunkport is taken as an idealized representation for the actual path of the trip.

P2: A point moving at an average speed of 49 miles per hour from the point representing Brooklyn to the point representing Kennebunkport along the line segment is taken as an idealized representation of the car trip itself.

P. Factors other than those considered by Andy and me to make the trip are negligible.

Rasa's Theorem

T1: The time required to make the trip is 6 hours and 44 minutes.

Rasa's theorem is in close agreement with her experience, which is evidence in favor of her math model as a realistic portrait of travel from Brooklyn to Kennebunkport on Labor Day weekends under current conditions.

NINO vs. RIRO

Whether Rasa's math model will serve as a suitable travel portrait in the future depends on the extent to which **current conditions** are maintained. Andy's model and Rasa's modified version of it took into account traffic delays due to road repairs that were in progress. When these repairs are completed, current conditions will have changed in a significant way and a suitable modification of both Andy's and Rasa's models would be in order to keep them current.

As we have seen,

> One way to get a grip on the question of a math model's realism is
> through its theorems. The other way is through its assumptions,
> formulated as postulates. Are the assumptions realistic?

We should always keep in mind that in addition to the explicitly stated
assumptions there is the implicit assumption that the factors not explicitly
addressed by the assumptions are not being addressed or are being viewed as
negligible to the focus of the situation under study. If the answer is *yes,* the
assumptions are realistic, and this is a correct assessment, then the theorems
will be realistic as well. The RIRO principle operates; if realistic input, then
realistic output in terms of theorems. If the answer is *no* to at least some of the
assumptions, and this is a correct assessment, then we cannot be sure about the
realism of the theorems. Some might be right on target, whereas others might
be considerably off reality's mark. The NINO principle operates; if nonsense
in, there is a good chance of nonsense out, even though some theorems might
be realistic.

Rasa was caught by the NINO machine. She took over Andy's math model
without carefully examining its assumptions and paid the price in terms of an
unrealistic theorem about travel time.

> None of these considerations have any affect on the status of
> Rasa's theorem as a theorem.

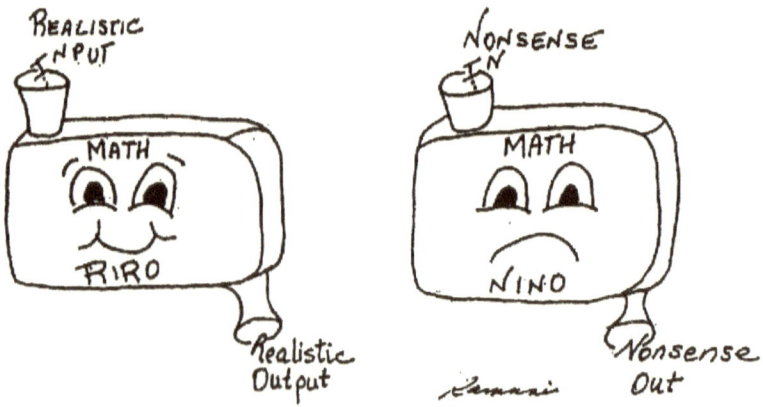

EXERCISES

6. Rasa's cousin Asta was planning to take a vacation trip from home in Woodhaven, Queens, to Kennebunkport, Maine, at the end of July. Based on data she collected, Asta formulated the following postulates:

Asta's Postulates

P1. A line segment of length 300 miles joining points representing Woodhaven and Kennebunkport is to be taken as an idealized representation for the actual path of the trip.

P2. An object moving at an average speed of 60 miles per hour from the point representing Woodhaven to the point representing Kennebunkport along the line segment is to be taken as an idealized representation of the car trip itself.

P. Factors not considered in the formulation of P1 and P2 have a negligible impact on the vacation trip.

(a) What theorem can be deduced from Asta's postulates concerning the time it would take to make the trip?

(b) Asta took the trip at the end of July as planned, but the journey took her six hours. Does the discrepancy between the projected trip time obtained from her postulates and the time it took her to make the trip establish that her conclusion is not valid? Explain.

(c) If your answer to (b) is no, what does this discrepancy establish? Explain.

7. Ann was planning to take a vacation trip from home in Brooklyn, New York, to Kennebunkport at the end of July with a major stop at Putnam, Connecticut for the annual picnic held there. Based on data she collected, Ann formulated the following postulates:

Ann's Postulates

P1. A line segment of length 250 miles joining points representing Brooklyn and Putnam and one of length 110 miles joining points representing

Putnam and Kennebunkport is to be taken as an idealized representation for the actual path of the trip.

P2. An object moving at an average speed of 50 miles an hour from the point representing Brooklyn to the point representing Putnam is to be taken as an idealized representation of the first leg of the car trip. For the second leg it's an object moving at an average speed of 55 miles per hour.

P3. Time at the Putnam picnic: 2 hours.

P. Factors not considered in the formulation of P1, P2, and P3 have a negligible impact on the vacation trip.

(a) What theorem can be deduced from Ann's postulates concerning the time it would take to make the (i) first leg of the trip?, (ii) the second leg?, (iii) the trip in total from Brooklyn to Kennebunkport?

(b) In making the trip it took Ann 5 hours and 35 minutes to cover the first leg, 1 hour and 50 minutes to cover the second leg, and 9 hours and ten minutes to make the entire trip from Brooklyn to Kennebunkport. Does this establish that the conclusion you obtained in answer to (a), part (i), is not valid, but that the result you obtained in answer to (a), Part (iii), is valid? Explain.

(c) If your answer to either or both parts of (b) is no, then what do the given data establish? Explain.

(d) Does the validity of the theorems obtained from the postulates of a math model portrait of a situation under study mean that all or none of theorems must be realistic? Explain.

5.3   THE MATH MODEL BUILDING PROCESS

The development of math models for the cost of smoking and Andy and Rasa's math model experiences reflect in miniature the general nature of math model building. It is useful to identify in general terms the factors involved in math modeling.

The development of a math model for a situation consists of the following steps.

> *1.   Specify the situation (process or problem) to be studied.*
>
> *2.   Collect data about the situation, make "suitable" assumptions about the factors involved, and formulate an idealized representation or portrait, if you will, for it.*

This is the **hypothesis of the math model,** and the individual statements that make it up are its **postulates.** They are based on the data collected and assumptions made from studying the situation.

Reflecting on Andy's development of a model for his trip, we see that he collected data about distances of various parts of the route he intended to take, travel times, and made assumptions concerning departure time, weather conditions, number of rest stops and their duration, and a number of other factors. Andy's postulates are a crystallization of these data and assumptions.

> *3.   Apply "math proof" to obtain valid conclusions, the theorems, with respect to the hypothesis of the model.*

In principle, it matters not whether the math methods employed to carry out math proof are simple, as in the cost of smoking and Andy's trip situations, or the ultimate in technical sophistication.

> 4. *Test the accuracy of the model as a portrait of the situation under study by comparing what the theorems say about the behavior of the situation with its actual real-world behavior.*

> "Close" agreement between what the theorems say about reality's behavior and reality itself is viewed as evidence in support of the math model as a realistic, though idealized, portrait of the situation. It does not establish that the model is in any sense a "perfect" description.

Just as a person's portrait may be sketched in many ways, so too may a math model portrait be developed in many ways. The math artist must exercise judgment and discretion as to which features of the situation are of paramount importance and must be taken into account in the portrait developed in the model, and which features are negligible and need not be reflected by the model. This calls for insight and sensitivity which, needless to say, cannot be pinned down in a sharp, unequivocal fashion. We can expect different math models to emerge from math artists who see a situation in different terms.

The models formulated by Jules and Janet for the cost of smoking and those formulated by Andy and Henry for a vacation trip time are illustration of this most important point.

Suppose two or more math models become available for a situation. How do we choose between them? The decisive verdict is rendered by the behavior of reality. The fundamental question is: Are the statements made by a math model's theorems in close agreement with the behavior of realty? When a math model theorem is found to be in disagreement with reality, this tells us that some part of the model's hypothesis is incomplete or unrealistic as a portrait of the situation at hand.

A theorem found to be false does not lose its status as a theorem. It remains a valid consequence of the hypothesis of the model, correct in the sense of validity, incorrect in the sense of reality.

5. *Refine the math model portrait of the situation under study when observations of the situation's actual behavior differ "significantly" from the behavior predicted by the model's theorems.*

Rasa's experience with the travel time theorem of Andy's model led her to modify his model to make it realistic for Labor Day weekend travel from Brooklyn to Kennebunkport.

Figure 5.2 summarizes the basic steps in the development of a math model in diagrammatic form.

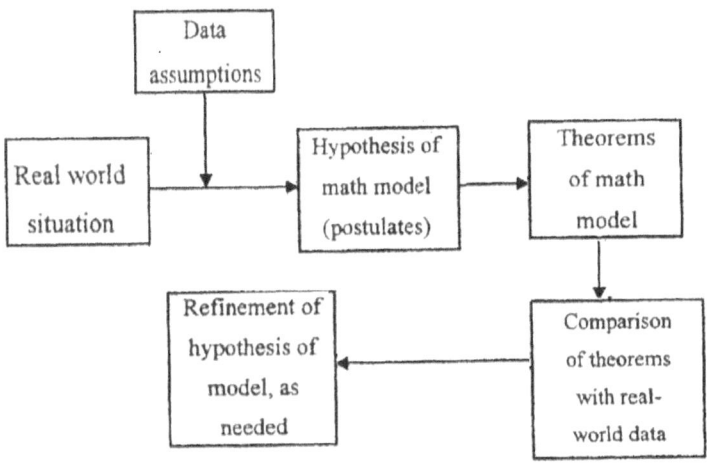

Figure 5.2 Development of a Math Model.

A Second Dimension of Math Modeling

Another dimension of math modeling arises when the theorems obtained from the model either cannot be compared with the behavior of reality or doing so might have dire consequences. The Axel Company, a producer of television sets, is eager to introduce two new models, TV-1 and TV-2, into the market. A theorem of a math model M1 developed for the company by its operations research department says that a monthly production schedule of 500 TV-1 and 600 TV-2 units would generate the largest possible monthly profit of $300,000. One way to determine if this theorem is realistic is to implement the production schedule and see what happens. If what happens is that a substantial monthly loss is incurred, the Axel Company could be forced out of business. Understandably, it is hesitant to take the plunge.

> **The alternative is to carefully, very carefully, examine the hypothesis of the production model. If it is indeed agreed that it is realistic, and this is a correct assessment, RIRO (if realistic input, realistic output) will prevail and it would make sense to implement the production schedule.**

If there are reservations about the realism of the assumptions, beware of NINO (if nonsense in, nonsense out). This is the best advice that can be followed in general. Figure 5.3 summarizes this dimension of math modeling.

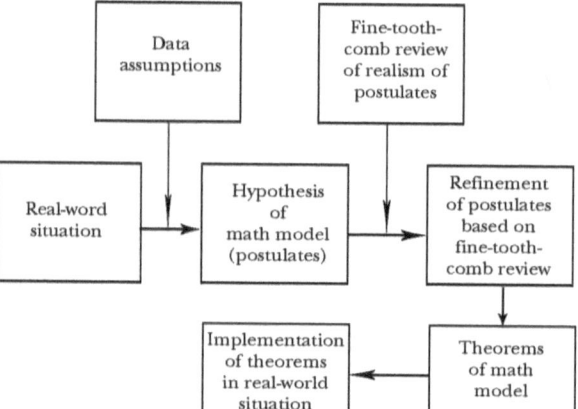

Figure 5.3 Development and implementation of a math model.

5.4   MODELING VS. REALITY AS SEEN BY ART BUCKWALD

Some Call It Art

The large doses of economic news we are all getting are not educating most Americans, but only confusing them.

Very few people understand them. Fortunately, I know someone who does. He is an economist named Alfred Daffy, and he endeared himself to the Reagan people with his economic theory that you can solve any problem if you throw enough Trojan horses at it.

When I first met Daffy he had constructed an economic model for unbelievable prosperity, full employment, and a surplus in the Treasury. It was a work of art, done in smooth clay without a line out of place. People from all over the country came to admire it; there was even talk Alfred might end up with a Nobel Prize.

I went to see Daffy at his studio the other day, and he had the model all torn apart.

"What are you doing?" I asked.

"I have to rework it," he said. "There are a few things I hadn't counted on." He took an enormous glob of clay and threw it at the side of the model.

"What's that?"

"The recession. On my original model I only allowed a little clay for a mild recession. Now we're in a real one and that puts my whole model out of kilter." He took another glob and put it on the other side. "There, that should balance it."

"What does that glob represent?"

"Unemployment. You can't have a large recession without large unemployment." He studied his model for a few moments, and then took some clay from the bottom and put it on the top.

"In my original model, I had interest rates down here. I never figured on them being up here."

"But they're falling," I said.

"Not for long," he said, grabbing a glob of clay in both hands and dumping it on top of the model. He took another glob and dumped that on top of the first one.

"What are you doing?" I cried.

Daffy said, "I'm adding a hundred-million-dollar deficit that wasn't in the first model."

"Why wasn't it there?"

"Because in my original model, everyone was going to get a tax cut which would spur the economy, and with more people working there would be more money going into the Treasury than the government was paying out, and we would have a surplus."

"What went wrong?"

Daffy kept throwing clay at his model indiscriminately. "The savings in government spending weren't there and the military budget jumped to over two hundred billion."

"That model is starting to look a mess," I said.

"I'm not through with it yet. Consumer spending is nil, our balance of payments is way out of whack, and the Gross National Product is down to zero."

"Are you sure you have enough clay?" I asked him.

There were tears in his eyes. "I created a masterpiece. Everything in the model was supposed to work. They were going to put it up in Rockefeller Center in place of the Christmas tree."

I tried to console him. "Albert, you're being too hard on yourself. Economists aren't scientists—they're dreamers. And they translate their dreams into beautiful works of art such as your original model. President Reagan may not know much about art, but he knows what he likes. And he wouldn't have bought the other model if he didn't like it."

"Yeah, but what is he going to think of this one?"

"Well, to be honest, it may not be to his taste. But he paid for it, so he's going to have to live with it."*

EXERCISES

1. The Gaja Company makes shoes. The Company plans to introduce two new styles, designated by A-18 and A-21, for the fall season and management wants to know how to set its monthly production schedule so as to maximize profit. Two consulting firms were hired to study this problem and to make recommendations. Each consulting firm formulated a mathematical model, designated by M1 and M2 for convenience. A conclusion obtained from M1 states that to maximize profit 50,000 A-18 pairs and 35,000 A-21 pairs should be produced monthly with an anticipated monthly profit of $230,000. A conclusion obtained from M2 states that to maximize profit 40,000 A-18 and 50,000 A-21 pairs should be produced monthly with an anticipated profit of $200,000.

    (a) If mathematics is the precise subject it is reputed to be, should there not be one solution to the problem of developing a production schedule to maximize profit instead of two? Explain.

---

* Art Buckwald, *While Reagan Slept* (New York: G. P. Putnam's Sons, 1983), 115-16.

(b) Since two solutions emerge, does it follow that not both are valid? Explain.

(c) Would you implement model M1 because it projects a larger monthly profit than M2? Explain.

(d) It came to pass that M1 was implemented. Subsequently, the Gaja Company sustained a consistent monthly loss on the A-18 and A-21 styles and discontinued them. How can one explain this?

The Gaja Company's Problem vs. Andy's Vacation Trip Problem

Outer appearances make these situations seem very different, but structurally speaking they are much the same. With Andy's trip problem we have Andy's vs. Henry's model with valid conclusions 6 hours vs. 7 hours for the trip time. With the Gaja Company's production problem we have math models M1 and M2 with valid conclusions $230,000 monthly profit vs. $200,000 monthly profit.

A trip time of 6 hours and monthly profit of $230,000 are more desirable than the alternatives, 7 hours and $200,000, but should the bait of what is desirable prompt them to swallow Andy's model and model M1 whole without careful thought? The answer that discussion of these scenarios hopefully make clear is, **NO, NO, NO**.

Serious thought should be given to the realism of these models if the potential for relatively mild inconvenience on the one hand and financial catastrophe on the other are to be avoided.

The same situation arises with linear program modeling considered in the next chapter. The difference between models M1 and M2 for the Gaja Company and LP-1 and LP-2 for the Austin Company considered in chapter 6 is that the specific assumptions behind LP-1 and LP-2 are given.

More information concerning appearances, but the same situation structurally speaking.

CHAPTER 6

Linear Program Models

6.1 THE BIRTH AND REBIRTH OF LINEAR PROGRAMMING

The seed from which the discipline called linear programming first germinated was planted in the late 1930's when the Leningrad Plywood Trust approached the Mathematics and Mechanics Department of Leningard University for help in solving a production scheduling problem of the following nature. The Plywood Trust had different machines for peeling logs for the manufacture of plywood. Various kinds of logs were handled and the productivity of each kind of machine (that is, the number of logs peeled per unit of time) depended on the wood being worked on. The problem was to determine how much work time each kind of machine should be assigned to each kind of log so that the number of peeled logs produced is maximized. A basic condition which had to be satisfied is that if logs of a given type of wood make up a specified percent of the input, then peeled logs of that type would also constitute that percent of the output.

The germination of this seed is due to Leonid Kantorovich who saw that it together with a wide variety of economic planning problems can be formulated in terms of what are today called linear program models. These problems involved the optimum distribution of worktime of machines, minimization of scrap in manufacturing processes, best utilization of raw materials, optimum distribution of arable land, optimal fulfillment of a construction plan with given construction materials, and the minimal cost plan for shipping freight from given sources to given destinations.

A simple example of a linear programming problem is a linear program involving two variables, x and y, let us say. Find the largest value of $P = 150x + 82y$, where x and are required to be non-negative and satisfy the linear constraint $2x + 3y \leq 120$.

More generally, a linear program, which may involve hundreds and even thousands of variables, is a problem requiring that we find the largest or smallest value of a linear function where the variables must satisfy linear constraints.

In 1939 Kantorovich published a report on his discoveries that included a method sufficient for solving all the linear program models he had formulated for the aforenoted problems. The chaos of the Second World War and the postwar intellectual climate in the Soviet Union were highly unfavorable for the development and implementation of Kantorovich's linear programming methods in the Soviet economic scene. Independently of the Soviet scene, linear programming methods were developed in the United States and Western Europe in the late 1940's, and the 1950's and 60's saw the development of a wide variety of linear program models for problems arising in such areas as economic planning, accounting, banking, finance, industrial engineering, and marketing. The thaw in the Soviet Union's intellectual climate which followed Joseph Stalin's death in 1953 saw the development and implementation of Kantorovich's ideas in the economic life of the U.S.S.R.

In 1975 Kantorovich was a co-recipient of the Nobel Prize in economics for his development of linear programming methods and their application to economics.

6.2 WHICH, IF EITHER, IS THE "RIGHT" LINEAR PROGRAM MODEL?

The Austin Company, a producer of high quality electronic home entertainment equipment, has decided to enter the digital tape player market by introducing two models, DT-1 and DT-2 (to be termed ultra and supreme when the marketing campaign is put into operation). Their problem is to determine the number of units of each model to be produced to maximize profit.

The Company's operations research department was asked to study the situation and make recommendations. The OR department began their analysis by collecting data. They divided the manufacturing process into three phases; construction, assembly, and finishing. The data collected and their analysis led them to introduce the following assumptions/postulates:

P1 In the construction phase each DT-1 unit requires 2 hours of labor and each DT-2 unit requires 3 hours of labor. At most 1,100 hours of construction time are available per week.

P2 In the assembly phase each DT-1 unit requires 5 hours of labor and each DT-2 unit requires 3 hours of labor. At most 1,400 hours of assembly time are available per week.

P3 In the finishing phase each DT-1 unit requires 4 hours of labor and each DT-2 unit requires 1 hour of labor. At most 756 hours of finishing time are available per week.

P4 After taking cost and revenue factors into consideration the anticipated profit for each DT-1 unit is $150 and the anticipated profit for each DT-2 unit is $120. In order for these unit profit values to be realistic the Company must produce at least 25 DT-1 and 40 DT-2 units per week.

P5 There is an unlimited market for the DT-1 and DT-2 models.

P   All factors other than the ones considered in the analysis of the production of the DT-1 and DT-2 models are negligible.

Its next task was to translate these assumptions into mathematical form, being careful to include everything stated in the assumptions and not go beyond them.

The OR department began by introducing variables for the quantities it sought to determine; let x denote the number of DT-1 and y the number of DT-2 units to be made weekly. There is a fair amount of data contained in the assumptions and it is useful to make it available at a glance in tabular form. This is done in Table 6.1.

Table 6.1

	No. of units to be made per week	Profit per unit	Construction time per unit (hrs)	Assembly time per unit (hrs)	Finishing time per unit (hrs)
DT-1	x	$150	2	5	4
DT-2	y	$120	3	3	1
			≤ 1100	≤ 1400	≤ 756

The key to expressing profit and the conditions that have emerged in terms of x and y is the information stated on unit profits and unit construction, assembly and finishing times for DT-1 and DT-2. We have:

$$profit = \begin{bmatrix} profit\ on \\ DT-1 \end{bmatrix} + \begin{bmatrix} profit\ on \\ DT-2 \end{bmatrix}$$

$$= \begin{bmatrix} profit\ on \\ one\ DT-1 \\ unit \end{bmatrix} \cdot \begin{bmatrix} no.\ of \\ units \\ made \end{bmatrix} + \begin{bmatrix} profit\ on \\ one\ DT-2 \\ unit \end{bmatrix} \cdot \begin{bmatrix} no.\ of \\ units \\ made \end{bmatrix}$$

$$= 150x + 120y$$

The profit obtained by making x DT-1 and y DT-2 units per week is expressed by the linear function:

$$P(x, y) = 150x + 120y$$

As to the conditions that x and y must satisfy, since the number of units made must be non-negative, we have:

$$x \geq 0$$
$$y \geq 0$$

The construction time condition is that

$$(\text{total construction time used}) \leq 1100.$$

In terms of unit construction times, 2 hours are needed for one unit of DT-1 and 3 hours are needed for one unit of DT-2, $2x$ hours are needed for x DT-1 units and $3y$ hours are needed for y DT 2 units. The total construction time utilized is expressed by $2x + 3y$. Thus, the construction time utilized is expressed by $2x + 3y$. Thus, the construction time condition is:

$$2x + 3y \leq 1100$$

Similarly, the assembly and finishing time conditions are stated by the inequalities:

$$5x + 3y \leq 1400$$
$$4x + y \leq 756$$

The conditions that at least 25 DT-1 and 40 DT-2 units must be produced weekly are expressed by the inequalities:

$$x \geq 25$$
$$y \geq 40$$

We thus emerge with the following mathematical structure, called **linear program model LP-1**, as a translation of the assumptions/postulates made by the OR department of the Austin Company.

$$\text{Maximize } P(x, y) = 150x + 120y$$

subject to

$$x \geq 0, \quad y \geq 0$$
$$2x + 3y \leq 1100$$
$$5x + 3y \leq 1400$$
$$4x + y \leq 756$$
$$x \geq 25, \ y \geq 40$$

Here x represents the number of DT-1 and y the number of DT-2 units to be made weekly.

A **linear program** is a mathematical problem with the following structure: there is specified a linear function of a number of variables that are required to satisfy linear conditions described by some mixture of linear inequalities and linear equations, called **constraints**. The problem is to find values for these variables which satisfy the constraints and yield the maximum, or minimum, value of the function, which is called an **objective function.**

A linear program may or may not arise from a real world situation/ problem under study. If it does, the linear program together with the assumptions/postulates that led to it is called a **linear program model** for the situation/problem under study.

LP-1 is a 2-variable linear program model, but the same kind of problem may involve 200 variables, 2000 variables or even 200,000 or more variables.

The Austin Company also hired the Aleksa Company, a consulting operations research firm, to independently study the digital tape player situation and make recommendations. The Aleksa OR group viewed the manufacturing process in terms of two phases: construction (which included assembly) and finishing. The data collected and their analysis led them to introduce the following assumptions/postulates:

P1a   In the construction phase each DT-1 unit requires 8 hours of labor and each DT-2 unit requires 5 hours of labor. At most 2,210 hours of construction time are available per week.

P2a   In the finishing phase each DT-1 unit requires 3 hours of labor and each DT-2 unit requires 2 hours of labor. At most 860 hours of finishing time are available per week.

P3a   The anticipated profit for each DT-1 unit is $140 and the anticipated profit for each DT-2 unit is $150. In order for these unit profit values to be realistic the company must produce at least 50 DT-1 and 50 DT-2 units per week.

P4a   There is an unlimited market for the DT-1 and DT-2 models.

P   All factors other than the ones considered in the analysis of the production of the DT-1 and DT-2 models are negligible.

The same sort of analysis that leads to LP-1 from the assumptions/postulates made by the Austin Company's operation research department leads to the Aleksa OR group's **linear program model LP-2**:

$$\text{Maximize } P(x, y) = 140x + 150y$$
subject to

$$x \geq 0, \quad y \geq 0$$
$$8x + 5y \leq 2210$$
$$3x + 2y \leq 860$$
$$x \geq 50, \quad y \geq 50,$$

where x represents the number of DT-1 and y the number of DT-2 units to be made weekly.

We turn our attention to developing a systematic approach to solving such problems in the next section.

EXERCISES

1. Set up LP-2 from the postulates formulated by the Aleksa Company.

6.3   THE CORNER POINT SOLUTION METHOD

We next address the problem of solving LP-1 and LP-2. To do this we develop a simple method for solving linear programs called the **corner point method.** It has an appealing geometric flavor and is effective for 2-variable linear programs.

As a working vehicle consider the linear program:

$$\text{Maximize } F(x, y) = 5x + 8y$$
subject to
$$x \geq 0, \quad y \geq 0$$
$$x + 2y \leq 8$$
$$3x + y \leq 9.$$

The points that satisfy the constraints of a linear program, called **feasible points,** are the points that the objective function to be optimized may be applied to.

Our problem here is to determine that feasible point (or those feasible points, if there is more than one) which yields the maximum value for the objective function $F(x, y) = 5x + 8y$.

Our first step in solving this problem is to obtain a geometric representation of the feasible points by graphing the constraints. This is done in Example 2 of Section 4.8 (p. 214).

The intersection points of at least two boundary lines which come out of the equality conditions of the constraints of a 2-variable linear program are called **corner points.** The corner points of our linear program are $(0, 0)$, $(0, 4)$, $(3, 0)$ and $(2, 3)$—obtained by solving the system of equations $x + 2y = 8$ and $3x + y = 9$; they are shown in Figure 6.1.

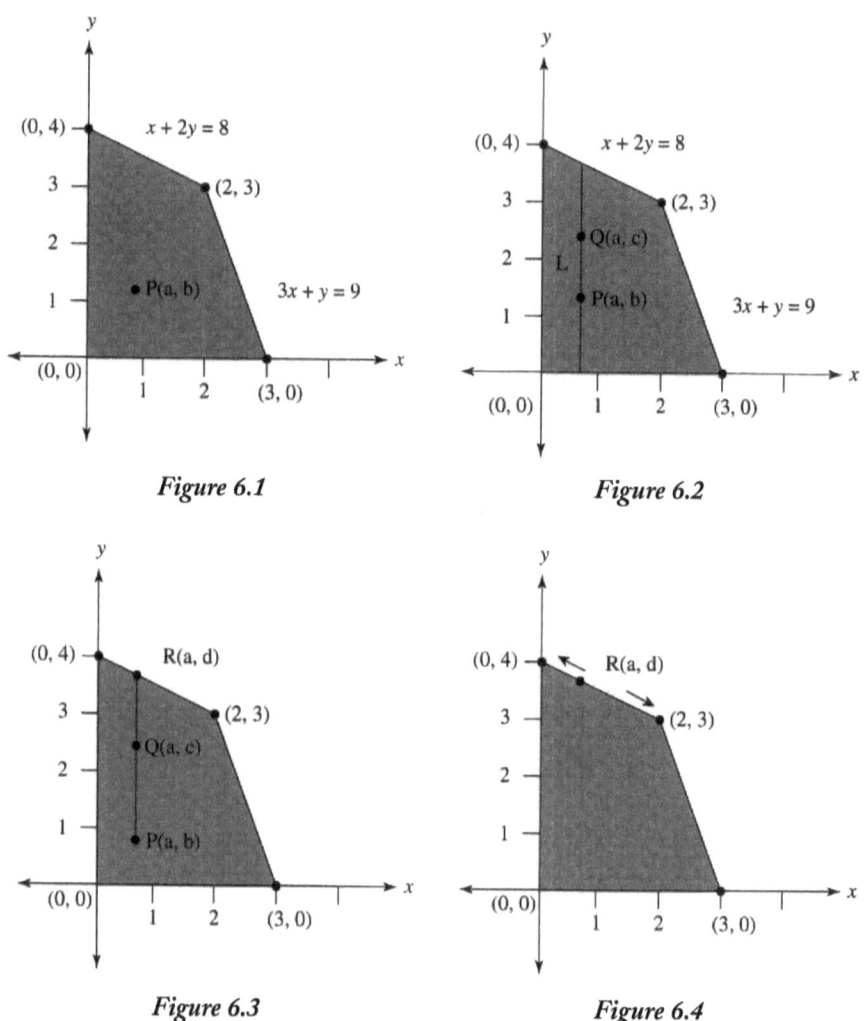

Figure 6.1

Figure 6.2

Figure 6.3

Figure 6.4

The significance of corner points is made clear by the following argument. As a starting point consider any feasible point $P(a, b)$ in the region of feasible points of our linear program; see Figure 6.1. If we move up the vertical line L passing through $P(a, b)$ to $Q(a, c)$, (see Figure 6.2), our objective function $F(x, y) = 5x + 8y$ increases in value from $5a + 8b$ to $5a + 8c$ since c is larger than b. By taking feasible points higher and higher on L we increase $F(x, y) = 5x + 8y$. Since we must remain within the set of feasible points in taking points higher and higher on L, we can go as far as $R(a, d)$ on the boundary line $x + 2y = 8$ (Figure 6.3). From there we can move in one of two directions on the boundary line until we come to a corner point (see Figure 6.4).

The question that this raises is, how does a linear function behave as we take points in one direction or the other along a boundary line? We shall not do so here, but one can prove that one of two things happens: (a) The linear function increases in value as we take points in one direction along the boundary line and decreases as we take points in the other direction, or (b) the linear function has the same value at all the points on the boundary line. In either case we are led to a corner point. This argument suggests the following theorem.

> **Corner Point Theorem:** If a 2-variable linear program has an optimal value (maximum or minimum value, depending on the nature of the linear program), then a solution yielding this optimal value can be found from among the corner points of the linear program.

The corner point theorem's hypothesis presupposes that the linear program under consideration has a solution. It does not say that a solution cannot occur at a feasible point which is not a corner point; this does happen with some linear programs. We are, however, assured of a solution at a corner point, assuming that the linear program has a solution to begin with.

Implementation of the corner point method to solve a 2-variable linear program involves the following sequence of steps:

> **Corner Point Method Steps**
>
> 1. Graph the feasible points of the linear program.
>
> 2. Locate its corner points on the graph.
>
> 3. Determine the coordinates of all corner points. For a corner point which is not on either of the coordinate axes this is done by solving the system of equations which describe a pair of boundary lines which intersect at the corner point.
>
> 4. Compute the value of the objective function at each corner point.
>
> 5. From these values pick out the largest or smallest value, depending on the nature of the linear program, and the solution(s) which yields it.

To illustrate these procedures we return to our vehicle:

$$\text{Maximize } F(x, y) = 5x + 8y$$

subject to

$$x \geq 0, \quad y \geq 0$$
$$x + 2y \leq 8$$
$$3x + y \leq 9$$

The graph of its feasible points is reproduced as Figure 6.5. The corner points, displayed in Figure 6.6, are $(0, 0), (3, 0), (2, 3)$ and $(0, 4)$; as was previously noted, $(2, 3)$ is obtained by solving the system of boundary line equations $3x + y = 9$ and $x + 2y = 8$.

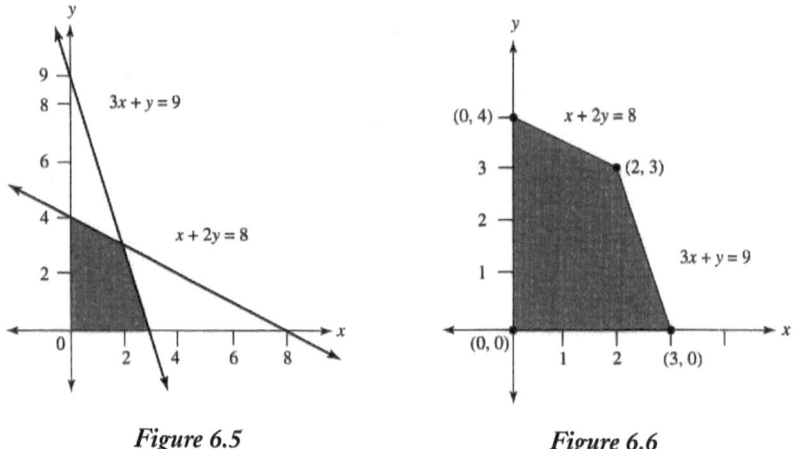

Figure 6.5   Figure 6.6

The computation of the value of the objective function $F(x, y) = 5x + 8y$ at the corner points yields the results summarized in Table 6.2, from which we see that 34 is the maximum value and $(2, 3)$ is the solution.

Table 6.2

Corner Point	$F(x, y) = 5x + 8y$
$(0, 0)$	0
$(3, 0)$	15
$(2, 3)$	34
$(0, 4)$	32

EXAMPLE 1   SOLUTION OF LP-1

Solve LP-1, the linear program model obtained by the operations research department of the Austin Company.

$$\text{Maximize } P(x,y) = 150x + 120y$$

subject to

$$x \geq 0, \quad y \geq 0$$

$$2x + 3y \leq 1100$$

$$5x + 3y \leq 1400$$

$$4x + \ y \leq \ 756$$

$$x \geq 25, \ y \geq 40$$

Our first step is to sketch the graph of the feasible points.

Locate the corner points on the graph and solve the appropriate systems of equations to determine their coordinates. There are five corner points, shown in Figure 6.7: $(25, 40), (25, 350), (100, 300)$—obtained by solving $2x + 3y = 1100$ and $5x + 3y = 1400$, $(124, 260)$—obtained by solving $4x + y = 756$ and $5x + 3y = 1400$ 1400, and $(179, 40)$.

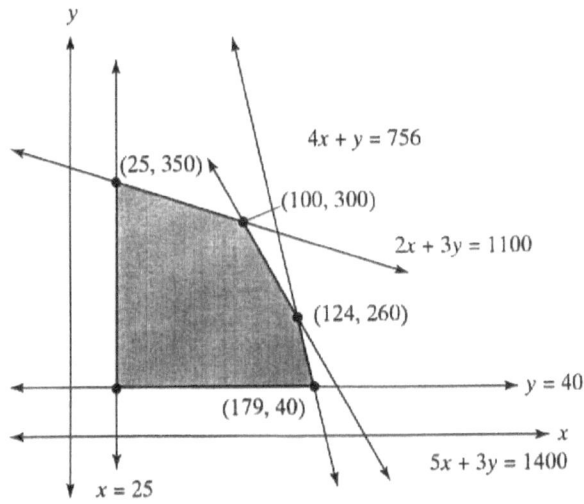

Figure 6.7

The computation of $P(x,y) = 150x + 120y$ at the corner points, summarized in Table 6.3, yields the solution $(100, 300)$ with maximum value 51,000.

Table 6.3

Corner Point	$P(x, y) = 150x + 120y$
(25, 40)	8,550
(25, 350	45,750
(100, 300)	51,000
(124, 260)	49,800
(179, 40)	31,650

EXAMPLE 2   SOLUTION OF LP-2

Solve LP-2, the linear program model obtained by the Aleksa Company for the Austin Company:

$$\text{Maximize } P(x, y) = 140x + 150y$$
subject to

$$x \geq 0, y \geq 0$$
$$8x + 5y \leq 2210$$
$$3x + 2y \leq 860$$
$$x \geq 50, \ y \geq 50$$

Our first step is to sketch the graph of the feasible points.

Locate the corner points on the graph and solve the appropriate systems of equations to determine coordinates. There are four corner points, shown in Figure 6.8: (50, 50), (245, 50), (120, 250), and (50, 355).

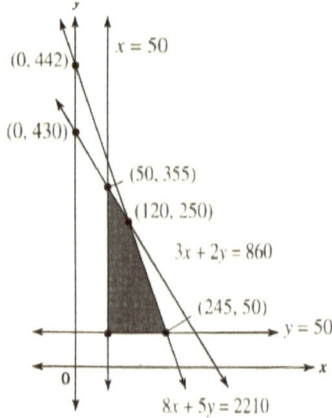

Figure 6.8

The computation of $P(x, y) = 140x + 150y$ at the corner points, summarized in Table 6.4, yields the solution (50, 355) with maximum value 60, 250.

Table 6.4

Corner Point	$P(x, y) = 140x + 150y$
(50, 50)	14,500
(245, 50)	41,800
(120, 250)	54,300
(50, 355)	60,250

Model LP-1 has solution (100, 300) with maximum value 51,000 whereas model LP-2 has solution (50, 355) with maximum value 60,250.

In terms of the Austin Company's situation, to implement LP-1's conclusion the production schedule would have to be set to manufacture 100 DT-1 and 300 DT-2 units per week with an anticipated weekly profit of $51,000; to implement LP-2's conclusion the production schedule would have to be set to manufacture 50 DT-1 and 355 DT-2 units per week with an anticipated weekly profit of $60,250.

Which solution should be implemented? It may seem foolish even to ask, but is it? We explore this issue in the next section.

EXERCISES

Solve, if possible, the following linear programs.

1. Max. $F(x, y) = 5x + 4y$
    subject to
    $x \geq 0, y \geq 0$
    $2x + 3y \leq 15$
    $x + y \leq 6$

6. Max. $F(x, y) = 3x + 2y$
    subject to
    $x \geq 0, y \geq 0$
    $x + 2y \geq 4$
    $x - 2y \geq 2$

2. Min. $G(x,y) = 10x + 12y$
    subject to
    $x \geq 0, y \geq 0$
    $3x + 5y \leq 12$
    $x + y \geq 3$

3. Min. $C(x,y) = 1.50x + 1.10y$
    subject to
    $x \geq 0, y \geq 0$
    $2x + y \geq 4$
    $x + y \geq 3$
    $x + 2y \geq 4$

4. Min. $C(x,y) = 8x + 12y$
    subject to
    $x \geq 0, y \geq 0$
    $x + y \geq 250,000$
    $2x + y \leq 400,000$

5. Max. $I(x,y) = 0.10x + 0.08y$
    subject to
    $x \geq 0, y \geq 0$
    $x + y \leq 60$
    $-x + 3y \geq 0$
    $x \geq 15$

7. Max. $I(x,y) = 110x + 70y$
    subject to
    $x \geq 0, y \geq 0$
    $x \leq 25$
    $2x + 4y \leq 120$
    $4x + 2y \leq 110$

8. Max. $C(x,y) = 39,600x + 54,000y$
    subject to
    $x \geq 0, y \geq 0$
    $2x + 3y \leq 50$
    $x + y \leq 30$

9. Max. $P(x,y) = 180x + 120y$
    subject to
    $x \geq 0, y \geq 0$
    $4x + 3y \leq 320$
    $5x + 2y \leq 330$

10. Max. $P(x,y) = 190x + 110y$
    subject to
    $x \geq 0, y \geq 0$
    $5x + 2y \leq 330$
    $3.25x + 2y \leq 225$
    $4x + 3y \leq 320$

6.4   WHICH SOLUTION, IF EITHER, SHOULD BE IMPLEMENTED?

It's always easier when you have a choice of one; take it, or leave it. But a choice of two is another matter. Bottom-line Bob, chairman of the ten member board charged with making a decision on how to implement the Company's entry into the digital tape player market, argued that it's obvious what we should do. "Implementation of LP-2 brings us a weekly profit of $60,250, whereas implementation of LP-1 brings us a weekly profit of $51,000. Since we want the largest possible return, we should go with LP-2. It's a no-brainer. The board voted nine to one to implement LP-2.

Alas, the $60,250 weekly profit was far from being realized after LP-2 was implemented, and two years later the Austin Company's venture into the digital tape player market had to be written off as a disaster.

Bottom-line Bob, who presided over this disaster was confused, upset, and out of a job. He went to Reflective Ramunė, the chair of the new board and the one person who had voted against implementation of LP-2, with some questions: "We were ultra-cautious and obtained the additional services of the Aleksa OR group to make

recommendations which, subsequently, had disastrous consequences for us; what went wrong? How is it that mathematics failed us? Why did you vote against implementation of LP-2?"

"Well Bob, as I pointed out at the board meeting, I voted against implementation of LP-2 because I was not convinced that its promise of a $60,250 weekly profit was realistic. As you yourself pointed out, 'the promise of LP-2 is $9,250 better than that promised by LP-1,' but promises may not be realizable if they are founded on unrealistic assumptions. The conclusion reached from LP-2 was indeed tempting, and in fact proved too tempting for my colleagues on the board, but since it came from a linear program founded on assumptions which I viewed as unrealistic, I resisted temptation. We have no quarrel with mathematics; mathematics gave us a valid conclusion from LP-2, which is all that we can legitimately expect. Unfortunately that conclusion proved to be unrealistic."

Is Mathematics Precise?

"Ramunė, I don't understand this. I always liked math in high school and college. Solving those equations, factoring those expressions, differentiating those functions, throwing the data into the computer and letting it do its thing, that was real fun. What I like most about math is its precision. You don't get ten sides to a story. You get one answer and that's that; no baloney.

"Bob, I think your math courses may have focused too much on technique and not enough on perspective. Technique can be fun to a point, but without a perspective

on its place in the over-all role of mathematics in applications, we see only a small tip of the mathematical iceberg. Mathematics is precise in the sense that it gives us valid conclusions based on the assumptions made, which is where technique—factoring, differentiating functions, and the like—plays its major role.

> Whether the assumptions made are realistic or not is another matter which technique can't help us with. The question of how to formulate these assumptions and reach a judgment on their realism may indeed yield ten sides to the story. I'm afraid that those who find mathematics attractive because of what they perceive to be its absolutist nature have misunderstood the meaning of mathematical precision."

6.5   HOW COULD IT BE WRONG? I USED A COMPUTER

"Ramunė, I still don't fully understand what went wrong. The company just spent millions to update its computer system. I had access to the latest and the best. Why didn't this save us from disaster." "Bob, Henry Clay's observation that 'statistics are no substitute for judgment' applies equally well to the computer.

We cannot expect the computer to employ technological alchemy and convert unrealistic assumptions into golden truths. Keep in mind the NINO principle; if nonsense in, then nonsense out. Indiscriminate use of computer technology has made possible the generation of more and more nonsense more quickly than ever before by more people having less and less understanding of what they are doing."

The Computer's Right of Way

"If what you say is true Ramunė, then what good is this super computer technology to us?" "For number crunching and delivering results quickly and efficiently, the computer is without equal, Bob.

In this dimension it is the undisputed master of the field. The mathematical model building process and computers have developed a symbiotic relationship in that computers have made it possible for us to solve previously unapproachable large scale problems that come out of mathematical models, while the accessibility of such

problems to computer solution has made possible the use of such complex models. Alas, none of this overrides the NINO principle."

EXERCISES

1. ZKB Electronics puts out two kinds of personal computers, model ZKB-47 and model ZKB-82. The management of ZKB called in the Rex Consulting Firm to determine how many units of each model should be made daily so as to maximize profit. The consulting firm set up a linear program model for the electronics company's production problem and, by applying the corner-point method, reached the conclusion that 300 ZKB-47 units and 250 ZKB-82 units should be made daily to maximize profit. Before implementing this conclusion, the management of ZKB put the following questions to the director of the consulting firm.

(a) Does use of the corner-point method guarantee that profit will be maximized when 300 ZKB-47 units and 250 ZKB-82 units are made daily and sold?

(b) What is your basis for recommending that we implement your conclusion?

How would you reply to these questions?

2. The Onutė Corporation plans to introduce two high resolution TV models, T20 and T24, to the market. Its own operations research group was led to introduce the following M1 model to determine the optimal production schedule for maximizing profit:

$$\text{Maximize } P(x, y) = 180x + 120y$$
subject to

$$x \geq 0, y \geq 0$$
$$4x + 3y \leq 320$$
$$5x + 2y \leq 330,$$

where, x and y denote the number of T20 and T24 units, respectively, to be made daily. Its solution is (50, 40) with maximum value 13,800 (see Exercise 9, sec. 6.3).

The Aleksa company was also hired to study the Onutė Corporation's production scheduling problem. It was led to introduce the following M2 model to determine the optimal production schedule for maximizing profit:

$$\text{Maximize } P(x, y) = 190x + 110y$$
subject to

$$x \geq 0, y \geq 0$$
$$5x + 2y \leq 330$$
$$3.25x + 2y \leq 225$$
$$4x + 3y \leq 320,$$

where x and y denote the number of T20 and T24 units, respectively, to be made daily. Its solution is $(60, 15)$ with maximum value 13,050 (see Exercise 10, sec. 6.3).

The following questions have arisen. How would you answer them?

(a) If mathematics is the precise subject that it is reputed to be, should there not be one solution to this problem rather than two?

(b) Since two solutions emerge, does it follow that not both are valid? Explain.

(c) Before making a decision about whether to implement M1 or M2, what questions would you put to the two operations research groups?

(d) Which model, if either, would you adopt and implement? Why? Is it possible that you would not adopt either model?

6.6 PROBLEMS LEADING TO LINEAR PROGRAM MODELS

Linear programming has turned out to have a wide spectrum of applications. To obtain some sense of this spectrum we look at four case studies. The case studies are all realistic, but are presented in miniature for the sake of manageability.

Actual real-life situations that emerge have the same structure and tone, but are more complex in that more factors are generally considered and more variables are required.

We view all of these situations through the eyes of others in much the same way that we see events through the eyes of a reporter or observer by reading his account of them in a newspaper, journal or book. Just as the reporter has selected what he believes are important features surrounding the events and has omitted those he considers unessential, we too are looking at features considered crucial to the situations we examine as seen by someone who has made such a selection. This selection reflects assumptions that have been made. To maintain a proper perspective on this it is important to keep in mind that other analysts, as other reporters, might see things in a different light and accordingly make other assumptions/postulates.

Case 1   Production Planning

The Austin Company's problem of determining the number of DT-1 and DT-2 digital tape players to be made per week so as to maximize profit, considered in section 6.2, is a production scheduling problem which we expressed in linear program terms under the assumptions introduced. The background that led to the Austin Company's linear program models is illustrative of situations with the following general features:

A firm makes a number of products or models of a product and utilizes a number of resources in their manufacture, such as raw materials, labor, capital, different machines, storage facilities. It is assumed that for each product made a fixed amount of each resource is required to make a unit of that product. Within the production time frame a fixed amount of each resource is available and cannot be exceeded. It is also assumed that for a range of possible output levels there is a fixed profit per unit of each product which does not depend on the number of units produced.

Under these conditions the problem of determining output levels of the products produced so as to maximize total profit can be formulated in terms of a linear program model.

The Veronika Company's Problem

As a further illustration of a production planning situation we turn to the Veronika Company's problem.

The Veronika Company makes stereo systems. Two new models, RA5 and RA9, are to be mass produced. Both models pass through assembly and finishing plants of the company. In the assembly plant an RA5 unit is worked on for 1 hour; an RA9 unit is worked on for 3 hours. In the finishing plant an RA5 unit is worked on for 2 hours; an RA9 unit is worked on for 1 hour. At most, 90 hours of assembly time and 80 hours of finishing time are available per week. The anticipated profit on an RA5 unit is $10 and the anticipated profit on an RA9 unit is $15.

The problem is to determine, with respect to the given assumptions/postulates, how many RA5 and RA9 units should be made per week so as to maximize profit.

We begin by introducing variables to stand for the quantities we wish to determine. Let x denote the number of RA5 units to be made and let y denote the number of RA9 units to be made. To make needed information available at a glance, we express the basic data in tabular form, as shown in Table 6.5

Table 6.5

	No. of units to be made per week	Profit per unit	Assembly time per unit (hours)	Finishing time per unit (hours)
Model RA5	x	$10	1	2
Model RA9	y	$15	3	1

Since profit is to be maximized, we must express profit in terms of x and y. It is useful to note that:

$$\text{profit} = \begin{bmatrix} \text{profit on} \\ \text{model RA5} \end{bmatrix} + \begin{bmatrix} \text{profit on} \\ \text{model RA9} \end{bmatrix}$$

$$\text{profit} = \begin{bmatrix} \text{profit on} \\ \text{one RA5} \\ \text{unit} \end{bmatrix} \cdot \begin{bmatrix} \text{no. of} \\ \text{units} \\ \text{made} \end{bmatrix} + \begin{bmatrix} \text{profit on} \\ \text{one RA9} \\ \text{unit} \end{bmatrix} \cdot \begin{bmatrix} \text{no. of} \\ \text{units} \\ \text{made} \end{bmatrix}$$

$$\text{profit} = 10x + 15y$$

The profit obtained by making x RA5 units and y RA9 units is expressed by the linear function

$$P(x, y) = 10x + 15y.$$

Our next task is to describe the conditions that x and y must satisfy. Since the number of units made is nonnegative, we have:

$$x \geq 0$$
$$y \geq 0$$

To express assembly plant operation time in terms of x and y, we note that:

$$\begin{bmatrix} \text{assembly} \\ \text{plant} \\ \text{time} \end{bmatrix} = \begin{bmatrix} \text{assembly} \\ \text{time on} \\ \text{RA5} \end{bmatrix} + \begin{bmatrix} \text{assembly} \\ \text{time on} \\ \text{RA9} \end{bmatrix}$$

$$\begin{bmatrix} \text{assembly} \\ \text{plant} \\ \text{time} \end{bmatrix} = \begin{bmatrix} \text{assembly} \\ \text{time for} \\ \text{1 RA5 unit} \end{bmatrix} \cdot \begin{bmatrix} \text{no. of} \\ \text{RA5 units} \\ \text{made} \end{bmatrix} + \begin{bmatrix} \text{assembly} \\ \text{time for} \\ \text{1 RA9 unit} \end{bmatrix} \cdot \begin{bmatrix} \text{no. of} \\ \text{RA9 units} \\ \text{made} \end{bmatrix}$$

$$\begin{bmatrix} \text{assembly} \\ \text{plant} \\ \text{time} \end{bmatrix} = 1 \cdot x + 3 \cdot y$$

Since at most 90 hours of assembly time per week is available, we have

$$x + 3y \leq 90.$$

Similarly, the condition that at most 80 hours of finishing time are available per week is expressed in terms of x and y by the inequality

$$2 + y \leq 80.$$

In summary, the postulates introduced lead to the linear program model

$$\text{Maximize } P(x, y) = 10x + 15y$$
$$\text{subject to}$$
$$x \geq 0$$
$$y \geq 0$$
$$x + 3y \leq 90$$
$$2x + y \leq 80,$$

where x and y denote the number of RA-5 and RA-9 units to be made per week, respectively.

To solve the Veronika Company's linear program we first sketch the graph of its feasible points, shown in Figure 6.9.

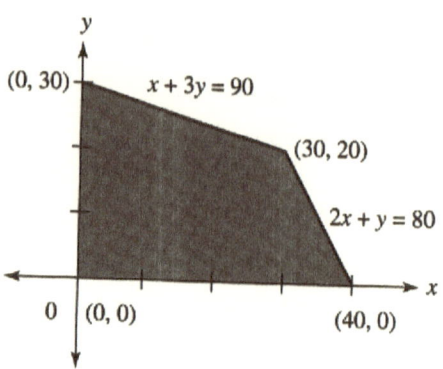

Figure 6.9

From Table 6.6 we see that (30, 20) is the solution of the Veronika Company's linear program model and 600 is its maximum value.

Table 6.6

Corner Point	$P(x, y) = 10x + 15y$
(0,0)	0
(0, 30)	450
(40, 0)	400
(30, 20)	600

Implementation of this solution by the Veronika Company calls for making 30 RA5 and 20 RA9 units per week. **Whether this solution should be implemented or not is, of course, another issue.**

EXERCISES

Should this solution be implemented? William Ganz, production manager of the Veronika Company gave thought to this question and asked for opinions.

1. "Yes," said Jim Turner, the resident computer "expert". "The solution was obtained by use of the latest computer technology available, and that's good enough for me." Would you agree with Jim? Explain

2. "Of course, no question about it," said June Carver, CEO of the Veronika Company. "The corner point method is the mathematical basis for this solution and its use ensures that profit will be maximized." Would you agree with June? Explain.

3. "Are there any factors that we should have taken into consideration in the formulation of this linear program model," Ganz asked Horace Black, head of the operations research department. "None that we we could come up with," replied Horace. "We took into account every factor that we judged relevant to maximization of profit." "What about the linear program models formulated for the Austin Company? Are there lessons that we should give thought to from these models?" "I don't think so," Would you agree with Horace's assessment? Explain.

4. If you were asked for your thoughts on whether the solution (30, 20) should be implemented, what would you say to Mr. Ganz? Explain.

Case 2   Diet Problems

A sack of animal feed is to be put together from linseed oil meal and hay. It is required that each sack of feed contain at least 2 pounds of protein, 3 pounds of fat, and 8 pounds of carbohydrate. It is estimated that each unit (a unit is 30 pounds) of linseed oil meal contains 1 pound of protein, 1 pound of fat, 2 pounds of carbohydrate, and that each unit of hay contains 1/2 pound of protein, 1 pound of fat, and 4 pounds of carbohydrate. Linseed oil meal costs $1.50 per unit and hay costs $1.10 per unit.

The problem is to determine how many units of linseed oil meal and hay should be used to make up a sack of animal feed that satisfies the nutritional requirements at minimal cost.

To translate this problem and the assumptions/postulates that underlie it into a linear program model, let x and y denote the number of units of linseed oil meal and hay, respectively, to be used in making up a sack of animal feed. The introduction of variables to represent the quantities we are seeking to determine is a humble but essential step in translating the background presented into mathematical form. These variables provide us with the crucial bridge linking the mathematical model with the background; if they are incorrectly or equivocally defined, the bridge they establish is in danger of collapsing.

The basic data are summarized in Table 6.7.

Table 6.7

	No. units used	Cost per unit	Protein per unit (lbs)	Fat per unit (lbs)	Carbohydrate per unit (lbs)
Linseed oil meal	x	$1.50	1	1	2
Hay	y	$1.10	1/2	1	4

Since the total cost equals the cost of linseed oil meal, 1.50x, plus the cost of hay, 1.10y, the cost function to be minimized is

$$C(x, y) = 1.50x + 1.10y.$$

The number of pounds of protein in the mixture equals the amount contributed by the linseed oil meal, x pounds, plus the amount contributed by the hay, $(1/2)y$ pounds. The sack of cattle feed must contain at least 2 pounds of protein, which translates to

$$x + (1/2)y \geq 2,$$

or equivalently

$$2x + y \geq 4.$$

Similarly, the fat and carbohydrate requirements are expressed by the constraints

$$x + y \geq 3$$

and

$$2x + 4y \geq 8,$$

or equivalently

$$x + 2y \geq 4.$$

We thus obtain the linear program

$$\text{Maximize } C(x, y) = 1.50x + 1.10y$$
$$\text{subject to}$$
$$x \geq 0, y \geq 0$$
$$2x + y \geq 4$$
$$x + y \geq 3$$
$$x + 2y \geq 4,$$

which has solution (1, 2) and minimum value 3.7 (see section 6.3. Exercise 3, p. 258).

To implement this **valid conclusion of the model** we would use 1 unit of linseed oil meal (30 pounds) and 2 units of hay (60 pounds). Based on the assumed costs of the ingredients, the anticipated cost of a sack of animal feed is $3.70.

The problem considered illustrates "diet problems" with the following general features: A diet, or food substance, is to be put together from a number of available foods. It is required that the diet be balanced in the sense that it must contain minimal amounts of stated nutrients—proteins, fats, carbohydrates, minerals, vitamins, etc. It is assumed that each food unit contains a known fixed amount of each nutritional unit and that the unit prices of the food items are known and fixed within

the time period considered. The problem is to determine the minimal cost diet which satisfies the prescribed nutritional requirements.

Case 3   Environmental Protection

The Saxon Company must produce at least 250 thousand tons of paper annually. From the current operating system 10 pounds of chemical residue is deposited into a neighboring water system for each ton of paper produced. The resulting pollution has become a problem of serious concern, and to remain eligible for state tax benefits the Saxon Company must restrict the chemical residue emitted into the state's water system to not exceed 200 tons per year. Two filtration systems, Delta and Beta, have emerged for consideration. It is estimated that the installation of the Delta system would reduce emissions to 2 pounds for each ton of paper produced, and installation of the Beta system would reduce emissions to 1 pound for each ton of paper produced. Capital and operating costs for the Delta and Beta systems have been estimated at $8 and $12, respectively, per ton of paper produced.

The problem is to determine how many tons of paper should be produced subject to the Delta system and how many should be produced subject to the Beta system so that the emissions standard is met at minimal cost.

Let x and y denote the number of tons of paper to be produced annually subject to use of the Delta and Beta systems, respectively.

The basic data are summarized in Table 6.8.

Table 6.8

	Nu. of tons produced	Emissions per ton (lbs)	Operating cost per ton
Delta	x	2	8
Beta	y	1	12
	$\geq 250,000$	$\leq 400,000$	

The cost function to be minimized is

$$C(x, y) = 8x + 12y.$$

The condition that the Saxon Company must produce at least 250 thousand tons of paper annually is expressed by

$$x + y \geq 250,000.$$

The total amount of chemical residue produced annually is the number of pounds produced through use of the Delta system, $2x$ pounds, plus the amount produced through use of the Beta system, y pounds. Since this cannot exceed 200 tons, we have

$$2x + y \leq 400,000 .$$

We thus emerge with the linear program

$$\text{Maximize } C(x, y) = 8x + 12y$$
$$\text{subject to}$$
$$x \geq 0, y \geq 0$$
$$x + y \geq 250,000$$
$$2x + y \leq 400,000,$$

which has solution $(150000, 100000)$ and minimum value 2,400,000 (see section 6.3, Exercise 4, p. 258).

> We should keep in mind that this conclusion is valid with respect to the assumptions made. The question of how close it is to reality's mark is another issue.

To implement this result the Saxon Company would have to produce 150,000 tons of paper annually subject to the Delta system and 100,000 tons of paper annually subject to the Beta system. The anticipated cost would be $2.4 million.

Case 4   Bank Portfolio Management

The Charles National Bank has assets in the form of loans and negotiable securities which, it is assumed, bring returns of 10 and 8 percent, respectively, in a certain time period. The bank has a total of $60 million to allocate between loans and securities. To meet unanticipated deposit withdrawals the bank maintains a securities balance greater than or equal to 25 percent of total assets. Lending is the bank's most important activity and to satisfy its clients it requires that at least $15 million be available for loans.

The bank wishes to determine, under these conditions, how funds should be allocated to maximize total investment income.

Let x and y denote the amount, in millions of dollars, to be allocated for loans and securities, respectively. The income function to be maximized is

$$I(x, y) = 0.10x + 0.08y.$$

The following constraints emerge:

$x + y \leq 60$: $60 million is available for investment in loans and securities.

$y \geq (1/4)(x + y)$,   or equivalently, $-x + 3y \geq 0$: A securities balance greater than or equal to 25% of total assets must be maintained. Note, total assets is defined as the sum of the amounts invested in loans and securities, which is $x + y$.

$x \geq 15$:   at least $15 million must be available for loans.

We thus obtain the linear program model

Maximize $I(x, y) = 0.10x + 0.08y$
subject to

$$x \geq 0, y \geq 0$$
$$x + y \leq 60$$
$$-x + 3y \geq 0$$
$$x \geq 15,$$

which has solution (45, 15) and maximum value 5.7 (see section 6.3, Exercise 5, p. 258).

> Again, let us keep in mind that what has been established is that this conclusion is valid with respect to the assumptions made. Its realism is another matter.

To implement this result the Charles Bank would have to allocate $45 million to loans and $15 million to securities. The anticipated interest on investment is $5.7 million.

Of Interest

A. Broaddus, "Linear programming: A New Approach to Bank Portfolio Management," *Federal Reserve Bank of Richmond: Monthly Review*, vol. 58, No. 11 (Nov. 1972), pp. 3-11. This article provides an introductory nontechnical discussion of linear programming for bank portfolio management.

K. J. Cohen and F. S. Hammer, "Linear Programming and Optimal Bank Asset Management Decisions," *Journal of Finance*, vol. 22 (May 1967), pp. 147-165. This paper describes a linear program model that had been used for several years by Bankers Trust Company in New York to assist in reaching portfolio decisions.

EXERCISES

The situations presented in the following exercises reflect assumptions made by an individual or group. Set up linear program models for the problems that arise, solve them, and state how the solutions obtained would be implemented. What concerns would you want to have satisfactorily addressed before implementing a solution obtained?

5. The Chuck Company makes radios. Two new portable stereo models, K15, and K31 are to be introduced. Both models pass through assembly and finishing plants of the company. In the assembly plant a K15 unit is worked on for 2 hours; a K31 unit is worked on for 3/2 hours. In the finishing plant a K15 unit is worked on for 5/2 hours; a K31 unit is worked on for 1 hour. At most 160 hours of assembly time and 165 hours of finishing time are available per week. The anticipated profit is $180 per K15 unit and $120 per K31 unit.

    The problem is to determine, under such assumptions/postulates, how many K15 and K31 units should be made weekly so as to maximize profit.

6. A fruit juice is to be made from orange juice concentrate and apricot juice concentrate. Particular attention is being paid to the vitamin A, C, and D content of the fruit juice. Each container of fruit juice is to contain at least 120 units of vitamin A, 150 units of vitamin C, and 55 units of vitamin D. One ounce of apricot juice concentrate contains 3 units of vitamin A, 2 units of vitamin C, and 1 unit of vitamin D. One ounce of orange juice concentrate contains 2 units of vitamin A, 3 units of vitamin C, and 1 unit of vitamin D. Orange juice concentrate costs 3¢ per ounce and apricot juice concentrate costs 2¢ per ounce.

The problem is to determine how many ounces of each concentrate should be used to make a least-cost container of juice that satisfies the vitamin requirements.

7. Cans of meat are to be mass produced and made available for distribution in emergency situations (such as earthquakes and floods). Each can of meat is to be a mixture of pork and beef and must contain at least 12 ounces of protein and 9 ounces of fat. It is estimated that a pound of the beef to be used contains 5 ounces of protein and 3 ounces of fat while a pound of pork contains 3 ounces of protein and 3 ounces of fat.

    If beef costs 50¢ a pound and pork costs 45¢ a pound, how many pounds of each should be used to make up a can of meat so that the nutritional requirements are met and the cost is minimal? What is the minimal cost?

8. The Saturn Company makes refrigerators and air-conditioners. Two plants, I and II, are used. The assembly work is done in plant I, and it is estimated that 5 labor-hours of work are required to produce a refrigerator and 2 labor-hours of work are required to produce an air-conditioner. The finishing work is done in plant II, and it is estimated that 3 labor-hours of work are needed to finish a refrigerator and 2 labor-hours are needed to finish an air-conditioner. Plant I has 220 labor-hours per week available and plant II has 180 labor-hours per week available. A market survey indicates that there is an unlimited market for these products.

    If the Saturn Company makes a profit of $50 on each refrigerator and $30 on each air-conditioner, how many of each should be produced weekly so as to maximize profit?

9. The Andrius Bank has assets in the form of loans and negotiable securities. For a certain time period it is assumed that loans and securities bring returns of 9 and 6 percent, respectively. The bank has a total of $25 million, provided by demand deposit accounts and time deposit accounts, to allocate between loans and securities. To meet unanticipated deposit withdrawals the bank always maintains a securities balance equal to or greater than 20 percent of total assets. Since lending is the bank's most important activity, it imposes certain restrictions on its loan balance to satisfy its principal clients. Specifically, it requires that at least $8 million be available for loans.

    Under the given assumptions/postulates, how should the bank allocate funds between loans and securities so that total investment income is maximized?

10. The Jay Toy Store plans to invest up to $2200 in buying and stocking two popular children's toys. The first toy costs $4 per unit and occupies 5 cubic feet of storage space; the second toy costs $6 per unit and occupies 3 cubic

feet of storage space. The store has 1400 cubic feet of storage space available. The owner expects to make a profit of $1.50 on each unit of the first toy he buys and stocks and a profit of $2.00 on each unit of the second toy.

How many units of each should be bought and stocked so that profit is maximized?

11. At Ecap University discussion has centered on determining the number of openings, called slots, to be made available in the forthcoming year at the associate and full-professor ranks. Each person promoted to associate professor is to receive a merit increment of $500, and each person promoted to full professor is to receive a merit increment of $1000. At most $15,000 is available for merit increments. A long-standing guideline is that the number of full-professor slots is not to exceed one fourth the number of associate professor slots. The university senate has recommended that at least 3 slots at the full-professor rank be established and that not more than 22 slots at the associate professor rank be established.

Of interest to the faculty council is the question of how many slots should be established at each rank so as to maximize the total number of promotions. Administration has raised the question of how many slots at each rank should be established so as to minimize the total cost of increments.

12. The Brooks and Darius mines of Lexington Mines, Inc., produce high-grade and medium-grade silver ore. The Brooks mine yields 1 ton of high-grade ore and 4 tons of medium-grade ore per hour. The Darius mine yields 2

tons of high-grade ore and 3 tons of medium-grade ore per hour. To meet its commitments the company needs at least 40 tons of high-grade and 100 tons of medium-grade ore per hour. It costs $500 per hour to operate the Brooks mine and $700 per hour to operate the Darius mine.

Lexington Mines, Inc., would like to determine how many hours per day each mine should be operated if their ore requirements are to be met at minimal cost.

13. The Petrovski Steel Company produces 2 million tons of steel annually. In the current operation of the blast and open-heart furnaces 50 pounds of particulate matter are emitted into the atmosphere for every ton of steel produced. The resulting air pollution has become a problem of serious concern and efforts are being directed at curbing the emissions. On the basis of studies that have been conducted, it is estimated that installation of the F14 filter system would reduce emissions to 20 pounds of particulate matter per ton of steel produced, and installation of the F24 filter system would reduce emissions to 18 pounds of particulate matter per ton of steel produced. Capital and operating costs for the F14 and F24 filter systems are estimated at $1.2 and $1.8, respectively, per ton of steel produced. It is desired that particulate emissions be reduced by 62,400,000 pounds or better per annum. At the same time cost is an important factor if the company is to remain competitive.

The problem is to determine how many tons of steel should be produced annually subject to the F14 system and how many tons should be produced subject to the F24 system so that the desired reduction in particulate emissions is achieved at minimal total cost.

CHAPTER 7

Further Food For Thought

7.1 A LINEAR PROGRAMMING SHORTFALL

Advertising Media Selection

To advertise its new best seller, the Brian Publishing Company is planning to buy morning and afternoon time on radio station GAJA. Morning time costs $2000 per minute and afternoon time costs $1000 per minute. The marketing department estimated that morning commercials reach 0.9 million listeners and afternoon commercials reach 0.6 million listeners. At most 16 minutes of morning time is available in the period in which the advertising campaign is to run. The marketing department feels that at least 8 minutes of morning time and at least 9 minutes of afternoon time should be purchased. The advertising budget for this campaign is $24,000.

The problem is to determine the number of minutes of morning and afternoon time that should be purchased to maximize total listener exposure time to the ads in the period in which the advertising campaign is to run.

Let x denote the number of minutes of morning time and y the number of minutes of afternoon time to be purchased.

Basic conditions are summarized in Table 7.1.

Table 7.1

	No. of minutes purchased	No. of listeners reached (millions)	Cost per minute (dollars)
Morning time	x	0.9	2000
Afternoon time	y	0.6	1000

$$x \geq 8, \ x \leq 16 \qquad\qquad\qquad \leq 24,000$$

$$y \geq 9$$

Total listener exposure time to the ads is expressed by

$$F(x, y) = 0.9x + 0.6y.$$

The inequalities

$$x \geq 8, \quad y \geq 9$$

express the requirement that at least 8 minutes of morning time and 9 minutes of afternoon time are to be purchased. That at most 16 minutes of morning time is available is expressed by

$$x \leq 16.$$

$2000x + 1000y$ expresses the cost of x minutes of morning time and y minutes of afternoon time, and since this cost cannot exceed $24,000, we have

$$2000x + 1000y \leq 24,000$$

which is equivalent to

$$2x + y \leq 24.$$

Therefore, we obtain the following linear program:

$$\text{Maximize } F(x, y) = 0.9x + 0.6y$$
subject to

$$x \geq 0$$
$$y \geq 0$$
$$x \geq 8$$
$$y \geq 9$$
$$x \leq 16$$
$$2x + y \leq 24$$

The afore problem is an example of a general class of problems called **advertising media-selection problems**.

The **general advertising media-selection problem** is to choose from various media capable of carrying an advertisement a selection that is, in some sense, best. Specific choices within a given medium as well as given media are included in the alternatives. The constraints in media selection include the size of the advertising budget, the minimum and maximum usages of specific media categories, and the desired minimum exposure rate to envisioned buyers.

A number of approaches have been developed for the media-selection problem,[1] and in the early 1960's hopes ran high in the world of advertising for the use of linear programming.

An early linear-program model for media selection was the one developed by James Engel and Martin Warshaw[2] for the McGraw-Edison Company. The Pennsylvania Transformer division of McGraw-Edison manufactures transformers for use by industrial plants, schools, hospitals, commercial construction projects, and so on. Ten trade publications were considered for advertising purposes, and $25,000 was allocated for industrial advertising for a period of 1 year. Since the purchase decision is usually made by the plant engineer, the objective posed was to maximize the number of plant engineers reached. The following linear program model was developed:

---

[1] See Dennis Gensch, "Different Approaches to Advertising Media Selection," *Operational Research Quarterly*, vol. 21, no. 2 (June 1970), pp. 193-219; Philip Kotler, *Marketing Management* (Englewood Cliffs, N.J.: Prentice-Hall, Inc., 1967), Chapter. 18.

[2] Allocating Advertising Dollars by Linear Programming," *Journal of Advertising Research*, vol. 4, no. 3 (September 1964), pp. 42-48.

$$\text{Maximize } f = 0X_1 + 15.15X_2 + 32.87X_3 + 49.44X_4 + 56.65X_5 + 17.54X_6$$
$$+ 58.20X_7 + 0X_8 + 23.53X_9 + 40.00X_{10}$$

subject to nonnegativity of the variables ($X_1 \geq 0, X_2 \geq 0$, etc.)

$$X_1 \leq 5.400$$
$$X_2 \leq 9,504$$
$$X_3 \leq 8,760$$
$$X_4 \leq 10,680$$
$$X_5 \leq 11,016$$
$$X_6 \leq 5,472$$
$$X_7 \leq 9,072$$
$$X_8 \leq 3,300$$
$$X_9 \leq 8,160$$
$$X_{10} \leq 6,900$$
$$X_1 + X_2 + X_3 + X_4 + X_5 + X_6 + X_7 + X_8 + X_9 + X_{10} \leq 25,000$$

where X_1 is the amount to be invested in media 1 (*Consulting Engineer Magazine*), X_2 is the amount to be invested in media 2 (*Electrical Construction Magazine*), and so on. The coefficients 0 of X_1, 15.15 of X_2, 32.87 of X_3, and so on, in the linear function f represent the number of plant engineers reached by each magazine per advertising dollar invested, so that f represents the total number of plant engineers reached. The last constraint expresses the condition that no more than $25,000 is to be spent on advertising and the other constraints are to prevent more dollars from being invested in any one monthly magazine than is necessary to buy 12 insertions.

Although linear program models were satisfactory for crude versions of the media-selection problem, it soon became clear that the features exhibited by more sophisticated versions of the problem could not be modeled in linear programming terms.

Frank Bass and Ronald Lonsdale[3] found linear-program models to be

---

[3] "An Exploration of Linear Programming in Media Selection," *Journal of Marketing Research*, vol. III, no. 2 (May 1966), pp. 179-188.

crude devices to apply to the media-selection problem. The linearity assumption itself, is the source of much of the difficulty. Justifying an assumption of linear response to advertising exposures on theoretical grounds would be difficult. Assumptions about the nature of response to advertising cause most difficulties in models of the type examined in this article.

Philip Kotler[4] noted the following limitations:

Linear programming assumes that repeat exposures have a constant marginal effect.

It assumes constant media costs (no discounts).

It cannot handle the problem of audience duplication.

It says nothing about when ads should be scheduled.

The Brian Company's linear program model reflects the conditions formulated by its marketing department, but in view of the reservations expressed by Bass, Lonsdale, and Kotler, the conditions formulated and the linear program model developed to express them must be considered a crude version of the advertising media-selection problem. Whether this version suffices for the Company's needs is an issue they would be well-advised to give serious thought to.

Another problem, a mathematical one, arises; the conditions formulated by the marketing department are not compatible. (Try to sketch the graph of the feasible points.) The linear program has no feasible points and thus no solution.

Although later linear programming approaches to the media-selection problem sought to overcome the criticisms that had been voiced, the message was clear: although linearity, as a mathematical tool, is too good not to be true, a linear-program model is not always a suitable fit for a media-selection problem; that is, the assumptions that must be made to force a fit are not always sufficiently realistic. When the model doesn't fit, don't use it.

---

[4] *Marketing Management* (Englewood Cliffs, N.J.: Prentice-Hall, Inc.) p. 478.

Exercises

1. How could the Brian Company modify its requirements so that it's media-selection linear program model will have a solution? Explain.

2. The automobile manufacturer Chuck Associates plans to advertise its new Chuck IV sports car model on cable station ACB. Late afternoon time costs $10,000 per minute and early evening time costs $20,000 per minute. Late afternoon commercials reach an estimated 1.5 million viewers and early evening commercials reach an estimated 2.3 million viewers. The marketing department of Chuck Associates believes that at least 12 minutes of late afternoon time and at least 10 minutes of early evening time are needed to effectively communicate the Company's message. At most 20 minutes of early afternoon time is available for the time period that Chuck Associates plans to run its advertising campaign. The budget for the campaign has been set at $100,000.

   The problem is to determine the number of minutes of late afternoon and early evening time to maximize the number of viewers reached.

   (a) Set up a linear program model for Chuck Associates's problem.

   (b) In terms of the points of view expressed by Bass, Lonsdale, and Kotler, should consideration be given to not using this linear program model? Explain.

   (c) If possible, solve the linear program of this L.P. model. If you have difficulty with this, what is the problem? Explain.

   (d) If it arises, how could this problem be overcome? Explain.

7.2 PROBLEMS REQUIRING SOLUTIONS IN INTEGERS

EXAMPLE 1 HOW MANY COATS AND DRESSES
SHOULD BE MADE?

The Hoffman Clothing Manufacturers, Inc., has available 120 square yards of
cotton and 100 square yards of wool for the manufacture of coats and dresses.

Two square yards of cotton and 4 square yards of wool are used in making a coat while 4 square yards of cotton are used in making a dress. Cotton costs $5 per square yard and wool cost $20 per square yard. Four hours of labor are needed to make a coat and 2 hours of labor are needed to make a dress. The cost of labor is $25 per hour. At most 110 hours of labor are available for the manufacture of the coats and dresses.

If a coat sells for $300 and a dress sells for $140, how many of each should be made if income is to be maximized?

Let x and y denote the number of coats and dresses to be made, respectively. The data given are summarized in Table 7.2.

Table 7.2

	Number made	Selling price	Cotton used per item	Wool used per item	Labor-hours per item
Coat	x	$300	2	4	4
Dress	y	$140	4	0	2
			$5 per sq yd; 120 sq yd available	$20 per sq yd; 100 sq yd available	$25 per labor-hour; 110 labor-hours available

Since income is to be maximized we turn our attention to expressing income in terms of x and y.

$$\text{Income} = \text{Amt from sales} - \text{Production cost}$$

$$I(x, y) = \underbrace{300x + 140y}_{\text{sales}} - \underbrace{[5(20x + 20(4)x]}_{\substack{\text{cost of coat} \\ \text{material}}} - \underbrace{5(4)y}_{\substack{\text{cost of} \\ \text{dress} \\ \text{material}}}$$

$$-\underbrace{[25(4)x + 25(2)y]}_{\text{labor cost}}$$

By multiplying and collecting terms we obtain:

$$I(x, y) = 110x + 70y$$

The constraint $4x \leq 100$ or, equivalently, $x \leq 25$, expresses the condition that the amount of wool used cannot exceed 100 square yards; $2x + 4y \leq 120$ expresses the condition that the amount of cotton used cannot exceed 120 square yards; $4x + 2y \leq 110$ expresses the condition that the number of labor-hours employed cannot exceed 110.

We thus emerge with the following linear program:

$$\text{Maximize } I(x, y) = 110x + 70y$$
subject to
$$x \geq 0, y \geq 0$$
$$x \leq 25$$
$$2x + 4y \leq 120$$
$$4x + 2y \leq 110$$

From section 6.3, Exercise 7 (p. 258) we see that $\left(\dfrac{50}{3}, \dfrac{65}{3}\right)$ yields the maximum value 3350.

What is actually required is a feasible point expressed in integers which maximizes $I(x, y)$. As one would expect, such linear programs are called **integer programs**. Sometimes, as in the case of the Austin Company's linear programs, it turns out that solutions in integers are obtained when a linear program solution method is applied. Sometimes not, as we have just seen.

We know that a solution of a linear program can be found among its corner points. Thus if an integer solution is desired and some of the corner points are not integers, the idea of modifying the given linear program by appending to it new constraints with the property that no integer feasible points are lost and the resulting corner points involve only integers is naturally suggested. Upon implementation of this idea, linear programming methods can then be applied to the modified linear program to obtain a solution in integers which will also be a solution of the given integer program. We illustrate the implementation of this idea in the case of two-variable integer programs by returning to the Hoffman Clothing Manufacturer's integer program:

Find integer values for x and y which

$$\text{Maximize } I(x, y) = 110x + 70y$$

subject to

$$x \geq 0, y \geq 0$$
$$x \leq 25$$
$$2x + 4y \leq 120$$
$$4x + 2y \leq 110.$$

The feasible points of this problem viewed as a linear program are shown in Figure 7.1.

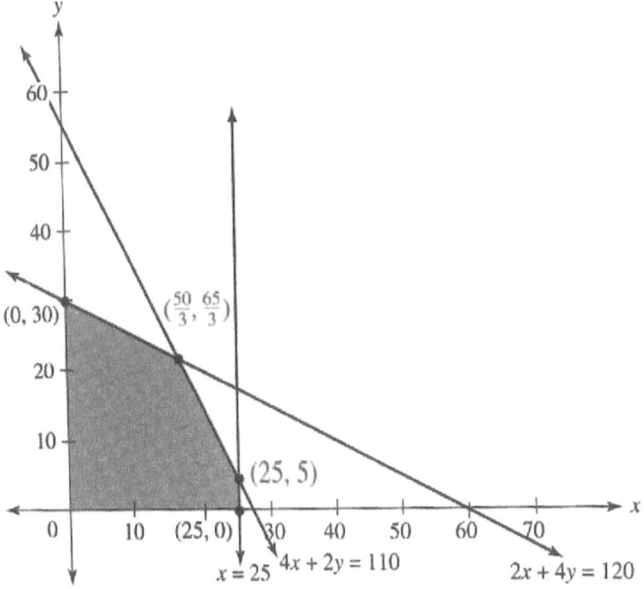

Figure 7.1

Figure 7.2 shows in detail the region surrounding $\left(\dfrac{50}{3}, \dfrac{65}{3}\right)$.

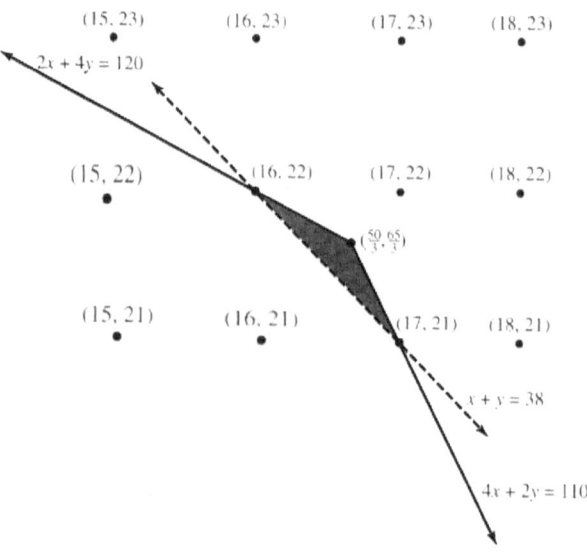

Figure 7.2

From Figure 7.2 we see that if the new constraints to be appended are to

1. eliminate $(\dfrac{50}{3}, \dfrac{65}{3})$ as a corner point.

2. not eliminate any integer feasible points,

3. yield corner points which are integers,

then we need only introduce the additional constraint $x + y \le 38$, based on the boundary line $(x + y = 38)$ passing through $(16, 22)$ and $(17, 21)$.

Thus the modified linear program is the following:

$$\text{Maximize } I(x, y) = 110x + 70y$$

subject to

$$x \ge 0, y \ge 0$$
$$x \le 25$$
$$2x + 4y \le 120$$
$$4x + 2y \le 110$$
$$x + y \le 38.$$

The feasible points of this modified program are shown in Figure 7.3. All of its corner points, $(0, 0)$, $(0, 30)$, $(16, 22)$, $(17, 21)$, $(25, 5)$, and $(25, 0)$, are expressed in

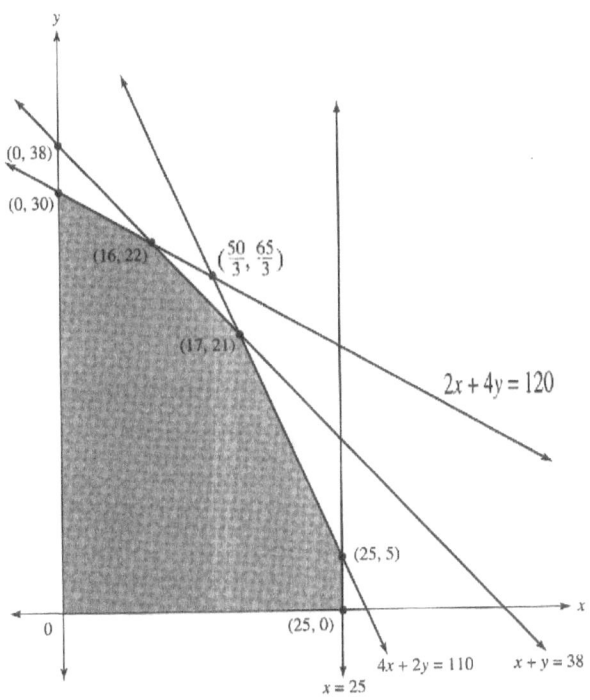

Figure 7.3

terms of integers. From Table 7.3

Table 7.3

Corner Point	$I(x, y) = 110x + 70y$
$(0, 0)$	0
$(0, 30)$	2100
$(16, 22)$	3300
$(17, 21)$	3340
$(25, 5)$	3100
$(25, 0)$	2750

we see that $(17, 21)$ is the solution and 3340 is the maximum value. Since this linear program differs from the original only in the loss of non-integer feasible points (those in the triangular region with vertices $(16, 22)$, $(17, 21)$, $(50/3, 65/3)$, above the line $x + y = 38$; see Figure 7.3, its solution $(17, 21)$ is also a solution of the Hoffman clothing manufacturer's integer program.

When the number of variables is greater than two, graphical techniques like the one employed to determine the new constraints for obtaining a solution in integers cannot be used and other methods must be sought. Algebraic techniques for the determination of the required new constraints which are computationally effective in many situations have been developed. Such techniques are beyond the scope of this book.

EXERCISES

1. Martins Plant Foods, Inc., plans to introduce Martins Miracle, hailed as a "blossom booster" for outdoors and indoors flowering plants. Martins Miracle is to be a mixture of bonemeal and processed vegetable matter. Each can of Martins Miracle is to contain at least 12 units of nitrogen and 9 units of phosphorus. Each pound of bonemeal contains 3 units of nitrogen and 1 unit of phosphorus; each pound of vegetable matter contains 2 units of nitrogen and 2 units of phosphorus.
    The problem is to determine the number of pounds of bonemeal and vegetable matter that should be used to make up a can of Martins Miracle so that the nutrient requirements are met at minimal cost.

    (a) Set up and solve the linear program model for this situation. What is the minimal cost in terms of this model and what mix would achieve it?

    (b) Management has been advised to obtain the best solution in integers to its component mixture problem. What additional constraint(s) must be added to the linear program obtained from (a) to guarantee a solution in integers?

    (c) Determine the solution to the integer program model that emerges. What does it mean to Martins Plant Foods, Inc.?

    For the integer programs given in Exercises 2 and 3, determine the additional constraints that must be imposed to guarantee a solution in integers.

2. Find nonnegative integers which minimize $F(x, y) = 3x + 2y$ subject to

$$3x + y \geq 6$$
$$x + y \geq 5.$$

3. Find nonnegative integers which maximize $P(x, y) = 16x + 6y$ subject to

$$3x + y \leq 6$$
$$x + y \leq 5.$$

SELF-TESTS FOR CHAPTERS 4-7

Allow 75 or so minutes for each self-test. Go over each one before proceeding to the next.

Self-Test 1

1. Write an equation for the line passing through (2, 5) with slope $\frac{1}{2}$.

2. Write an equation for the line passing through $(-1, 2)$ and $(3, 1)$.

3. Solve the system: $x + 2y = 40$
                              $4x + 3y = 100$

4. Solve the system: $5s + 7t = 18$
                              $3s + 2t = 2$

5. The following table shows employee age (x) in years and income (y) in thousands of dollars for a random sample of six employees of a very large corporation.

x	30	35	42	46	50	58
y	12	15	14	22	25	22

(a) Find the sample linear regression equation specifying the average value of y in terms of x.

(b) Find the predicted average income corresponding to age 45 and interpret the result obtained.

(c) Plot the original data as well as the sample regression line on one diagram.

6. Solve the system:

$$2x + y - 4z = 8$$
$$3x + 2y - 7z = 12$$
$$x - y + 3z = -4$$

7. Solve the system:

$$x + 6y + 4z = -1$$
$$x - y - 3z = -3$$
$$3x + 4y - 2z = -7$$

8. Is $f(x, y, z) = 3x + 2y + 2^z$ a linear function? Explain.

9. Sketch the graph of the following system:

$$x \geq 0, y \geq 0$$
$$4x + 5y \leq 70$$
$$x + y \leq 16$$

Self-Test 2

1. Answer TRUE or FALSE. Explain the basis for your answer.

   (a) Linear programming could be used to maximize $F(x, y) = 3x + y^2$.

   (b) A feasible point of a linear program satisfies all of the constraints.

   (c) A constraint is a mathematical statement guaranteed to precisely reflect reality.

   (d) A company should implement a linear program model's solution if they are certain that it is valid.

   (e) In real life it will always be the case that the solution of a linear program model will be the optimum one for the company.

2. Solve the system:

$$x + 3y - z = -2$$
$$-x - y + 2z = 7$$
$$3x - 2y + z = -5$$

3. Solve: Maximize $I(x,y) = 110x + 70y$
                  subject to

$$x \geq 0, y \geq 0$$
$$2x + 3y \leq 5$$
$$3x - y \leq 2$$

4. Set up the following problem in terms of a linear program model and solve.
    Audio Acoustics, Inc., plans to spend up to $12,000 in buying and stocking two speaker system models that are in great demand. The K17 model costs $50 per unit and occupies 3 cubic feet of storage space; the K24 model costs $60 per unit and occupies 4 cubic feet of storage space. The store has 760 cubic feet of storage space available. The management expects to make a profit of $2 on each K17 unit and $3 on each K24 unit.
    The problem is to determine, with respect to the given conditions, how many units of each model should be bought and stocked so that profit is maximized.

5. The Atlantic Company, a lamp manufacturer, began production of two models, A14 and A51. To determine the best production schedule, two consulting firms, the Aleksa Company and Veronika Consultants, were hired to analyze the company's operations and make recommendations. The Aleksa Company set up a linear program model for the production process, which when solved by a mathematical method called the simplex method yielded the solution (500, 280) with a maximum value of 3000. The Aleksa Company recommended that 500 A14 units and 280 A51 units be made per week, for an anticipated maximum profit of $3000 per week.
    Veronika Consultants set up a different linear program model for the production process, which when solved by the corner-point method yielded (450, 300) as the solution with a maximum value of 2500. Veronika Consultants recommended that 450 A14 units and 300 A51 units be made weekly to obtain a maximum profit of $2500 per week.

    The management of the Atlantic Company found these developments puzzling and raised the following questions. Answer these questions in appropriate detail.

    (a) How is it possible for different solutions to be obtained? After all, isn't mathematics a precise subject?

(b) Which solution is correct and in what sense is it correct?

(c) Which solution should be implemented and why?

6. "Math Will Rock Your World" by Stephen Baker, *Business Week*, Jan 23, 2006 (pp. 54-62) discusses the role of mathematics—math modeling in particular—in transforming the modern world. It would be difficult to find a more richly textured paean to mathematics.

   As a sample to give you an idea of this texture I quote one section, "How Mathematics Transforms Industries" (p. 57).

Mathematicians have long enjoyed celebrity status in Silicon Valley and on Wall Street. Now they're plying their trade throughout the U.S. economy.

Consulting:

IBM: Big Blue is building math profiles of 50,000 consultants so that computers can pick the perfect team for every assignment. Other tools eventually will be able to track their progress, hour by hour, and rate their performance. Workers will eventually labor in virtual assembly lines.

Food and Beverage:

ENOLOGIX: The goal of this California consultancy is to help vintners mimic the chemistry of wines highly ranked by leading critic Robert B. Parker. It employs algorithms to cull a database of 70,000 vintages and run analyses. Precise studies of customer data provide blueprints for new products.

Advertising:

EFFICIENT FRONTIER: The Silicon Valley startup provides mathematical optimization for online ad campaigns. It calculates response rates and return on investment for every advertisement. Broad shift from hunch-based campaigns to mathematical targeting.

Police and Intelligence:

NATIONAL SECURITY AGENCY: Mathematicians at the nation's top techno-spy agency build algorithms to trawl internet and phone traffic looking for pattern speech, subject and frequency that might point to the next attack. Investigators wade through rivers of data in search for would-be terrorists.

Marketing:

UMBRIA: Colorado startup assigns numeric values to picks and pans of products that pop up on blogs. Using vector graphics, it confirmed that raunchy Burger King ads online turned off nearly everyone, except for the target audience of young men. Math—based consultancies scour blogs and podcasts for market intelligence.

Media:

INFORM: This New York startup turns written articles into bits of geometry and organizes them in a virtually library. It can match the articles to readers' math-based profiles. Automatic systems threaten to supplant editors.

In the closing part of the article the author notes that midcareer managers "still must understand enough about math to question the assumptions behind the numbers."

Do you agree or disagree with this assessment? Explain.

Self-Test 3

1. Solve the following systems:

(a) $-3x + 6y = 12$
$\quad\ \ x - 2y = -4$

(b) $\ \ x - 2y + z = 4$
$\quad 2x + y - z = 2$
$\quad\ \ x + y + 2z = 6$
$\quad\ \ x - y - z = 2$

2. Sketch the graphs of the following systems:

(a) $x \geq 0, y \geq 0$ (b) $x \geq 0, y \geq 0$

        $x + 2y \geq 40$ $x + y \leq 15,000$

        $4x + 3y \geq 100$ $x + y \geq 2000$

                                    $x \leq 10,000, \ y \leq 12,000$

3. The Daniel Company is a producer of fine baked goods. To obtain more visibility for its products the Daniel Company plans to advertise on the food cable channel BAKE. Morning time costs $2000 per minute and afternoon time costs $1600 per minute. The marketing department estimates that morning time commercials reach 1.1 million viewers and afternoon time commercials reach 0.8 million viewers. At most 16 minutes of morning time is available in the period in which the Company plans to run its campaign. The marketing department recommends that at least 8 minutes of morning time and at least 6 minutes of afternoon time be purchased. The advertising budget has been set at $48,000.

    The problem is to determine the number of minutes of morning and afternoon time that should be purchased to maximize listener exposure time.

(a) Set up a linear program model for the Daniel Company's problem.

(b) Solve the linear program model.

(c) In view of the reservations expressed by Bass, Lonsdale, and Kotler concerning the suitability of linear program modeling for advertising media-selection, what thoughts would you express to the management of the Daniel Company on implementing the solution obtained in answer to (b)? Explain.

(d) On reconsideration the marketing department changed its recommendation that at least 8 minutes of morning time be purchased to at least 10 minutes of morning time be purchased.

(i) What modification would have to be made in the model that had been set up to accomodate this change.

(ii) Does the modified L.P. model have a solution? Explain.

(iii) If your answer to (ii) is no, what would the Daniel Company's management have to do to accomodate this recommendation? Explain.

4. Camera World, which makes VHS video cameras among other products, has three service departments—accounting, shipping and marketing, and two production departments. Each service department's total cost must be distributed to the service departments and to the production departments based on their respective usages of the services provided. For each service department listed in the left most column of Table 1

Table 1

Service	Service Depts			October
Dept.	Accounting	Shipping	Marketing	Overhead
Accounting	0.01	0.08	0.02	$30,000
Shipping	0.02	0	0.15	$20,000
Marketing	0.02	0.01	0	$95,000

the fraction of its total cost assigned to the service departments of the firm is given. The October overhead of the service departments is also given.

Set up the system of equations that describes the conditions to be satisfied by the total costs of the service departments. Be sure to clearly state what your variables represent.

5. The Hudson Furniture Manufacturing Company, Inc., has available 300 square yards of pine veneer and 228 square yards of walnut veneer for the manufacture of model D-1 desks and model B-14 bookcases. 8 square yards of pine veneer and 6 square yards of walnut veneer are used in making a desk; 10 square yards of pine veneer and 7 square yards of walnut veneer are used in making a bookcase. Pine veneer costs $2 per square yard and walnut veneer costs $4 per square yard. 5 labor-hours are needed to manufacture a desk and 4 labor-hours are needed to manufacture a bookcase. The cost of labor is $4 per labor-hour. At most 150 labor-hours are available for the manufacture of the desks and bookcases. A desk sells for $100 and a bookcase sells for $108.

The problem is to determine the number of each that should be made and sold to maximize profit.

(a) Set up and solve the linear program model for this problem.

(b) Determine the additional constraint(s) that must be imposed to guarantee a solution in integers.

(c) Solve the integer program model that emerges.

(d) What advice would you give to the Hudson Company concerning the implementation of the afore integer program model.

Self-Test 4

1. Solve:
$$2x + 3y + z = -3$$
$$x - 4y - z = 7$$
$$x + 7y + 2z = 8$$

2. Sketch the graph of the system:
$$x \geq 0, \; y \geq 0$$
$$x + y \geq 3$$
$$x + y \leq 9$$
$$-x + y \leq 5$$

3. Set up the following problem in terms of a linear program model and solve.

The Borg Company makes desks. Two models, B7 and B9, are currently in high demand. Both models pass through finishing and assembly plants of the company. In the finishing plant, a B7 unit is worked on for 1 hour; a B9 unit is worked on for 2 hours. In the assembly plant, a B7 unit is worked on for $\frac{3}{2}$ hours; a B9 unit is worked on for 1 hour. At most 220 work hours of finishing time and 270 work hours of assembly time are available per week. The profit on a B7 unit is $12; the profit on a B9 unit is $10.

The problem is to determine, with respect to the given assumptions/postulates, how many units of each model should be made per week so as to maximize profit.

(a) What advice would you give the Borg Company concerning the implementation of the afore linear program model?

4. Do you agree or disagree with the following point of view. So state and explain in appropriate detail.

   "Mathematical methods have the advantage of certitude. No qualified person can resist the truth of a mathematical conclusion properly communicated. The job of communication may be difficult if the solution is complex, but when the communication is competent, agreement is inevitable. If anyone doubts a solution, he can recalculate the equations and check the steps in the derivation. Then he must either demonstrate that there has been an error or acknowledge the truth of the solution."

5. Which of the following might be an objective function for a linear program? Explain.

   (a) $F(x, y) = 2x - 0.33y$ (b) $F(x, y) = 2x + xy - 3y$

   (c) $F(x, y, z) = \dfrac{2x + 3y}{z}$ (d) $F(x, y) = 2x^2 + 3y$

6. Which of the following might be a constraint for a linear program? Explain.

   (a) $3x - 2y + z \le -4$ (b) $x^2 + y^2 \le 4$
   (c) $0.25x + 2y \le 8$ (d) $3x - xy \ge 10$

7. Does the following linear program have a solution? Explain.

$$\text{Maximize } F(x, y) = 3x + 2y$$
   subject to

$$x \ge 0, y \ge 0$$
$$x + 3y \ge 9$$
$$2x + y \ge 8$$

CHAPTER 8

Topics in Algebra with Consideration of Realism

8.1 OPERATIONS WITH POLYNOMIALS AND FACTORING

Algebraic expressions such as

$$2x+1, \qquad 2x^2 + \frac{1}{2}x - 1$$

$$x^3 - x^2 + 4, \qquad 2x^8 - \frac{1}{3}x^6 - 4$$

illustrate the nature of polynomials in one variable. The **general polynomial in x of degree n,** where n is a positive integer, is an expression that is, or can, be put into the form

$$ax^n + bx^{n-1} + cx^{n-2} + \cdots + hx + k$$

where $a, b, c, \ldots, h$ and k are constants, and $a \neq 0$. Polynomials of degree 1, illustrated by

$$2x+1, \qquad \frac{1}{4}x - 2, \qquad 3x$$

are called **linear,** and polynomials of degree 2, illustrated by

$$2x^2 + \frac{1}{2}x - 1, \qquad 3x^2 - x, \qquad x^2 - 9$$

are called **quadratic.** Our main concern is with linear and quadratic polynomials.

Addition and Subtraction of Polynomials

Addition and subtraction of polynomials is no different from addition and subtraction of algebraic expressions in general; but since we shall have occasion to employ such operations, we pause to refresh our memories at this point. To add two polynomials, we write like terms under each other and then add.

EXAMPLE 1

Add $2x^3 + 4x^2 + 2$ and $-2x^2 + x - 6$.

$$
\begin{array}{r}
2x^3 + 4x^2 + 2 \\
-2x^2 + x - 6 \\
\hline
2x^3 + 2x^2 + x - 4
\end{array}
$$

Thus the sum is $2x^3 + 2x^2 + x - 4$.

To subtract a first polynomial from a second, write the first under the second, placing like terms under each other, change the signs of the terms in the lower polynomial, and add.

EXAMPLE 2

Subtract $4x^3 - 2x + 1$ from $3x^2 - 3x + 4$.

We obtain

$$
\begin{array}{cc}
3x^2 - 3x + 4 & \qquad 3x^2 - 3x + 4 \\
4x^3 \qquad -2x + 1 \xrightarrow[\substack{\text{change} \\ \text{signs}}]{} & -4x^3 \qquad + 2x - 1 \\
& \qquad \overline{-4x^3 + 3x^2 - x + 3}
\end{array}
$$

Thus the difference is $-4x^3 + 3x^2 - x + 3$.

The basis for this procedure is the use of the distributive property for removing parentheses. The problem is to find $(3x^2 - 3x + 4) - (4x^3 - 2x + 1)$. Removing parentheses yields $3x^2 - 3x + 4 - 4x^3 + 2x - 1$. The follow-through involves combining like terms.

EXERCISES

1. Add $3x^2 + 4x - 3$ and $x^3 + 5x^2 - 1$.

2. Add $2x^4 - 4x^3 + 6$ and $5x^3 - x^2 + 3$.

3. Add $-4x^2h + 2xh - 1$ and $-4x^2h + 3xh + h^2$.

4. Add $-x^3h - 3x^2h + 4h$ and $2x^3 - 4x^2h - xh^2$.

5. Add $3x^2y - 3y + 2$ and $x^2y - 3xy + 5$.

6. Subtract $2x^2 + x - 1$ from $4x^2 - 3x + 5$.

7. Subtract $-x^2 + 4x - 3$ from $x^2 + 3x - 8$.

8. Subtract $x^2h - 3xh + h$ from $3x^2h + 4xh + 7$.

9. Subtract $4x^3h + 2x^2h + 3xh$ from $4x^3h - 3x^2h + 2x + 1$.

10. Subtract $3x^2h^2 + 2xh - 4$ from $4x^2h^2 + 3x^2h + 7xh$.

Multiplication of Polynomials

The distributive property for the real-number system states

$$a(b - c + d - e + \cdots + z) = ab - ac + ad - ae + \cdots + az.$$

For example:

$$-5(3x^2 + x - 1) = -15x^2 - 5x + 5$$

$$3x^2(4x^3 - 3x^2 - 2) = 3x^2(4x^3) + 3x^2(-3x^2) + 3x^2(-2)$$
$$= 12x^5 - 9x^4 - 6x^2$$

More generally, to multiply a first polynomial by a second polynomial, multiply each term of the first polynomial by each term of the second, and take the algebraic sum of these products.

EXAMPLE 3

Multiply $(x-5)(2x^2+x-1)$.

By the distributive property, we have:

$$\begin{aligned}
(x-5)(2x^2+x-1) &= (x-5)2x^2+(x-5)x+(x-5)(-1)\\
&= (2x^3-10x^2)+(x^2-5x)+(-x+5)\\
&= 2x^3-10x^2+x^2-5x-x+5\\
&= 2x^3-9x^2-6x+5
\end{aligned}$$

This work may be more conveniently carried out by means of the following vertical arrangement.

$$\begin{array}{ll}
2x^2+\quad x-1 & \\
\underline{x\ -\quad 5} & \\
2x^3+\quad x^2-\ x & \rightarrow\quad \text{product } x(2x^2+x-1)\\
\underline{\quad -10x^2-5x+5} & \rightarrow\quad \text{product } -5(2x^2+x-1)\\
2x^3-\ 9x^2-6x+5 & \rightarrow\quad \text{sum of like terms}
\end{array}$$

Multiplying each term of $2x^2+x-1$ by each term of $x-5$ and combining similar terms is what this comes down to in practice.

EXAMPLE 4

Find $(x+h)^3$.

Multiplying $(x+h)$ by itself yields:

$$\begin{array}{l}
x+h\\
\underline{x+h}\\
x^2+hx\\
\underline{\quad +hx+h^2}\\
x^2+2hx+h^2
\end{array}$$

Multiplying $(x+h)^2 = x^2 + 2hx + h^2$ by $(x+h)$ yields:

$$
\begin{array}{l}
x^2 + 2hx + h^2 \\
x+h \\
\hline
x^3 + 2hx^2 + h^2 x \\
\quad + hx^2 + 2h^2 x + h^3 \\
\hline
x^3 + 3hx^2 + 3h^2 x + h^3
\end{array}
$$

Thus $(x+h)^3 = x^3 + 3hx^2 + 3h^2 x + h^3$.

EXAMPLE 5

Find $(2ax+b)^2$.

$$
\begin{array}{l}
2ax + b \\
2ax + b \\
\hline
4a^2 x^2 + 2abx \\
\quad + 2abx + b^2 \\
\hline
4a^2 x^2 + 4abx + b^2
\end{array}
$$

Thus $(2ax+b)^2 = 4a^2 x^2 + 4abx + b^2$.

EXERCISES

Find the following products.

11. $3x(x+4)$ 12. $2x(2x^2 +1)$

13. $-x(x^2 - x + 1)$ 14. $2x(3x^2 - 3x + 4)$

15. $-3x(x^3 - 4x^2 + 7)$ 16. $-2x^2(3x^2 + 4x - 9)$

17. $2xh(3x^2 h + 4h - 1)$ 18. $3xh(5x^2 - 3xh + 4)$

19. $(x-1)(x+2)$ 20. $(x-3)(x+3)$

21. $(x-h)(x+h)$ 22. $(x-h)^2$

23. $(2x-3)(x+1)$

24. $(5x+1)(x-3)$

25. $(2x-1)(3x+2)$

26. $(5x-3)(2x-4)$

27. $(x-h)^3$

28. $(x-1)(x^2+3x+4)$

29. $(x-2)(x^3+3x^2+4)$

30. $(x+h)(2hx^2+4hx-3)$

31. $(x+2)(x^4-3x^2+2x+1)$

32. $2(x-3)(4x+5)$

33. $(3x-2)(3x^3+4x^2+7)$

34. $(2x^2+h)(x^2h+3hx-4)$

35. $(4x+3)(4x^3+3x+7)$

36. $4x(x-2)(x+5)$

37. $5x^2(x-3)(2x^2+1)$

38. $(x-h)^4$

39. $(x-3)(x+4)(x-6)$

40. $(3x+1)(x+3)(2x+6)$

41. $(x+h)^5$

42. $(x-h)^5$

Factoring

The distributive property lies at the center of basic factoring since it justifies taking out a factor common to all members of a sum of terms and expressing the sum as a product of the term taken out and the remaining sum. Thus, for example, from

$$2x^3 - 2x^2 - 12x$$

we may factor out $2x$ and write:

$$2x^3 - 2x^2 - 12x = 2x(x^2 - x - 6)$$

Can we do better? That is, can $x^2 - x - 6$ be expressed as a product? To explore this question, let us consider how the factors, if there are such, would multiply back to yield $x^2 - x - 6$.

Consider possible factors of the form

$$x+a \quad \text{and} \quad x+b.$$

Multiplying these factors yields:

$$x + a$$
$$\underline{x + b}$$
$$x^2 + ax$$
$$\underline{+bx + ab}$$
$$x^2 + ax + bx + ab$$
$$= x^2 + (a + b)x + ab$$

For $x^2 - x - 6$, $a + b = -1$, and $ab = -6$. Thus to yield a negative product, -6, a and b must be of opposite sign and add up to -1. This suggests $a = -3$ and $b = 2$. To verify that $x^2 - x - 6 = (x - 3)(x + 2)$, we check by multiplying $(x - 3)$ and $(x + 2)$.

$$(x - 3)$$
$$\underline{(x + 2)}$$
$$x^2 - 3x$$
$$\underline{+2x - 6}$$
$$x^2 - x - 6$$

Thus we have:

$$2x^3 - 2x^2 - 12x = 2x(x^2 - x - 6) = 2x(x - 3)(x + 2)$$

Factoring is useful because it often permits simplifications through application of the cancellation principle and is useful in solving certain kinds of equations.

Much can be said about factoring a variety of special forms, but for our purposes it suffices to focus on factoring quadratic polynomials of the form

$$ax^2 + bx + c$$

where $a \neq 0$.

EXAMPLE 6

Factor, if possible, $2x^2 - 4x - 30$.

The first order of business is to see if any terms that are common to all members of the sum can be factored out. From $2x^2 - 4x - 30$ we can factor out 2.

$$2x^2 - 4x - 30 = 2(x^2 - 2x - 15)$$

To factor $x^2 - 2x - 15$, we need factors that must be of the form

$$(x + a)(x + b)$$

to obtain the term x^2. The numbers a and b must yield a product of -15 (and thus be of opposite sign) and a sum of -2. If we try $a = 5$ and $b = -3$, we obtain a product of -15, but a sum of 2, so that this possibility washes out.

On the other hand, $a = -5$ and $b = 3$ seem to work. Verification via multiplication confirms this.

$$
\begin{array}{r}
x - 5 \\
x + 3 \\
\hline
x^2 - 5x \\
+3x - 15 \\
\hline
x^2 - 2x - 15
\end{array}
$$

Thus we have:

$$2x^2 - 4x - 30 = 2(x^2 - 2x - 15) = 2(x - 5)(x + 3)$$

EXAMPLE 7

Factor, if possible, $x^2 + 4$.

To factor $x^2 + 4$, we consider factors of the form

$$(x + a)(x + b)$$

which yield the term $x^2 + 4$. The numbers a and b must yield a product of 4 and a sum of 0. To yield a product of 4, a and b must be of the same sign (both positive or both negative), and therefore cannot yield a sum of 0.

Thus it is not possible to factor $x^2 + 4$.

EXAMPLE 8

Factor, if possible, $6x^2 + x - 15$.

This problem is a bit more troublesome because the number of possibilities to be considered is larger; but our basic approach still pertains. We need linear factors that yield a product of $6x^2$ rather than x^2. To begin, consider factors of the form

$$(6x+a)(x+b).$$

The product ab must be -15, and thus a and b must be of opposite sign.

The following possibilities are suggested, and we only stop to determine the middle term since the other requirements are met.

$$
\begin{array}{ccc}
6x+15 & 6x+5 & 6x-5 \\
\dfrac{x-1}{15x} & \dfrac{x-3}{5x} & \dfrac{x+3}{-5x} \\
\dfrac{-6x}{9x} & \dfrac{-18x}{-13x} & \dfrac{18x}{13x}
\end{array}
$$

None of these products work, so let's consider factors of the form:

$$(2x+a)(3x+b)$$

Since we must have $ab = -15$, the following possibilities are suggested.

$$
\begin{array}{cc}
2x+5 & 2x-3 \\
\dfrac{3x-3}{15x} & \dfrac{3x+5}{-9x} \\
\dfrac{-6x}{9x} & \dfrac{10x}{x}
\end{array}
$$

Thus $(2x-3)(3x+5) = 6x^2 + x - 15$.

There is often a fair amount of trial and error in this kind of analysis, and in the end the factors we are searching for may not exist. One may well ask, what kind of factors are we searching for? As the examples considered suggest, we shall restrict our attention to factors with integer values. Thus, for example, $x^2 - 5$ cannot be factored further from this point of view.

If irrational numbers are permitted in the factors, then $x^2 - 5$ can be expressed as the product $(x-\sqrt{5})(x+\sqrt{5})$.

EXERCISES

Factor each of the following as completely as possible and check by multiplication.

43. $5x + 3xy + xz$ 44. $3a^2 + 2ab + a$ 45. $6ax - 4a^2 y$

46. $3b + 6b^2 + 12b^3$ 47. $2xy + x^2 y - xy^2$ 48. $2ab - 3a^2 b^2 + a^3 b^2$

49. $x^2 + 5x - 6$ 50. $x^2 + 4x + 3$ 51. $x^2 + 12x + 11$

52. $x^2 - 9$ 53. $t^2 - 6t + 5$ 54. $y^2 + 16$

55. $x^2 - 10x + 21$ 56. $2x^2 - 18x + 16$ 57. $x^2 + 4x + 5$

58. $2y^3 - 12y^2 + 16y$ 59. $3x^3 - 27x^2 + 42x$ 60. $x^2 - 18x + 72$

61. $t^3 - 16t^2 + 60t$ 62. $x^2 - 16x + 64$ 63. $4x^3 - 8x^2 - 32x$

64. $t^2 + 11t - 60$ 65. $-x^2 - 10x - 25$ 66. $x^2 + 6x + 12$

67. $x^2 - a^2$ 68. $2y^3 - 12y^2 - 54y$ 69. $2x^2 - 7x + 15$

70. $4x^2 - 12x + 5$ 71. $2x^2 - 3x + 1$ 72. $5x^2 - 3x + 8$

73. $6x^2 + 5x - 6$ 74. $10t^2 + 49t - 5$

8.2 QUADRATIC EQUATIONS AND QUADRATIC FUNCTIONS

A sugar refinery has revenue and cost functions defined by

$$R(x) = 405x - 2x^2, \qquad c(x) = \frac{1}{2}x^2 + 5x + 3000$$

where x is output in tons per week, and $R(x)$ and $c(x)$ are revenue and cost in dollars. Where does the firm break even? That is, for what output x does revenue equal cost? Setting $R(x)$ equal to $c(x)$ yields

$$405x - 2x^2 = \frac{1}{2}x^2 + 5x + 3000,$$

from which we obtain:

$$-\frac{5}{2}x^2 + 400x - 3000 = 0 \qquad (8.1)$$

Determination of the break-even point reduces to solving Equation (8.1) which is an equation of the form

$$ax^2 + bx + c = 0 \qquad (8.2)$$

where a, b, and c are constants and $a \neq 0$.

Any equation that can be written in this **standard form** is called a **quadratic equation**. We next turn our attention to developing methods for solving quadratic equations and then return to the previous break-even problem.

Solution by Factoring

EXAMPLE 1

Solve $2x^2 - 16 = 0$.

To solve an equation of this form, where the term involving x is missing, is relatively easy. We have:

$$2x^2 - 16 = 0, \quad x^2 - 8 = 0, \quad x^2 = 8, \quad x = \sqrt{8} \quad \text{and} \quad -\sqrt{8}$$

It is easily verified that these values are solutions.

EXAMPLE 2

Solve $6x^2 + x - 5 = 10$.

We first express this equation in standard form by transposing 10. This yields:

$$6x^2 + x - 15 = 0$$

$6x^2 + x - 15$ can be expressed as the product $(2x - 3)((3x + 5)$. Thus we have:

$$(2x - 3)(3x + 5) = 0$$

Since the product of two factors can be zero only when at least one of the factors is zero, our problem reduces to the consideration of two cases.

Case 1: $2x - 3 = 0$ Case 2: $3x + 5 = 0$

$\qquad\qquad 2x = 3$ $\qquad\qquad\qquad\qquad\quad 3x = -5$

$\qquad\qquad x = \dfrac{3}{2}$ $\qquad\qquad\qquad\qquad\quad x = -\dfrac{5}{3}$

Since the operations used are operations of equivalence, which take us from an equation to an equivalent equation, we may conclude that if our equation has a solution, then $\dfrac{3}{2}$ and $-\dfrac{5}{3}$ are solutions of $6x^2 + x - 5 = 10$.

It is easily verified that these values are solutions.

EXAMPLE 3

Solve $2x^3 - 2x^2 - 12x = 0$.

We begin by factoring out $2x$. This yields:

$$2x(x^2 - x - 6) = 0$$

Since $x^2 - x - 6$ factors into $(x - 3)(x + 2)$, we obtain:

$$2x(x - 3)(x + 2) = 0$$

Case 1: $2x = 0$ $\qquad$ Case 2: $x - 3 = 0$ $\qquad$ Case 3: $x + 2 = 0$

$\qquad\quad x = 0$ $\qquad\qquad\qquad\quad x = 3$ $\qquad\qquad\qquad\quad x = -2$

By substitution, it is easily verified that 0, 3, and −2 are solutions of $2x^3 - 2x^2 - 12x = 0$.

Since many quadratic equations cannot be easily factored, and, indeed, most cannot be factored at all, a more general approach to solving quadratic equations is needed.

The Quadratic Formula

The solutions of the quadratic equation $ax^2 + bx + c = 0$ are given by

$$x = \frac{-b + \sqrt{b^2 - 4ac}}{2a} \quad \text{and} \quad x = \frac{-b - \sqrt{b^2 - 4ac}}{2a}$$

which is usually condensed to:

$$x = \frac{-b \pm \sqrt{b^2 - 4ac}}{2a}$$

This presupposes that $ax^2 + bx + c = 0$ has a solution in the real-number system, which is not the case if $b^2 - 4ac < 0$.

Before deriving this result, which is called the **quadratic formula,** we shall illustrate its application to the quadratic equation

$$6x^2 + x - 15 = 0$$

which we solved in Example 2 by factoring. In this situation $a = 6$, $b = 1$, and $c = -15$. We have:

$$x = \frac{-1 \pm \sqrt{(1)^2 - 4(6)(-15)}}{2(6)} = \frac{-1 \pm \sqrt{1 + 360}}{12}$$

$$= \frac{-1 \pm \sqrt{361}}{12} = \frac{-1 \pm 19}{12} = \frac{3}{2} \quad \text{and} \quad -\frac{5}{3}$$

Thus the solutions of our equation are $\frac{3}{2}$ and $-\frac{5}{3}$, which, as we would expect, agrees with the results obtained in Example 2.

To derive the quadratic formula, we begin with the quadratic equation

$$ax^2 + bx + c = 0$$

and multiply both sides by $4a$, obtaining:

$$4a(ax^2 + bx + c) = 4a \cdot 0$$
$$4a^2x^2 + 4abx + 4ac = 0 \tag{8.3}$$
$$4a^2x^2 + 4abx \quad\quad = -4ac$$

From section 8.1, Example 5 (p. 306), $(2ax + b)^2 = 4a^2x^2 + 4abx + b^2$. Thus to make the left side of Equation (8.3) the perfect square $(2ax + b)^2$, we must add b^2 to it. Adding b^2 to both sides yields:

$$4a^2x^2 + 4abx + b^2 = b^2 - 4ac$$
$$(2ax + b)^2 = b^2 - 4ac$$

Taking square roots gives us:

$$2ax + b = \pm\sqrt{b^2 - 4ac}$$

Transposing b yields:

$$2ax = -b \pm \sqrt{b^2 - 4ac}$$

Finally, dividing both sides by $2a$ gives us the desired result:

$$x = \frac{-b \pm \sqrt{b^2 - 4ac}}{2a}$$

Substitution of these values for x in $ax^2 + bx + c = 0$ and carrying out the computations verifies that they are solutions of the general quadratic equation.

If $b^2 - 4ac = 0$, the quadratic equation has one solution, $x = -b/2a$; if $b^2 - 4ac > 0$, there are two solutions; and if $b^2 - 4ac < 0$, there are no solutions within the real-number system.

To further illustrate the use of the quadratic formula, we return to the break-even problem considered at the beginning of the section, which led to the quadratic equation

$$-\frac{5}{2}x^2 + 400x - 3000 = 0 \tag{8.4}$$

which we simplify to

$$x^2 - 160x + 1200 = 0$$

by multiplying both sides of (8.4) by $-\dfrac{2}{5}$. In this situation $a = 1$, $b = -160$, and $c = 1200$. We have:

$$x = \frac{-(-160) \pm \sqrt{(-160)^2 - 4(1)(1200)}}{2(1)} = \frac{160) \pm \sqrt{25,600 - 4800}}{2}$$

$$= \frac{160) \pm \sqrt{20,800}}{2} = \frac{160 \pm 144.22}{2} = 152.1 \quad \text{and} \quad 7.9$$

For the sugar refinery, **with respect to the given conditions,** the revenue function equals the cost function for output levels of 7.9 and 152.1 tons per week.

> We should keep in mind that whether these output levels are in **reality** the break-even output levels of the sugar refinery depends on how realistic the given revenue and cost functions are.

Quadratic Functions

A function of the form

$$y = ax^2 + bx + c$$

where a, b, and c are constants and $a \neq 0$, is said to be a **quadratic function.** The revenue and cost functions of a sugar refinery, defined by

$$R(x) = 405x - 2x^2, \qquad c(x) = \frac{1}{2}x^2 + 5x + 3000$$

and considered at the beginning of this section, are quadratic functions.

Geometrically speaking, quadratic functions are **parabolas,** upright if $a > 0$ (shown in Figure 8.1), and downward if $a < 0$ (shown in Figure 8.2).

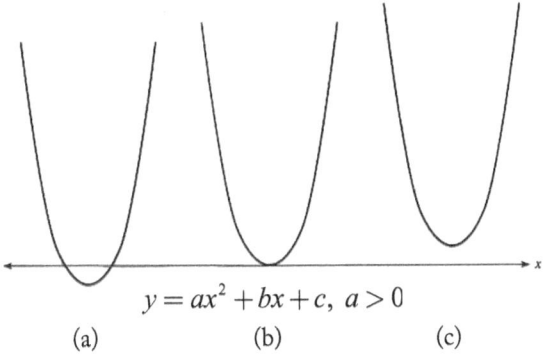

$$y = ax^2 + bx + c, \ a > 0$$

(a)   (b)   (c)

Figure 8.1

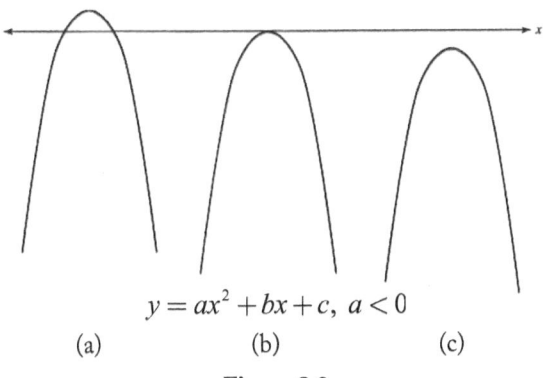

$$y = ax^2 + bx + c, \ a < 0$$

(a)   (b)   (c)

Figure 8.2

From these figures we see that a parabola may intersect the x-axis ($y = 0$) at two distinct points, one point, or no points. For the quadratic equation

$$0 = ax^2 + bx + c$$

associated with $y = ax^2 + bx + c$ these possibilities translate to the equation having two distinct solutions, one distinct solution, or no solution within the real-number system.

To plot the graph of a quadratic function it suffices to give x a number of values, calculate the corresponding values of y, plot the points (x, y) obtained, and join them by a smooth curve. In doing this in applied settings we must be sensitive to the domain of definition of the function, which is often not the set of all real-numbers.

In plotting the graph of a quadratic function it is useful to know where its minimum, or maximum, point is. To determine the location of the extreme point, we consider

$$y = ax^2 + bx + c \tag{8.5}$$

and employ, with a suitable variation, the technique used to establish the quadratic formula. Multiplying numerator and denominator of the right side of (8.5) by $4a$ yields:

$$y = \frac{4a^2x^2 + 4abx + 4ac}{4a} \tag{8.6}$$

Adding and subtracting b^2 to the numerator of (8.6) gives us

$$y = \frac{(4a^2x^2 + 4abx + b^2) - b^2 + 4ac}{4a} \tag{8.7}$$

Substitution of $(2ax + b)^2$ for $4a^2x^2 + 4abx + b^2$ in the numerator of (8.7) yields

$$y = \frac{(2ax + b)^2 - b^2 + 4ac}{4a}$$

$$= \frac{(2ax + b)^2}{4a} + \frac{-b^2 + 4ac}{4a}$$

The first component, $(2ax + b)^2 / 4a$, varies as x varies and the second component, $(-b^2 + 4ac)/4a$, does not involve x and is constant for a given function.

Consider two cases: suppose that $a > 0$, so that the parabola opens upward and has a minimum point. Since $(2ax+b)^2 \geq 0$ for all x and $4a > 0$, $(2ax+b)^2/4a \geq 0$, and the minimum value of y is attained when $(2ax+b)^2/4a = 0$. This requires that $2ax+b = 0$, or $x = -b/2a$.

Suppose that $a < 0$, so that the parabola opens downward and has a maximum point. Then $(2ax+b)^2/4a \leq 0$ (since $4a$ is negative), and the maximum value of y is obtained when $(2ax+b)^2/4a = 0$. This requires that $2ax+b = 0$, or $x = -b/2a$.

> In summary, then, the minimum, or maximum, value of a quadratic function is attained at $x = -b/2a$.

Example 4

Returning to the revenue and cost functions of the sugar refinery,

$$R(x) = 405x - 2x^2, \qquad c(x) = \frac{1}{2}x^2 + 5x + 3000$$

we have that the maximum value of $R(x)$ is attained at
$x = -b/2a = -405/(-4) = 101.3$.

$$R(101.3) = 405(101.3) - 2(101.3)^2 = 20,503$$

Thus the maximum point of the graph of the revenue function $R(x)$ is (101.3, 20,503); the revenue function is maximized at the output level of 101.3 tons of sugar per week.

Values of $R(x)$ for selected values of x are given in Table 8.1, and the graph of $R(x)$ is shown in Figure 8.3.

Table 8.1

x	0	50	101.3	150	200
$R(x)$	0	15,250	20,503	15,750	1000

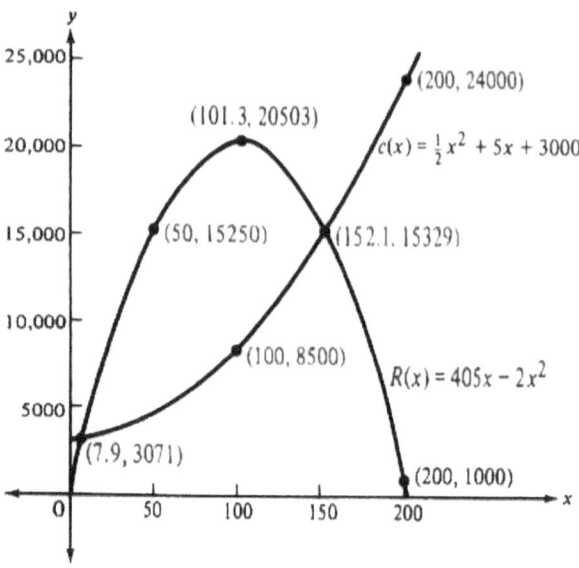

Figure 8.3

The line $x = -b/2a$ serves as a line of symmetry for the graph of $y = ax^2 + bx + c$. For the sugar refinery's revenue function, the line of symmetry is $x = 101.3$; the behavior of the graph on one side of this line is reproduced on the other side.

The minimum value of the cost function $c(x) = \frac{1}{2}x^2 + 5x + 3000$ is attained at

$x = -b/2a = -5/1 = -5$, which is outside of the domain of definition of $c(x)$. Values of $c(x)$ for selected values of x are given in Table 8.2, and the graph of $c(x)$ is shown in Figure 8.3. Also shown in Figure 8.3 are the break-even points (7.9, 3071) and (152.1, 15,329).

Table 8.2

x	0	7.9	100	152.1	200
$c(x)$	3000	3071	8500	15,329	24,000

The profit function $P(x)$ of the sugar refinery is obtained by subtracting cost from revenue:

$$P(x) = R(x) - c(x)$$

$$= 405x - 2x^2 - (\frac{1}{2}x^2 + 5x + 3000)$$

$$= -\frac{5}{2}x^2 + 400x - 3000$$

The maximum value of $P(x)$ is attained at $x = -b/2a = -400/(-5) = 80$. The profit function is maximized at the output level of 80 tons of sugar per week.

Mathematical Precision vs. Real World Accuracy

Does the afore analysis establish that it is true in the real world sense that profit will be maximized at the output level of 80 tons per week? No; per the discussion of real world accuracy (realism) vs. mathematical precision (validity) in sec. 3.3 of ch. 3, the math analysis establishes the validity of this conclusion with respect to our starting point, namely the revenue and cost functions

$R(x) = 405x - 2x^2$ and $c(x) = \dfrac{1}{2}x^2 + 5x + 3000$ obtained for the sugar refinery.

The real world accuracy of this valid conclusion is another issue.

May I suggest a review of sec. 3.3 at this point for perspective on these most important issues.

EXAMPLE 5   A RETURN TO THE THOMAS COMPANY'S SITUATION

This takes us back to Example 3 (p. 126) of sec. 3.3, Functions from Applications, where we obtained

$$I(x) = (20,000 - 200x)(60 + x),$$

x = 0, 1, . . . perhaps extending as far as 99, as the function expressing the total average income of all restaurants in the chain as a function of the number of new restaurants opened.

The Thomas Company requires an answer to the question, how many additional restaurants should be opened in order to maximize the total daily average income of all the restaurants in its chain.

To address this question consider the mathematical model counterpart of $I(x)$, which is

$$I(x) = (20,000 - 200x)(60 + x)$$
$$= -200x^2 + 8000x + 1,200,000$$

where $0 \le x \le 99$.

Here $a = -200$ and $b = 8000$, so that $\dfrac{-b}{2a} = 20$. Mathematically speaking, we can tell the Thomas Company that $I(x)$ is maximized for $x = 20$.

What Does this Mean to the Thomas Company?

Does it mean that if the Thomas Company were to implement this finding by adding 20 additional restaurants to its chain the total daily average income of all the restaurants in its chain will be maximized? We cannot say for sure. Mathematically speaking, we have a valid conclusion with respect to the assumptions made. How close it is to reality's mark is another issue. This takes us back to the question of how realistic are the two starting points of the mathematical analysis, the explicity stated assumption,

P1: The average amount of business done by each restaurant in the chain will drop by $200 per day for each new restaurant opened.

And the not explicitly stated assumption, but implicit in all such analyses,

P: Other factors that bear on the situation are not being considered. This may be the case because other factors that may bear on the situation are viewed as negligible, or it is not understood how to come to grips with them, or a combination of both.

If the management of the Thomas Company is satisfied that these assumptions/postulates are realistic, then it would be reasonable to implement the mathematical conclusion by adding 20 additional restaurants to its chain.

EXERCISES

Solve the following equations by factoring, if possible, or by use of the quadratic formula.

1. $x^2 + 5x - 6 = 0$ 2. $x^2 - 25 = 0$ 3. $x^2 + 4x + 3 = 0$

4. $x^2 + 4x = 0$ 5. $x^2 + 12x + 11 = 0$ 6. $2x^2 - 18x + 16 = 0$

7. $x^2 - 18x + 72 = 0$ 8. $3x^3 - 27x^2 + 42x = 0$ 9. $4x^2 - 12x + 5 = 0$

10. $6x^2 + 5x - 6 = 0$ 11. $x^2 + 7x + 2 = 0$ 12. $(x-3)^2 = 64$

13. $2x^2 - 8x + 1 = 0$ 14. $x^2 - x - 5 = 0$ 15. $x^2 - 2x - 7 = 0$

16. $10t^2 + 49t - 5 = 0$ 17. $x^2 + 4 = 0$ 18. $3x^2 - 2x - 5 = 0$

19. $4x^3 + 5x^2 - 7x = 0$ 20. $5x^2 - 3x - 8 = 0$

Sketch the graphs of each of the following functions, and find the coordinates of the maximum, or minimum, point, as the case may be.

21. $y = x^2 + 5x - 8$ 22. $y = x^2 + 4x + 3$ 23. $y = 4x^2 - 12x + 5$

24. $y = 2x^2 - 8x + 1$ 25. $y = x^2 + 12x + 11$ 26. $y = 5x^2 - 3x - 8$

27. A coffee producer has revenue and cost functions defined by $R(x) = 306x - 5x^2$ and $c(x) = \frac{1}{2}x^2 + 20x + 900$, where x is output in tons per week, and $R(x)$ and $c(x)$ are revenue and cost in dollars.

  (a) Determine the break-even output levels of the firm.

  (b) Does this analysis establish the real world accuracy of the break-even output levels of the firm? Explain

  (c) Sketch the graphs of the math model functions of the revenue and cost functions on the same coordinate system.

  (d) Determine the profit function of the firm and the output level at which the profit function is maximized.

  (e) Is the output level for which the profit function is maximized necessarily the profit level for which profit is maximized? Explain

28. The revenue and cost functions of a tea producer are $R(x) = 310x - 2x^2$ and $c(x) = \frac{1}{2}x^2 + 10x + 1500$, where x is output in tons per day, and $R(x)$ and $c(x)$ are revenue and cost in dollars.

  (a) Determine the break-even output levels of the firm.

  (b) Does this analysis establish the real-world accuracy of the break-even output levels of the firm? Explain

  (c) Sketch the graphs of the math model functions of the revenue and cost functions on the same coordinate system.

(d) Determine the profit function of the firm and the output level at which the profit function is maximized.

(e) Is the output level for which the profit function is maximized necessarily the profit level for which profit is maximized? Explain.

29. The demand and supply functions for chocolate in a certain region are $D = 5000/x$ and $S = 2x + 10$, where x is the price of chocolate in dollars per 100 pounds, D is demand in hundreds of pounds per day, and S is supply in hundreds of pounds per day.

(a) Determine the price at which market equilibrium is achieved. (For further discussion of market equilibrium, see sec. 3.6.)

(b) Is the price at which market equilibrium achieved, mathematically speaking, necessarily the price at which market equilibrium is achieved, realistically speaking? Explain.

8.3   THE LANGUAGE OF VARIATION

The terminology of variation is useful for describing a number of situations that arise in business, economics, and science.

If x and y are variables, then **y is said to vary directly as x**, or **y is said to be directly proportional to x**, if there is a nonzero constant k such that

$$y = kx$$

The constant k is called the **constant of variation** or **constant of proportionality.** Thus, for example, the circumference C of a circle varies directly as the radius r, since $C = 2\pi r$; the constant of proportionality is 2ϖ.

EXAMPLE 1

Suppose that the monthly profit P of a firm is directly proportional to sales s, and that $P = \$1000$ for $s = \$20,000$. Write P as a function of s, and find the profit that corresponds to a sales volume of $\$30,000$ per month.

$P = ks$. For $s = 20,000$, $P = 1000$. Thus $1000 = k(20,000)$,

$k = \dfrac{1000}{20,000} = 0.05$, and we have

$$P = 0.05s .$$

For $s = 30,000$, $P = 0.05(30,000) = 1500$ or $\$1500$.

The statement **y varies directly as the square of x**, or **y is directly proportional to the square of x**, means that

$$y = kx^2$$

for some nonzero constant k.

To say, for example, that the distance d an object falls (in feet) varies directly as the square of the time of fall t (in seconds) means that

$$d = kt^2 .$$

If $d = 16$ feet when $t = 1$ second, then $16 = k(1)^2$, $k = 16$, and we have

$$d = 16t^2 .$$

We say that **y varies inversely as x**, or **y is inversely proportional to x**, if

$$y = \frac{k}{x}$$

for some nonzero constant k.

EXAMPLE 2

Boyle's law says that, for an enclosed gas at a constant temperature, the pressure p varies inversely at the volume v. If $v = 10$ cubic inches when $p = 8$ pounds per square inch, express p as a function of v, and find v when $p = 12$ pounds per square inch.

$p = k/v$, for $v = 10$, $p = 8$. Thus $8 = k/10$, $k = 80$, and we have:

$$p = \frac{80}{v}$$

For $p = 12$, we have:

$$12 = \frac{80}{v}, \qquad 12v = 80$$

$$v = \frac{80}{12} = 6.67 \text{ or } 6.67 \text{ cubic inches}$$

To say that **y varies inversely as the square of x** means that

$$y = \frac{k}{x^2}$$

for some nonzero constant k.

　　To say that **y varies jointly as s and t** means that

$$y = kst$$

for some nonzero constant k.

　　To say that **y varies jointly as s and t and inversely as u** means that

$$y = k \cdot \frac{st}{u}$$

for some nonzero constant k.

EXERCISES

1.　If y varies directly as x, and $y = 32$ when $x = 8$, (a) express y as a function of x; (b) find the value of y when $x = 12$.

2.　If y varies inversely as x, and $y = 4$ when $x = 2$, (a) express y as a function of x; (b) find the value of y when $x = 5$.

3. If s varies jointly as u and v and inversely as t, and $s = 12$ when $u = 2$, $v = 3$, and $t = 6$, (a) express s as a function of u, v, and t; (b) find s when $u = 4$, $v = 6$, $t = 8$.

4. If r varies directly as the cube of t and inversely as the square of s, and $r = 9$ when $t = 3$ and $s = 2$, (a) express r as a function of t and s; (b) find r when $t = 5$ and $s = 3$.

5. If the demand D for a certain product varies inversely as its price p, and the demand is 1000 units per week when the price is \$4 per unit, (a) express D as a function of p; (b) find D when p is \$5 per unit.

6. The quantity S supplied to the market of a certain commodity varies directly as the square of its price p, and the quantity supplied is 5000 units per week when the price is \$5 per unit. (a) Express quantity supplied as a function of price; (b) find the quantity supplied when the price is \$3 per unit.

7. The area A of a circle is directly proportional to the square of its radius r. If $A = 4\pi$ when $r = 2$, (a) express A as a function of r; (b) find A when $r = 4$.

8. The cost of labor varies jointly as the number of workers and the number of days they work. If 5 persons working 10 days are paid 2500, (a) express cost of labor as a function of the number of workers and the number of days worked; (b) find the cost of 6 persons working 8 days.

9. The illumination I from a source of light varies inversely as the square of the distance d from the source. If $I = 10$ units when $d = 2$ feet, express I as a function of d.

CHAPTER 9

Math Modeling for Money

9.1 ARITHMETIC AND GEOMETRIC PROGRESSIONS

The sequence of numbers 2, 5, 8, 11, 14, 17 has the property that each term is obtained by adding the same constant, 3, to the preceding term and, as such, illustrates the nature of an arithmetic progression. In general, a sequence of numbers in which each term is obtained by adding the same constant d to the preceding term is called an **arithmetic progression.** The number d is called the **common difference.** If a is the first term of an arithmetic progression with common difference d, the first n terms are:

$$a, \; a+d, \; a+2d, \; a+3d, \ldots, a+(n-1)d \qquad (9.1)$$

The *n*ᵗʰ *term* l is:

$$l = a+(n-1)d \qquad (9.2)$$

Let S_n denote the sum of the fist n terms of the arithmetic progression described by (9.1). Then:

$$S_n = a+(a+d)+(a+2d)+\cdots+(l-d)+l \qquad (9.3)$$

Also,

$$S_n = l+(l-d)+(l-2d)+\cdots+(a+d)+a \qquad (9.4)$$

where the same terms are written in reverse order. Adding terms that appear under each other in (9.3) and (9.4) yields:

$$2S_n = \underbrace{(a+l)+(a+l)+\cdots+(a+l)}_{n \text{ terms}} = n(a+l)$$

Thus $2S_n = n(a+l)$, which yields:

$$S_n = \frac{n}{2}[a+l]$$

Since $l = a+(n-1)d$, substitution for l in the preceding gives us

$$S_n = \frac{n}{2}[2a+(n-1)d] \qquad (9.5)$$

as the sum of the first n terms of an arithmetic progression with first term a and common difference d.

EXAMPLE 1

Find the sum of the first 200 positive integers 1, 2, ..., 200.

The first 200 positive integers can be thought of as an arithmetic progression with $a = 1$, $d = 1$, $n = 200$. Thus:

$$S_{200} = \frac{200}{2}[2(1)+199(1)] = 100(201) = 20{,}100$$

EXAMPLE 2

The Johnston Company had sales of $100,000 during its first year of operation, and sales increased by $20,000 per year for each year after the first. Find the sales for the tenth year of operation and the total sales of the company during its first 10 years.

Since the sales behavior of the Johnston Company is described by an arithmetic progression with $a = 100{,}000$ and $d = 20{,}000$, the sales for the tenth year of operation is the tenth term in the progression. From (9.2), with $n - 1 = 9$ we obtain

$$l = 100,000 + 9(20,000) = 280,000$$

so that sales amounted to $280,000 for the tenth year of operation.

The total sales of the company over its first 10 years is given by:

$$S_{10} = \frac{10}{2}[2(100,000)???9(20,000)] = 1,900,000$$

or $1,900,000

Geometric Progressions

A sequence of numbers in which each term is obtained by multiplying the preceding term by a constant r is called a **geometric progression**. The constant r is called the **common ratio**. If a is the first term of a geometric progression with common ratio r, the first n terms are:

$$a, \; ar, \; ar^2, \; ar^3, ..., \; ar^{n-1} \tag{9.6}$$

The **n^{th} term** is:

$$l = ar^{n-1} \tag{9.7}$$

Thus, for example, the first four terms of the geometric progression whose first term is 9 and whose common ratio is $\frac{1}{3}$ are 9, 3, 1, and $\frac{1}{3}$.

Let S_n denote the sum of the first n terms of the geometric progression described by (9.6). Then:

$$S_n = a + ar + ar^2 + ar^3 + \cdots + ar^{n-1} \tag{9.8}$$

Multiplying by r yields:

$$rS_n = ar + ar^2 + ar^3 + \cdots + ar^{n-1} + ar^n \tag{9.9}$$

Subtracting (9.9) from (9.8) gives us $S_n - rS_n$ on the left side and $a - ar^n$ on the right side, since the $ar, ar^2, \ldots, ar^{n-1}$ terms cancel out. Thus:

$$S_n - rS_n = a - ar^n$$
$$S_n(1-r) = a - ar^n$$

Dividing both sides of this result by $1-r$ (assuming that $r \neq 1$) yields

$$S_n = \frac{a - ar^n}{1-r} = \frac{a(1-r^n)}{1-r} \qquad (9.10)$$

as the sum of the first n terms of a geometric progression with first term a and common ratio $r \neq 1$. When $r = 1$, $S_n = na$.

EXAMPLE 3

Find the sum of the first eight terms of a geometric progression whose first three terms are 2, 4, 8.

Here $a = 2$ and $r = \dfrac{4}{2} = 2$. Thus:

$$S_8 = \frac{2(1-2^8)}{1-2} = \frac{2(-255)}{-1} = 510$$

EXAMPLE 4

Over a certain time interval the number of bacteria in a culture doubles every hour. If there are initially 100 bacteria in the culture, determine the function that describes the number of bacteria in the culture at the end of t hours. What are the number of bacteria in the culture at the end of 4 hours?

The number of bacteria in the culture initially $(t = 0)$, at the end of 1 hour $(t = 1)$, at the end of 2 hours $(t = 2)$, at the end of 3 hours $(t = 3)$, and, more generally, at the end of k hours $(t = k)$ is described by the successive terms of the following geometric progression:

$$100, \; 100(2)^1, \; 100(2)^2, \; 100(2)^3, \ldots, \; 100(2)^k$$

Thus

$$f(t) = 100(2)^t$$

describes the number of bacteria in the culture at the end of t hours. The domain of definition of $f(t)$ consists of nonnegative integer values ($t = 0$, 1, 2, etc.) in the time interval in question.

The number of bacteria in the culture at the end of 4 hours is $f(4) = 100(2)^4 = 1600$.

EXAMPLE 5   BEWARE THE WIND

The common ratio $r = 2$ describes the procedure of doubling each term of a geometric progression to obtain the next term. Sums of geometric progressions increase modestly at first, but then take off. This was clear to Senator Lloyd L. Wind in his dealings with his colleagues who wanted to keep him quiet for a while.

The agreement was that "beginning next Monday we will donate one dollar to the charity you designate and double the size of the donation every week thereafter for as many weeks as you remain silent in the Senate." Wind stayed the course for 20 weeks, which generated a total of 21 payments, each payment being twice that of its predecessor.

The sum of this geometric progression, with $a = 1$, $r = 2$ and $n = 21$, is

$$S_{21} = \frac{1(1 - 2^{21})}{1 - 2}$$
$$= 2^{21} - 1$$
$$= 2,097,151,$$

which required Wind's colleagues to pay his favorite charity $2,097,151.

Wind could easily had stayed the course another week, which would have payment given by:

$$S_{22} = 2^{22} - 1, \text{ which is } 4,194,303$$
$$\text{or } \$4,194,303$$

Staying silent an additional week would have brought payment of $8,388,607. But Senator Wind, while longwinded, was also considerate of his colleagues. He wanted to teach them a lesson for being "so smart," but one that they could afford.

EXAMPLE 6   DECLINING BALANCE METHOD

Certain assets lose value rapidly at fist and then less rapidly as time passes. To depreciate such assets, the **declining balance method of depreciation** is sometimes used. If A is the original cost of the asset, then the first year's depreciation is $2A/n$, where n is the number of years of useful life of the asset. The depreciation is successive years after the first is obtained by multiplying the preceding year's depreciation by the constant factor $1-(2/n)$.

Thus the depreciation of the asset from year to year forms a geometric progression with first term $2A/n$ and common ratio $1-(2/n)$. The depreciation in years 1, 2, 3, and, more generally, r (where $1 \le r \le n$) is:

$$D_1 = \frac{2A}{n} \qquad\qquad D_2 = \frac{2A}{n}\left(1-\frac{2}{n}\right)$$

$$D_3 = \frac{2A}{n}\left(1-\frac{2}{n}\right)^2 \qquad D_r = \frac{2A}{n}\left(1-\frac{2}{n}\right)^{r-1}$$

For example, if an asset costing $20,000 with a lifetime of 10 years is depreciated by use of the declining balance method, the depreciation in years 1, 2, 3, and r is given by:

$$D_1 = \frac{2(20,000)}{10} = 4000 \qquad D_2 = 4000\left(\frac{8}{10}\right) = 3200$$

$$\text{or } \$4000 \qquad\qquad \text{or } \$3,200$$

$$D_3 = 4000\left(\frac{8}{10}\right)^2 = 2560 \qquad D_r = 4000\left(\frac{8}{10}\right)^{r-1}$$

$$\text{or } \$2560$$

From (9.10), with $a = 4000$ and $r = \frac{8}{10}$ or $\frac{4}{5}$, the total declining balance depreciation over the 10-year life of the asset is:

$$S_{10} = \frac{4000[1-(\frac{4}{5})^{10}]}{1-\frac{4}{5}} = \frac{4000\left(\frac{8,717,049}{9,765,625}\right)}{\frac{1}{5}} = 17,853 \text{ or } \$17,853$$

Let us observe that the total depreciation, $17,853, is less than the original cost, $20,000. This result also presupposes that the salvage value of the asset does not exceed $20,000 - \$17,853 = \$147$; otherwise, an adjustment must be made.

EXERCISES

1. An arithmetic progression with 20 terms has first term 3 and common difference 4. Find the last term and the sum.

2. An arithmetic progression with 25 terms has first term -25 and common difference 5. Find the last term and the sum.

3. Find the sum of the first 300 positive integers.

4. Find the sum of the first 200 even integers.

5. Bob Ford was hired with an initial salary of $16,000, with annual increases of $800 for 7 years. What will Bob's salary be at the end of 5 years? What will his total income be during 7 years with the company?

6. For the geometric progression 1, 3, 9, . . . , with six terms, find the common ratio, the last term, and the sum.

7. For the geometric progression 2, $\dfrac{4}{3}$, $\dfrac{8}{9}$, . . . , with six terms, find the common ratio, last term, and the sum.

8. If each bacterium in a culture divides into two bacteria at the end of 2 hours, how many bacteria will be in the culture at the end of 10 hours if there were 500 bacteria initially?

9. Suppose that a car costing $50,000 depreciates 5 percent in value each year. What is the car worth at the end of 5 years?

10. **Do you have enough funds for this?** Amy Allen, teenage daughter of Fred and Gail Allen, consistently practiced one basic principle of life: shop 'till you drop. Her parents believed that it was long overdue for Amy to come to grips with the "realities" of money and institute a savings plan.

    To help encourage her to get this going Fred offered Amy the following deal. "If you deposit $1 into your account each week, then I will deposit $1 into your account for your first deposit and double the size of my deposit every week thereafter for a year, provided that you maintain your deposits without interruption. If you miss a deposit, you forfeit the entire amount that I will have deposited to that point."

    If Amy agrees to this arrangement, how much will her father be committed to depositing into her account?

11. **Reward Thy Servant.** Legend has it that Prince John of the Sea served his master Emperor Ronald the Astrologer faithfully at great risk to himself for many, many years. Finding himself an old man with not much to show for his years of devoted service, John requested an audience with the emperor.

The emperor was delighted to see him and offered to secure for him a place in the Senate where many before had made their fortunes. But this was not what John wanted. He wanted a grant that would allow him to purchase an estate and live out his remaining years in comfort. "What sum would you consider sufficient for your needs," asked the emperor. "Ten million dollars," answered John of the Sea.

"I want you to receive your just reward," replied the emperor, "and this is what I suggest." "Go to my Keeper of the Coin, who will give you one coin of silver weighing 1 ounce; bring it here. Go to him tomorrow and he will give you another coin worth twice the first; bring it here and place it beside the first. On the day after tomorrow he will give you a coin worth four times the first, on the day after that he will give you one worth eight times the first, and so on. Bring them here. You must do this yourself. When you can no longer lift the coin our agreement will have ended, but you may keep all the coins that you will have taken from the Keeper of the Coin; that will be your reward." John of the Sea was delighted, for he imagined riches beyond his wildest expectations. He thanked Ronald the Astrologer profusely for his generosity and set about his task.

But John no longer had the strength of his youth and, although driven by desire, the best he could manage was a coin weighing 512 pounds.

Geometric progressions are more reliable than astrology!

(a) How many coins will John take from the Keeper of the Coin?

(b) If silver is worth $10 an ounce, what is the value of John's reward?

(c) How does John's reward compare with his request for $10 million?

12. A company buys an item having a useful life of 10 years for $30,000. If the company depreciates the item by the declining balance method, (a) what is the depreciation for the first year? (b) What is the depreciation for the second year? (c) What is the total depreciation for the 10-year life of the asset?

13. Show that the total declining balance depreciation of an asset that costs A dollars and has a useful life of n years is $A[1-(1-2/n)^n]$, assuming that a salvage value adjustment does not have to be made.

9.2 APPLICATIONS TO THE MATHEMATICS OF MONEY*

Suppose that $1000 is invested 6 months from now and in the following three successive 6-month periods at the rate of 6 percent per annum compounded semiannually. The time at which each sum of $1000 is invested coincides with the time at which compound interest is added to the existing amount. At the end of the investment period (2 years from now) these respective amounts will have the following values:

$$1000(1.03)^3 = 1000(1.09273) = 1092.73 \text{ or } \$1092.73$$
$$1000(1.03)^2 = 1000(1.06090) = 1060.90 \text{ or } \$1060.90$$
$$1000(1.03)^1 = 1000(1.03000) = 1030.00 \text{ or } \$1030.00$$
$$1000(1.03)^0 = 1000(1.00000) = 1000.00 \text{ or } \$1000.00$$

The growth exhibited by this investment situation is shown in Figure 9.1.

---

* Please review section 2.3 at this point.

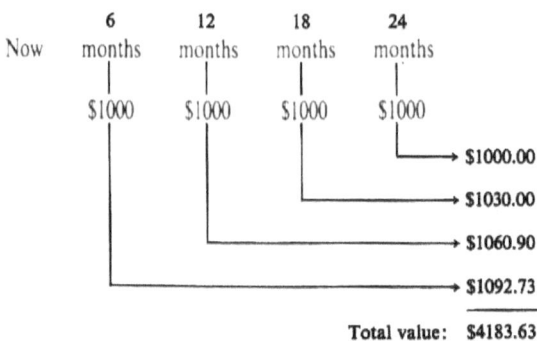

Figure 9.1

Thus the total future value of this investment stream at the end of the investment period is the sum $\$1092.73 + \$1060.90 + \$1030.00 + \$1000.00 = \$4183.63$.

This example illustrates a structural feature common to many financial situations.

A Math Model for the Future and Present Values of an Annuity

This model is based on the following assumptions/postulates:

P1. Equal payments R are made at regular intervals of time (in our example, $R = \$1000$). Such a series of payments is called an **annuity.**

P2. Each payment R, called **rent,** is made at the end of the time interval in which it is due. Each such time period is termed a **payment period** (in our example, the payment period is 6 months). Although the term annuity suggests annual payments, the payment period of an annuity may be any length of time.

P3. The interest period of the compound interest rate coincides with the payment period of the annuity.

The **term** of an annuity is the time between the beginning of the first payment period and the end of the last payment period. The term of the annuity considered in the preceding example is 2 years.

The **amount**, or **future value**, of an annuity is the total amount that would be accumulated at the end of the term of the annuity if each payment is invested at a given rate of compound interest at the time of payment. The future value of the annuity discussed in the preceding example is $4183.

More generally, suppose that a rent of R dollars is paid at the end[*] of each payment period for n payment periods, where the compound interest rate is i per period. The R dollars paid at the end of the first payment period will earn interest for $n-1$ periods, and will amount to $R(1+i)^{n-1}$ dollars at the end of the term. The R dollars paid at the end of the second payment period will earn interest for $n-2$ periods, and will amount to $R(1+i)^{n-2}$ dollars. And so on; the R dollars paid at the end of the $(n-1)$st (next to last) payment period will earn interest for one period, and will amount to $R(1+i)$ dollars. The R dollars paid at the end of the nth payment period will earn no interest at all, and will amount to R dollars.

In summary, we obtain the time diagram shown in Figure 9.2. Letting F denote the future value of this annuity

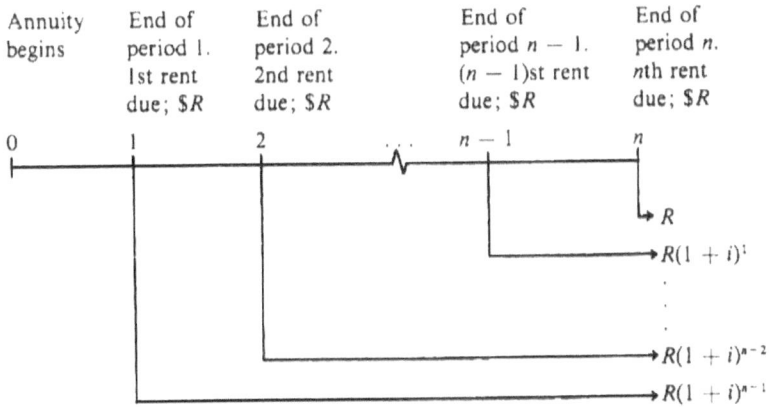

Figure 9.2

(the total amount on deposit after n payment periods), we have

$$F = R + R(1+i) + R(1+i)^2 + \cdots + R(1+i)^{n-2} + R(1+i)^{n-1}$$

---

[*] If payments are made at the beginnings of the payment periods, slight adjustments must be made in the results that will be derived. The same general principles, however, apply. See Exercise 18.

the sum of the first n terms of a geometric progression with first term R and common ratio $(1+i)$. From the formula for the sum of the first n terms of a geometric progression,

$$S_n = \frac{a(1-r^n)}{1-r}$$

with $a = R$ and $r = (1+i)$, we obtain:

$$F = \frac{R[1-(1+i)^n]}{1-(1+i)} = \frac{R[1-(1+i)^n}{-i}$$

Multiplying numerator and denominator by -1 and regrouping terms yields the valid conclusion:

$$F = R\left[\frac{(1+i)^n - 1}{i}\right]$$

The quantity $[(1+i)^n - 1]/i$ is denoted by $s_{\overline{n}|i}$, read **"s angle n with respect to i,"** and expresses the future value of an annuity in which \$1 is paid at the end of each payment period for n payment periods, where the compound interest rate is i per period. Thus we have:

$$F = R \cdot s_{\overline{n}|i}$$

Values of $s_{\overline{n}|i}$ for various n and i are given in Appendix Table 4. More extensive tables are available.

EXAMPLE 1

To save up enough money for a downpayment on a house, the Roberts family plans to deposit \$300 at the end of every 3 months into a fund that pays 8 percent per annum compounded quarterly. How much will be on deposit at the end of 5 years?

This is an annuity with rent $R = \$300$, compound interest of 2 percent per payment period, and 20 payment periods. Thus the amount that will be on deposit at the end of 5 years is $F = 300 \cdot s_{\overline{20}|0.02}$. From Appendix Table 4, $s_{\overline{20}|0.02} = 24.29737$. Thus:

$$F = 300(24.29737) = 7289.21 \text{ or } \$7289.21$$

A Sinking Fund Problem

EXAMPLE 2

The Riccardi Company anticipates an expenditure of $6000 to replace a company car in 5 years. How much should be deposited into a fund paying 8 percent per annum compounded semiannually if $6000 is to be available in 5 years?

The interest rate is 4 percent per payment period, the number of payment periods is 10, the amount F is $6000, and the rent R is to be determined. We have $6000 = R \cdot s_{\overline{10}|0.04}$, from which we obtain:

$$R = \frac{6000}{s_{\overline{10}|0.04}}$$

From Appendix Table 5, $1/s_{\overline{10}|0.04} = 0.08329$. Thus:

$$R = 6000(0.08329) = 499.74 \text{ or } \$499.74$$

Problems of this sort, for which the amount F and $s_{\overline{n}|i}$ are known and the rent R is to be determined, are often referred to as **sinking fund** problems.

EXERCISES

1. Find the amount in a fund at the end of 7 years if the fund pays interest at the rate of 4 percent per annum (a) compounded annually and $50 is deposited at the end of each year; (b) compounded semiannually and $50 is deposited at the end of each 6 months; (c) compounded quarterly and $50 is deposited at the end of each 3 months.

2. At the end of each year Professor Janet Reed deposits 10 percent of her
    $50,000-a-year teaching salary into a fund paying 5 percent per annum
    compounded annually. How much money will the professor have in the fund
    at the end of 30 years?

3. Oliver Lukas deposits $500 at the end of each year into a fund that pays 4
    percent per annum compounded annually. Construct a schedule showing the
    growth of the fund during the first 5 years.

4. Find the rent required to develop a sinking fund of $2500 in 9 years if
    deposits are made (a) annually and the interest rate is 5 percent per annum
    compounded annually; (b) semiannually and the interest rate is 4 percent
    per annum compounded semiannually; (c) quarterly and the interest rate is
    4 percent per annum compounded quarterly.

5. Jack Bryan buys a new car every 2 years. The list price of the new car is
    $14,000 and the trade-in value of the old car is $7000. If Jack can deposit
    money into a fund that pays 8 percent per annum compounded quarterly,
    how much should he deposit each quarter to take care of depreciation on
    the car?

Present Value of an Annuity

Consider an annuity in which $R is paid at the end of each payment period,
where the compound interest rate is i per period. One concern, which we have
already explored, is with the **future value** of the annuity after a certain number
of payment periods have elapsed.

Another feature, which we now examine, is concerned with the present.
What is the annuity worth now? More precisely, what sum initially invested will
yield $R at the end of each payment period, where the compound interest rate
is i per period? To determine this sum, which, naturally enough, is called the
present value of the annuity, we determine the present value of each rent R of
the annuity and add.

The present value of the first rent of $R, which is due at the end of the first
period, is $R(1+i)^{-1}$. The present value of the second rent is $R(1+i)^{-2}$, and so
on. The present value of the nth rent (which is the last payment) is $R(1+i)^{-n}$.
In summary, we obtain the time diagram shown in Figure 9.3. Letting P denote
the present value of this annuity, we have:

$$P = R(1+i)^{-1} + R(1+i)^{-2} + R(1+i)^{-3} + \cdots + R(1+i)^{-n}$$

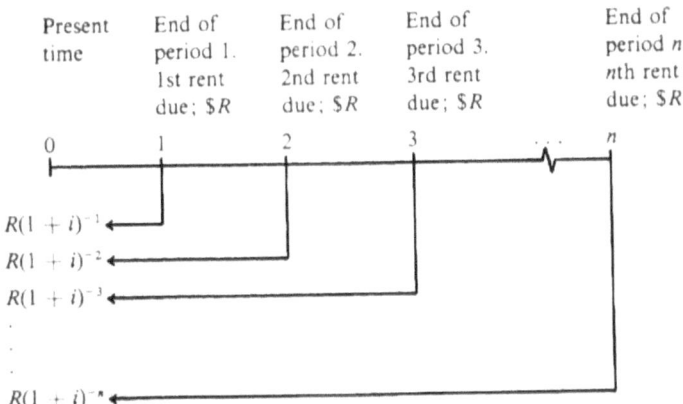

Figure 9.3

the sum of the first n terms of a geometric progression with first term $R(1+i)^{-1}$ and common ratio $(1+i)^{-1}$. From the formula for the sum of the first n terms of a geometric progression,

$$S_n = \frac{a(1-r^n)}{1-r}$$

with $a = R(1+i)^{-1}$ and $r = (1+i)^{-1}$, we obtain:

$$P = \frac{R(1+i)^{-1}[1-[(1+i)^{-1}]^n]}{1-(1+i)^{-1}}$$

$$= \frac{R(1+i)^{-1}[1-(1+i)^{-n}]}{1-(1+i)^{-1}}$$

Multiplying numerator and denominator by $(1+i)$ neutralizes $(1+i)^{-1}$ [since $(1+i)(1+i)^{-1} = (1+i)^0 = 1$], and yields the following valid conclusion:

$$P = \frac{R[1-(1+i)^{-n}]}{(1+i)-1}$$

$$P = R\left[\frac{1-(1+i)^{-n}}{i}\right]$$

The quantity $[1-(1+i)^{-n}]/i$ is denoted by $a_{\overline{n}|i}$, read **"a angle n with respect to i,"** and expresses the present value of an annuity in which \$1 is paid at the end of each payment period for n payment periods, where the compound interest rate is i per period. Thus we have:

$$P = R \cdot a_{\overline{n}|i}$$

Values of $a_{\overline{n}|i}$ for various n and i are given in Appendix Table 6. More extensive tables are available.

EXAMPLE 3

An annuity pays \$600 every 4 months for 10 years, where the interest rate is 9 percent per annum compounded 3 times a year. Find the present value of this annuity.

The interest rate i is 3 percent per payment period, the number of payment periods is $n = 30$, and the rent is $R = \$600$. From Appendix Table 6, $a_{\overline{30}|0.03} = 19.60044$. Thus we have:

$$P = 600 \cdot a_{\overline{30}|0.03} = 600(19.60044) = 11,760.26 \text{ or } \$11,760.26$$

\$11,760.26 should be deposited initially into a fund paying interest at the rate of 9 percent per annum compounded 3 times a year if \$600 is to be paid every 4 months for 10 years.

An Amortization Problem

EXAMPLE 4

The Blyden Company has a \$10,000 mortgage, which is to be paid off in equal installments to be made quarterly for 5 years. If the interest rate is 8 percent per annum compounded quarterly, what is the amount due in each installment?

The interest rate is 2 percent per payment period, and the number of payment periods is 20. The present value of the annuity is $P = \$10,000$, the amount that is due now but is to be paid off in equal installments. The amount of each installment R is the rent to be determined. From $P = R \cdot a_{\overline{n}|i}$, we obtain:

$$R = \frac{P}{a_{\overline{n}|i}} = \frac{10,000}{a_{\overline{20}|0.02}}$$

From Appendix Table 7, $1/a_{\overline{20}|0.02} = 0.06116$. Thus:

$$R = 10,000(0.06116) = 611.60 \text{ or } \$611.60$$

Problems of this sort for which P and $a_{\overline{n}|i}$ are known and R is to be determined are called **amortization problems.**

Let's take another look at the structure underlying the Blyden Company's problem, but cast it in terms of a different setting.

EXAMPLE 5

Jennifer Rice purchased a new car for $14,000. She paid $4000 down and obtained financing for the $10,000 balance (mortgage, if you will) which required the debt to be paid off quarterly in equal installments over five years. The interest rate is 8 percent per annum compounded quarterly. How much is each installment (rent, if you will)?

Since Jennifer's situation, structurally speaking, is the same as the Blyden Company's, the answer is the same: $611.60.

EXAMPLE 6

Apart from taxes and miscellaneous fees, how much did the car actually cost Jennifer? What interest did she incur?

As we saw in Example 5, Jennifer's loan of $10,000 is to be paid in 20 installments of $611.60, which comes to:

$$20(611.60) = 12,232 \text{ or } \$12,232$$

Thus, the cost of the car comes to $16,232 with $2,232 being the interest paid on the loan.

EXAMPLE 7

What percent of the total cost of the car is interest?

$$\frac{\text{interest}}{\text{total cost}} \cdot 100 = \frac{2,232}{16,232} \cdot 100 = 0.1375 \text{ or } 13.75\%$$

EXAMPLE 8

Let us suppose that on reconsideration Jennifer decides not to take out the $10,000 loan, but to put the $611.60 installments that she would have had to pay into a fund paying the same rate of 8 percent per annum compounded quarterly.

How long would it take her to accumulate the additional $10,000 to purchase the car?

To answer this question consider a sequence of $611.60 payments from the point of view of an annuity. The basic relationship giving the amount (or future value) of this sequence of payments is:

$$F = R \cdot s_{\overline{n}|i}$$

In this situation $F = \$10,000$, $R = \$611.60$, $i = 0.02$ and n, the number of 3-month payment periods, is to be determined. We have:

$$10,000 = (611.60)s_{\overline{n}|0.02}$$

$$s_{\overline{n}|0.02} = \frac{10,000}{611.60}$$

$$= 16.3506$$

Our next step is to run our finger down the 2% column of Appendix Table 4 for $s_{\overline{n}|i}$ until we locate the smallest value exceeding 16.3506. This value is 17.29342 for which $n = 15$.

Thus, it will take Jennifer 15 3-month periods or 3 years and 9 months to accumulate the additional $10,000 needed to purchase the car, assuming that the price hasn't changed.

EXERCISES

6. An annuity pays $400 every 3 months for 10 years, where the interest rate is 12 percent per annum compounded quarterly. Find the present value of this annuity.

7. An annuity pays $1000 every 6 months for 8 years, where the interest rate is 8 percent per annum compounded semiannually. Find the present value of this annuity.

8. An annuity pays $100 every month for 3 years, where the interest rate is 12 percent per annum compounded monthly. Find the present value of this annuity.

9. If money is worth 12 percent per annum compounded monthly, which television set is cheaper and by how much?

   (a) Set A is paid for with a $55 downpayment followed by 12 monthly payments of $10 each.

   (b) Set B is paid for with a $25 downpayment followed by 15 monthly payments of $10 each.

10. A car costing $850 is paid for with a downpayment followed by 12 monthly payments of $40 each. How much is the downpayment if the interest rate on the unpaid balance is 12 percent per annum compounded monthly?

11. An $8000 mortgage is to be paid off in equal installments to be made at the end of the year for 4 years.

    (a) If the interest rate is 5 percent per annum compounded annually, find the amount paid in each installment.

    (b) Make a schedule showing the part of each installment that goes for interest and principal.

12. To pay off a mortgage of $6000 in 5 years, what should the rent be if the installments are payable (a) annually at 4 percent per annum compounded annually; (b) semiannually at 4 percent per annum compounded semiannually; (c) quarterly at 4 percent per annum compounded quarterly?

13. Amy Ito will need college expenses of $1500 on each of her 18th, 19th, 20th, and 21st birthdays. What single sum of money should be deposited on the girl's 14th birthday in a fund paying 4 percent per annum compounded annually to provide the necessary college funds?

14. William Baxter, who is 55, plans to deposit equal sums of money in a retirement fund every 6 months until he is 65 years old, when the fund has $10,000 in it. He arranges for the money to be paid to him (or his heirs) in equal amounts at the end of every 6 months for 10 years. Money is worth 6 percent per annum compounded semiannually.

(a) Before retirement, how much must be deposited into the fund at the end of each 6-month period?

(b) Upon retirement, how much does he receive at the end of each 6-month period?

15. At the end of each year for 10 years, Susan Rey deposits $200 in a bank that pays interest at the rate of 5 percent per annum compounded annually. The money continues to draw interest until the 15th year, at which time the sum on deposit is withdrawn. How much is withdrawn?

16. Karen Duran borrows $2000 at 12 percent per annum compounded monthly and agrees to pay off the debt in equal monthly payments for 2 years. After she makes her 15th payment, she decides to pay off the balance of the debt. How much should she pay?

17. Joseph Carter purchases an article and agrees to pay for it in 24 equal monthly payments of $50 each. Money is worth 12 percent per annum compounded monthly. After he makes his 20th payment, he decides to pay off the balance of the debt. How much should he then pay?

18. Consider an annuity for which a rent of R is paid at the beginning of each payment period for n payment periods, where the interest rate is i per period. Show that the future value, or amount, of this annuity is $F = (1+i)R \cdot s_{\overline{n}|i}$, and that its present value is $P = (1+i)R \cdot a_{\overline{n}|i}$.

SELF-TESTS FOR CHAPTERS 8-9

Allow 100 or so minutes for each self-test. Go over the first one before proceeding to the second.

Self-Test 1

1. Add $5x^2 - 6x + 1$ and $-2x^2 + 4x - 3$.

2. Add $3x^2 y^2 + 4x^2 y - 1$ and $2x^2 y^2 + 6x^2 y + xy$.

3. Subtract $4x^2 h - 11xh + 2$ from $5x^2 h - 3xh + h^2$.

Find the following products.

4. $4x(x^2 + 5x - 1)$

5. $(5x + 2)(3x - 1)$

6. $(2x^2 + 1)(3x^2 + 5)$

7. $3(3x + 2)(4x - 1)$

8. $3x(x + y)^3$

If possible, factor each of the following as completely as possible and check by multiplication.

9. $t^2 + 4t - 21$

10. $4x^2 - 25$

11. $4x^3 + 8x^2 - 140$

12. $2x^2 + 11x + 5$

13. $2y^2 + 7y + 6$

Solve the following equations by factoring, if possible, or by use of the quadratic formula.

14. $t^2 - 6t + 5 = 0$

15. $2y^3 - 12y^2 - 54y = 0$

16. $3x^2 - 3x - 5 = 0$

17. Sketch the graph of $y = -2x^2 + 4x + 9$, and find the coordinates of its extreme point.

18. If z varies directly as t and $z = 12$ when $t = 2$, (a) express z as a function of t; (b) find the value of z when $t = 4$.

19. Find the sum of the first 100 odd integers.

20. A company buys an item that has a useful life of 15 years for $20,000. If the company depreciates the item by the declining balance method, (a) what is the depreciation for the first year? (b) What is the depreciation for the second year? (c) What is the total depreciation for the 15-year life of the item?

21. A company anticipates an expenditure of $6000 for equipment replacement in 10 years. How much should be deposited semiannually into a fund paying 6 percent per annum compounded semiannually if $6000 is to be available in 10 years?

22. An $8000 mortgage is to be paid off in equal installments to be paid at the end of each 6 months for 4 years. If the interest rate is 6 percent per annum compounded semiannually, find the amount to be paid in each installment.

Self-Test 2

1. Add $3x^2 + 8x - 5$ and $2x^2 + 8x - 4$.

2. Subtract $-2x^2 + x - 11$ from $3x^2 + 6x - 14$.

3. Subtract $5x^2t^2 - 6xt^2 + 11$ from $3x^2t^2 - 2xt - 12$.

Find the following products.

4. $2x^2h(3x^2 + 4xh - 8)$

5. $(4x+1)(3x^2+2)$

6. $(3x-2)(x^4+2x^2-3x+1)$

7. $(x-2)(x+3)(x-5)$

8. $4xy(x+2y)^2$

If possible, factor each of the following as completely as possible and check by multiplication.

9. $2x^3-6x^2-8x$

10. $x^2-12x-13$

11. $3x^3-18x^2-21x$

12. $6t^2+5t-4$

13. $3x^2+10x+8$

Solve the following equations by factoring, if possible, or by use of the quadratic formula.

14. $6x^2+x-15=0$

15. $2x^2-5x+3=0$

16. $-2x^2+4x+9=0$

17. The revenue and cost functions of a coffee producer are $R(x)=204x-x^2$ and $c(x)=x^2+4x+1000$, where x is output in tons per day, and $R(x)$ and $c(x)$ are revenue and cost in dollars.

   (a) Determine the break-even output levels of the firm.

   (b) Does this analysis establish the real world accuracy of the break-even output levels? Explain.

   (c) Sketch the graphs of the math model functions of the revenue and cost functions on the same coordinate system.

(d) Determine the profit function of the firm and the output level at which the profit function is maximized.

(e) Is the output level for which the profit function is maximized necessarily the profit level for which profit is maximized? Explain.

18. If v varies jointly as w and x and inversely as s, and $v = 8$ when $s = 3$, $w = 6$, and $x = 2$, (a) express v as a function of w, x, and s; (b) find v when $s = 2$, $w = 3$, and $x = 4$.

19. For the geometric progression $\dfrac{1}{2}, -\dfrac{1}{6}, \dfrac{1}{18}, \ldots$, with seven terms, find the common ratio, the last term, and the sum.

20. A family, wishing to provide for the college education of a newborn child, plans to deposit $200 at the end of every 6 months into a fund that pays 8 percent per annum compounded semiannually. How much will be on deposit after 18 years if this plan is carried out?

21. An annuity pays $200 every 3 months for 5 years, where the interest rate is 8 percent per annum compounded quarterly. Find the present value of this annuity.

22. Julio Monteiro purchased a new car for $22,000. He paid $5000 down and obtained financing for the $17,000 balance. The agreement called for the debt to be paid off semi-annually in equal installments over eight years. The interest rate is 10 percent per annum compounded semi-annually.

(a) How much is each installment?

(b) How much did the car actually cost Julio, apart from taxes and miscellaneous fees?

(c) What interest did he incur?

(d) What percent of the total cost of the car is interest?

(e) Suppose that on reconsideration Julio decides not to take out the $17,000 loan, but to put the installments that he would have had to pay into a fund paying the same rate of 10 percent per annum compounded semi-annually. How long would it take him to accumulate the additional $17,000 to purchase the car, assuming that its price hasn't changed?

CHAPTER 10

The World of
Complex-Numbers

10.1 INTRODUCTION

At one time positive integers and fractions such as 1/10, 2/3, 1, 5/4, 2, . . . and the like were sufficient for internal mathematical needs as they were perceived and real world applications as well. With the passage of time this ceased to be the case. As a simple illustration, consider the problem faced by a business in recording profits and loses. Positive integers are quite naturally brought into service to record profits, but without the availability of negative integers the recording of loses is not at all a simple matter. We have become so accustomed to negative numbers today that we do not give them a second thought. But at one time they were regarded as controversial and viewed with great suspicion.

For most of us the real-number system occupies the same position today that the positive integers and fractions once held for our ancestors. We have become accustomed to them through exposure and use and it is difficult to see how "other" numbers would be useful and even possible. In the evolution of mathematics and its applications the real-number system came to be insufficient for internal mathematical needs and the needs of applied situations as well. The outcome was the development of another system, called the complex-number system, which contains the real-number system as a special case, but goes beyond it.

> As to "real world" applications that touch our daily lives, it is noteworthy that modern electrical engineering and the theory of flight with their rich variety of very practical applications ranging from turning on a TV to catching a flight from New York to New Orleans would not be possible without the complex-number system.

A few observations about nomenclature are, perhaps, in order at this point. The term "real" attached to number, yielding real number, is unfortunate in that its use in this way may suggest that some numbers are "real" in a tangible application oriented sense whereas others are not real in this sense. This is most certainly not the case; the prefix real in real number should not be thought of as an adjective modifying the noun number. Real number is one term and to emphasize this it would be better to put a dash between them, real-number, to emphasize the unity. (For further discussion see **Words and Wording** in sec. 1.1.) What gives mathematical life to the real-number system are for the most part, the properties of closure, commutativity, and so on, discussed in section 1.2 *(Properties of Addition and Multiplication)* and properties of order discussed in section 1.3 *(Properties of an Order Relation)*.

Much the same can be said about the complex-number system. We introduce complex-numbers and the basic operations of addition and multiplication in the next section. These operations and their properties give mathematical life to the complex-number system, which, in a strange and fascinating way, has led to so many very tangible benefits.

As to the terms complex-number and imaginary-number, which we shall also encounter, we should not think of complex as signifying complicated and imaginary as suggesting a mathematical fantasy land. These are different numbers in that they belong to different systems which give them a different mathematical life. If we had the freedom to do it all over again and introduce more neutral language, it might be preferable to call complex-numbers different-numbers and imaginary-numbers unusual-numbers. Alas, . . .

10.2 DEFINITION AND PROPERTIES OF THE COMPLEX-NUMBER SYSTEM

The **complex-number system** C is the set of all ordered pairs of real numbers (x, y), called **complex-numbers,** together with two operations called addition, +, and multiplication, $\cdot$, defined as follows: If $z_1 = (x_1, y_1)$ and $z_2 = (x_2, y_2)$:

$$z_1 + z_2 = (x_1, y_1) + (x_2, y_2) = (x_1 + x_2, y_1 + y_2)$$

$$z_1 \cdot z_2 = (x_1, y_1) \cdot (x_2, y_2) = (x_1 x_2 - y_1 y_2, x_1 y_2 + x_2 y_1)$$

The product $z_1 \cdot z_2$ is also written $z_1 z_2$. Two complex-numbers (x_1, y_1) and (x_2, y_2) are said to be **equal** if $x_1 = x_2$ and $y_1 = y_2$; we write $(x_1, y_1) = (x_2, y_2)$. Thus, for example, if $z_1 = (1,3)$ and $z_2 = (2,5)$, $z_1 + z_2 = (3,8)$ and $z_1 + z_2 = (1 \cdot 2 - 3 \cdot 5, \ 1 \cdot 5 + 3 \cdot 2) = (-13,11)$.

EXERCISES

Find the sum and product of the following complex-numbers.

1. $z_1 = (2,3);\ z_2 = (-1,3)$

2. $z_1 = (-3,4);\ z_2 = (2,3)$

3. $z_1 = (5,1);\ z_2 = (-1,4)$

4. $z_1 = (4,-2);\ z_2 = (\frac{1}{2},3)$

5. $z_1 = (5,4);\ z_2 = (0,0)$

6. $z_1 = (x,y);\ z_2 = (0,0)$

7. $z_1 = (2,3);\ z_2 = (-2,-3)$

8. $z_1 = (x,y);\ z_2 = (-x,-y)$

9. $z_1 = (x,y);\ z_2 = (1,0)$

10. $z_1 = (2,3);\ z_2 = (\frac{2}{13}, \frac{-3}{13})$

11. $z_1 = (-3,4);\ z_2 = (\frac{-3}{25}, \frac{-4}{25})$

12. $z_1 = (x,y);\ x \ne 0,\ y \ne 0;;$

$$z_2 = \left(\frac{x}{x^2 + y^2}, \ \frac{-y}{x^2 + y^2} \right)$$

13. $i = (0,1);\ i = (0,1)$

14. $i = (0,1);\ (0,y)$

15. $(x,0) + (y,0)$

Find and compare $z_1 z_2$ and $z_2 z_1$ for the following z_1 and z_2.

16. $z_1 = (2,3);\ z_2 = (-1,3)$ 17. $z_1 = (5,1);\ z_2 = (-1,4)$

Find and compare $(z_1 z_2) z_3$ and $z_1 (z_2 z_3)$ for the following z_1, z_2, and z_3.

18. $z_1 = (2,3);\ z_2 = (-1,3);\ z_3 = (3,-2)$

19. $z_1 = (5,1);\ z_2 = (-1,4);\ z_3 = (-2,-3)$

Find and compare $z_1(z_2 + z_3)$ *and* $z_1z_2 + z_1z_3$ *for the following* z_1, z_2, *and* z_3.

20. $z_1 = (2,3)$; $z_2 = (-1,3)$; $z_3 = (3,-2)$

21. $z_1 = (5,1)$; $z_2 = (-1,4)$; $z_3 = (-2,-3)$

Properties of Addition and Multiplication

The complex-number system has the same properties with respect to addition and multiplication—closure, commutativity, existence of identity elements *et al*—as those exhibited by the real-number system (see section 1.2).

Closure property of addition and multiplication. Whenever two complex-numbers are added or multiplied, we obtain the same kind of mathematical creature, a complex-number. Formally stated, we have the following:

A1. If z_1 and z_2 are any numbers in the complex-number system C, then the sum of z_1 and z_2, denoted by $z_1 + z_2$, is in C.

M1. If z_1 and z_2 are any numbers in the complex-number system C, then the product of z_1 and z_2, denoted by $z_1 \cdot z_2$ or z_1z_2 is in C.

Commutative property of addition and multiplication. The order in which two complex-numbers are added or multiplied does not affect the result obtained. Illustrations of this are seen in Exercises 16 and 17. Formally stated, we have the following:

A2. If z_1 and z_2 are any numbers in the complex-number system C, then $z_1 + z_2 = z_2 + z_1$.

M2. If z_1 and z_2 are any numbers in the complex-number system C, then $z_1z_2 = z_2z_1$.

Associative property of addition and multiplication. Parentheses, brackets, and the like, we recall are used in algebra to group together whatever terms are within them. Thus $2 + (3 + 4)$ means 2 is to be added to the sum of 3 and 4, yielding $2+7$ or 9, whereas $(2+3)+4$ means that the sum of 2 and 3, or 5, is to be added to 4, which also yields 9. Real-number addition and multiplication are associative operations.

Complex-number addition is clearly associative and the results of Exercises 18 and 19, which are indeed special cases, might suggest that complex-number multiplication is as well. Formally stated, we have the following:

A3. If z_1, z_2, and z_3 are any numbers in the complex-number system C, then

$$z_1 + (z_2 + z_3) = (z_1 + z_2) + z_3$$

M3. If z_1, z_2, and z_3 are any numbers in the complex-number system C, then

$$z_1(z_2 z_3) = (z_1 z_2) z_3.$$

The complex-numbers zero and one. As we see from Exercises 6 and 9, the complex-number system has counterparts to zero and one in the real-number system; namely, $0 = (0,0)$ and $1 = (1,0)$. Formally stated, we have the following:

A4. There is an unique complex-number, namely $0 = (0,0)$, called **zero**, with the property that $z_1 + 0 = 0 + z_1 = z_1$, where z_1 is any complex-number.

M4. There is an unique complex-number, namely $1 = (1,0)$, called **one**, with the property that $z_1 \cdot 1 = 1 \cdot z_1 = z_1$, where z_1 is any complex-number.

Additive and multiplicative inverses of a complex-number. From Exercise 8 we have that for any complex-number $z_1 = (x, y)$ there is $z_2 = (-x, -y)$ such that their sum is 0. Moreover, it can be shown that z_2 is unique. z_1 and z_2 are said to be **additive inverses** of each other, or **negatives of** each other. Each is called the **additive inverse** or **negative of** the other.

There is a major difference between the **negative of** a complex-number and **negative** complex-number. As we shall establish in the next section, there are **no negative complex numbers** (or positive ones either, for that matter).

> The concept of negative number is **not viable for complex numbers**; the little "**of**" makes all the difference in the world.

From Exercise 12 we have:

> For any nonzero complex-number $z_1 = (x, y)$ there is
>
> $$z_2 = (\frac{x}{x^2 + y^2},\ \frac{-y}{x^2 + y^2})$$
>
> such that $z_1 z_2 = z_2 z_1 = 1$.

It can also be shown that z_2 is unique. z_1 and z_2 are said to be **multiplicative inverses** or **reciprocals** of each other. Each is said to be the **multiplicative inverse of** or **reciprocal of** the other.

In summary, we have:

> **A5.** If z_1 is any complex-number, then there is an unique complex-number called the **additive inverse of** z_1 or **negative of** z, denoted by $-z_1$, such that $z_1 + (-z_1) = (-z_1) + z_1 = 0$.
>
> **M5.** If z_1 is any nonzero complex-number, then there is an unique complex-number called the **multiplicative inverse of** z_1 or **reciprocal of** z_1, denoted by z_1^{-1} or $1/z_1$, such that $z_1 \cdot \dfrac{1}{z_1} = \dfrac{1}{z_1} \cdot z_1 = 1$.

Distributive property. Real-numbers have the distributive property of multiplication over addition, more simply called the distributive property; that is, for all real-numbers a, b, and c, $a(b + c) = ab + ac$. Exercises 20 and 21 suggest the possibility that it holds for complex-numbers as well. It does.

> **D. If** z_1, z_2, **and** z_3 **are any complex-numbers, then**
> $$z_1(z_2 + z_3) = z_1 z_2 + z_1 z_3.$$

More generally, the distributive property holds for z_1 multiplying the sum of any number of terms.

Subtraction and Division of Complex-Numbers

Subtraction and division of complex-numbers are defined in terms of the basic operations of addition and multiplication, respectively, as their real-number cousins.

> **Subtraction.** If z_1 and z_2 are any complex-numbers, the difference $z_1 - z_2$ is defined by $z_1 - z_2 = z_1 + (-z_2)$.
>
> **Division.** The quotient $z_1 \div z_2$, also written $\dfrac{z_1}{z_2}$, where $z_2 \neq 0$, is defined by $z_1 \div z_2 = z_1 \cdot \dfrac{1}{z_2}$.

Standard or Rectangular Form of a Complex-Number

> In enlarging the real-number system to obtain the complex-number system we are in the fortunate position of not having lost the real-numbers in the sense that complex-numbers of the form $(x, 0)$ are **algebraically indistinguishable** from the real-numbers.

If we set up a correspondence between $(x, 0)$ and x we have that the sum $(x, 0) + (y, 0) = (x + y, 0)$ corresponds to $x + y$ and the product $(x, 0)(y, 0) = (xy, 0)$ corresponds to xy. Thus, for the addition and multiplication of these complex-numbers with each other the results are the same as if $(x, 0)$ were replaced by the real-number x. For this reason

> The real-number x and the ordered pair $(x, 0)$ are used interchangeably.

At the same time the complex-numbers have properties not possessed by the real-numbers. For example, there is a complex-number whose square is the "real-number" $(-1,0)$, namely, $(0, 1)$, since $(0,1)^2 = (-1,0)$ (see Exercise 13). **The complex-number (0, 1) is denoted by i,** and we have:

$$i^2 = (0,1)(0,1) = (-1,0) = -1$$

That is, i^2 is the negative **of** 1 (**of** makes all the difference). It is useful to observe that a complex-number (x, y) can be expressed by:

$$(x, y) = (x, 0) + (0, y)$$
$$= (x, 0) + (0, 1)(y, 0)$$

$$(x, y) = x + iy = x + yi$$

Thus, (x, y) may be denoted by $x + iy$, or $x + yi$, with the understanding that x is $(x, 0)$ and y is $(y, 0)$. When $z = (x, y)$ is written in the form $x + iy$ it is said to be expressed in **standard** or **rectangular form.** When written in this way x is called the **real-part** and yi the **imaginary-part** of the complex-number. Complex-numbers of the form $(0, y) = 0 + iy$ or iy are called **imaginary-complex-numbers** or, more simply, **imaginary-numbers** (a most unfortunate term, to be sure).

The term rectangular form for $x + iy$ comes from the rectangular coordinate representation shown in Figure 10.1. The complex-number $(x, y) = x + iy$ is represented graphically by the point with coordinates (x, y).

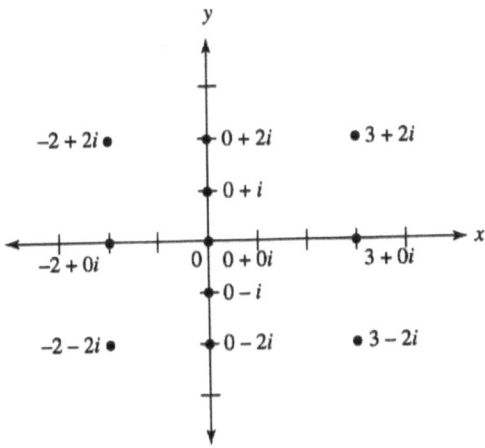

Figure 10.1

Addition, subtraction, multiplication and division of complex-numbers take on a particularly simple form when they are expressed in $x + yi$ form. If $z_1 = (x_1, y_1) = x_1 + y_1 i$ and $z_2 = (x_2, y_2) = x_2 + y_2 i$, we have:

$$z_1 + z_2 = (x_1 + y_1 i) + (x_2 + y_2 i) = (x_1 + x_2) + (y_1 + y_2)i$$

$$z_1 - z_2 = (x_1 + y_1 i) - (x_2 + y_2 i) = (x_1 - x_2) + (y_1 - y_2)i$$

That is, to add or subtract complex-numbers add or subtract their real and imaginary parts, respectively, and attach i to the imaginary part. For example, $(2+3i)+(4-2i)=6+i$ and $(2+3i)-(4-2i)=-2+5i$.

The product $z_1 z_2 = (x_1 + y_1 i)(x_2 + y_2 i)$ can be obtained by multiplying real and imaginary components without regard for the special meaning of i and replacing i^2 where it occurs by -1. Thus:

$$x_1 + y_1 i$$
$$\underline{x_2 + y_2 i}$$
$$x_1 x_2 + x_2 y_1 i$$
$$\underline{+x_1 y_2 i - y_1 y_2}$$
$$(x_1 x_2 - y_1 y_2) + (x_1 y_2 + x_2 y_1)i$$

For example, for $(2+3i)(4-2i)$ we have:

$$2+3i$$
$$\underline{4-2i}$$
$$8+12i$$
$$\underline{-4i-6i^2}$$
$$14+8i$$

To obtain the quotient $\dfrac{z_1}{z_2} = \dfrac{x_1 + y_1 i}{x_2 + y_2 i}$, multiply numerator and denominator of the quotient by $x_2 - y_2 i$ and combine terms as previously described. For example:

$$\frac{2+3i}{4-2i} = \frac{2+3i}{4-2i} \cdot \frac{4+2i}{4+2i} = \frac{2+16i}{20} = \frac{1}{10} + \frac{4}{5}i.$$

EXERCISES

22. Is $-1+2i$ the additive inverse of $1+3i$? Explain.

23. Is $4+3i$ the multiplicative inverse of $-1-2i$? Explain.

24. Find the additive inverses of the following:

(a) $7-2i$ (c) $2+3i$

(b) $-3+4i$ (d) $-1-4i$

25. Find the multiplicative inverses of the following:

(a) $7-2i$ (c) $2+3i$

(b) $-3+4i$ (d) $-1-4i$

26. Does $0 = (0,0)$ have a multiplicative inverse? Explain.

Find the sums, differences, products, and quotients indicated.

27. $(2+3i)+(4-2i)$

28. $(3-i)-(4+2i)$

29. $(3-2i)(4+i)$

30. $(5-3i)(2+3i)$

31. $\dfrac{5+4i}{2+3i}$

32. $\dfrac{6-2i}{1+i}$

33. $(4+i\sqrt{2})(4-2i\sqrt{2})$

34. $(5+2i)(3-i\sqrt{3})$

35. $(2-i\sqrt{2})-(3+2i\sqrt{2})$

36. $\dfrac{2+3i}{2-4i}$

37. $\dfrac{2+i\sqrt{2}}{3-i\sqrt{2}}$

38. $\dfrac{4+i}{3-2i}$

39. Can division by $0 = (0,0)$ be defined? Explain.

40. For what values of z, if any, is $\dfrac{z-1}{z^2+1}$ not defined? Explain.

10.3 CAN THE COMPLEX-NUMBER SYSTEM BE ORDERED?

Can we speak of one complex-number being smaller than or larger than another the way we do of the real-numbers? Is it not obvious that i sitting so comfortably above 0 on the y-axis (see Figure 10.1) is greater than 0?

To answer these questions we must travel back to section 1.3 and examine the conditions that give mathematical life to an order relation $<$. The conditions 01-04 noted there are the following in terms of any number system in general and the complex-number system in particular. Without them there is no order relation.

---

01. For any two z_1 and z_2 in C, one and only one of the relations $z_1 < z_2$, $z_1 = z_2$, $z_1 > z_2$ holds.

02. If $z_1 < z_2$, then $z_1 + z_3 < z_2 + z_3$. Adding the same complex-number to both sides of $z_1 < z_2$ preserves the sense of <.

03. If $z_1 < z_2$ and $0 < z_3$, then $z_1 \cdot z_3 < z_2 \cdot z_3$; if $z_3 < 0$, then $z_1 \cdot z_3 > z_1 \cdot z_3$.

04. If $z_1 < z_2$ and $z_2 < z_3$, then $z_1 < z_3$.

We have the following result for the complex-number system C:

It is not possible to define an order relation on the complex-number system C.

Proof. The proof is indirect; suppose it were possible to establish an order relation < on C which satisfies conditions 01 through 04. Let us consider i. Since $i \neq 0$, we have $i > 0$ or $i < 0$. If $i > 0$, then $i^2 > 0$, or $-1 > 0$, which implies $0 > 1$. Since $-1 > 0$, we have $(-1)(-1) > 0(0)$, or $1 > 0$, which contradicts $0 > 1$. Thus, $i > 0$ is not possible.

Suppose $i < 0$, or $0 > i$. Multiplying $0 > i$ by $i < 0$ gives us $0 < i^2$, or $0 < -1$. Adding 1 to both sides gives us $1 < 0$. Now $0 < -1$ means $-1 > 0$. Multiplying $-1 > 0$ by $-1 > 0$ yields $1 > 0$, which contradicts $1 < 0$.

Thus, i does not satisfy any of the conditions $i = 0$, $i < 0$, $i > 0$, which means that condition 01 cannot be satisfied.

Thus, as suggestive as the geometric representation of complex-numbers shown in Figure 10.1 might be, it is **meaningless** to talk about i being greater than 0 or, more generally, about one complex-number being smaller or larger than another.

While complex-numbers cannot be compared directly for size, there is an alternative which is useful. This alternative is also suggested by the rectangular coordinate representation of complex-numbers given by Figure 10.1 and involve what is called the **modulus** or **length** or **absolute value** of a complex-number.

Consider the complex-number $z = x + iy$ represented by the point P with coordinates (x, y) in Figure 10.2.

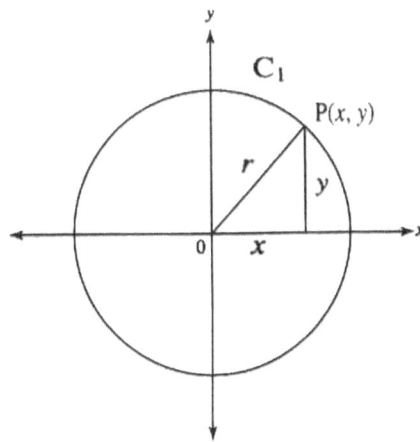

Figure 10.2

The length r of the line segment OP joining O and P is called the **modulus** or **length** or **absolute value** of z. r is also denoted by $|z| = |x + iy|$. We have:

$$|z_1| = \sqrt{x^2 + y^2}$$

While we cannot compare complex-numbers z_1 and z_2 for size directly we can compare them for size indirectly through their absolute values $|z_1|$ and $|z_2|$, which are nonnegative real-numbers. All complex-numbers that lie on the same circle C_1 (Figure 10.2) have the same modulus, whereas $|z_2| > |z_1|$ means that z_2 is on a circle with a larger radius than that of z_1.

EXERCISES

1. Determine the modulus of the following complex-numbers.

(a) $2 + 3i$
(b) $-4 + 2i$
(c) $-1 + 4i$

(d) $6 - 2i$
(e) $5 - 3i$
(f) $-3 - 7i$

2. Arrange the complex-numbers stated in 1(a) through 1(f) in order according to the size of their absolute values.

3. Is i greater than $-i$? Explain.

10.4 RETURN TO THE QUADRATIC EQUATION

Consider the quadratic function

$$y = x^2 + 1 \tag{10.1}$$

and the corresponding quadratic equation

$$x^2 + 1 = 0. \tag{10.2}$$

The graph of (10.1), is a parabola which lies above the x-axis (Figure 10.3).

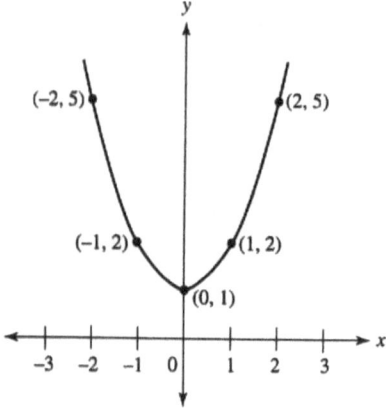

Figure 10.3

The equation (10.2) has no solution within the real-number system since there are no x for which $y = 0$, which, geometrically speaking, are x at which the parabola intersects the x-axis.

But if we extend the scene to

$$z^2 + 1 = 0 \tag{10.3}$$

by allowing for the possibility of complex-number solutions, then there are two solutions of (10.3), namely, i and $-i$, since $i^2 = (-i)^2 = -1$ (the complex-number $-1 = (-1, 0)$).

More generally, the quadratic equation

$$ax^2 + bx + c = 0 \qquad (10.4)$$

has solutions given by the quadratic formula

$$\frac{-b \pm \sqrt{b^2 - 4ac}}{2a}$$

even when $b^2 - 4ac < 0$ if we allow as options complex-number solutions.

There is no good reason why we should not do so. We may formally replace x by z in (10.4) or do so mentally in allowing for complex-number solutions.

EXAMPLE 1

Solve $x^2 - 2x + 4 = 0$. $\qquad (10.5)$

We have $a = 1$, $b = -2$, and $c = 4$. The quadratic formula yields:

$$x = \frac{2 \pm \sqrt{-12}}{2}$$
$$= \frac{2 \pm i\sqrt{12}}{2} = \frac{2 \pm 2i\sqrt{3}}{2}$$
$$= 1 + i\sqrt{3},\ 1 - i\sqrt{3}$$

EXERCISES

Solve the following quadratic equations.

1. $x^2 + x + 5 = 0$
2. $x^2 - 3x + 6 = 0$
3. $x^2 + 2x + 4 = 0$

4. $2x^2 + 2x + 1 = 0$
5. $-3x^2 - 2x - 1 = 0$
6. $-5x^2 + 2x - 2 = 0$

In the following three sections preliminaries are developed that will enable us to determine the nth roots of a number, real-number or complex-number.

10.5 SOME BASIC TRIGONOMETRY

Consider a ray which is anchored at the origin O and the angle A obtained by rotating this ray, called the **terminal ray,** from the positive x-axis, termed the **initial ray,** as shown in Figure 10.4(a).

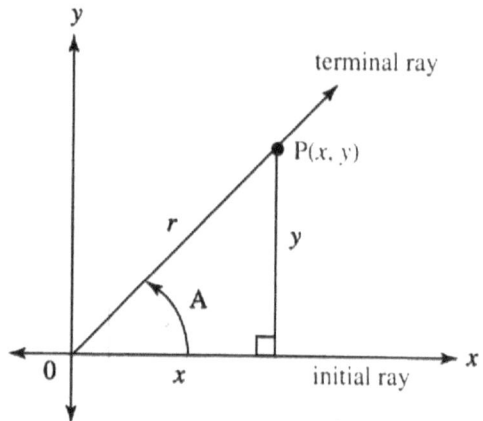

Figure 10.4(a)

Case 1: Choose any point $P(x, y)$ on the terminal ray. The length r of the segment OP is given by

$$r = \sqrt{x^2 + y^2}\ .$$

The sine and cosine of A, denoted by sin A and cos A, respectively, are defined by:

$$\sin A = \frac{\text{ordinate}}{r} = \frac{y}{r}$$

$$\cos A = \frac{\text{abscissa}}{r} = \frac{x}{r}$$

If $A = 45°$, for example (see Figure 10.4(b)),

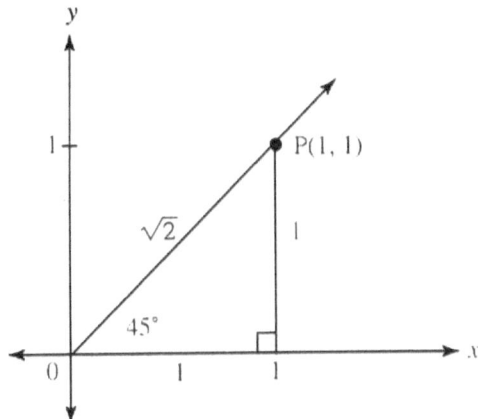

Figure 10.4(b)

and $P(1, 1)$ is the point chosen, then $r = \sqrt{2}$ and we have:

$$\sin 45° = \frac{1}{\sqrt{2}}, \quad \cos 45° = \frac{1}{\sqrt{2}} = 0.7071,$$

correct to four places.

It does not matter which point $P(x, y)$ is chosen on the terminal ray since the ratio y/r is the same for all $P(x, y)$ as is the ratio x/r.

Values of $\sin A$ and $\cos A$ to four places are given for A ranging from $0°$ to $90°$ in Appendix Table 8 (p. 437).

Case 2: Suppose angle A puts the terminal ray in the second quadrant, such as $135°$, shown in Figure 10.5.

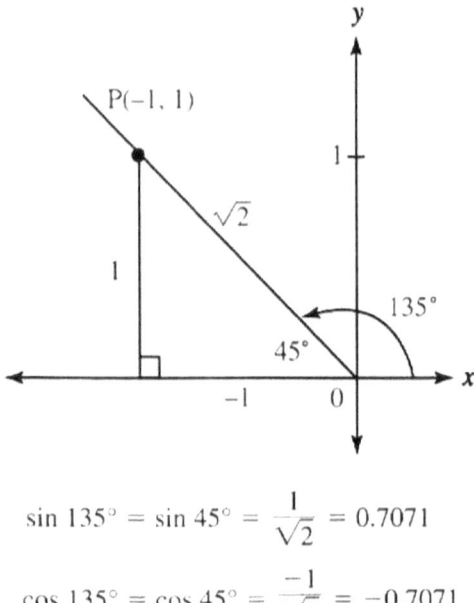

$$\sin 135° = \sin 45° = \frac{1}{\sqrt{2}} = 0.7071$$

$$\cos 135° = \cos 45° = \frac{-1}{\sqrt{2}} = -0.7071$$

Figure 10.5

Pick a point $P(x, y)$ on the terminal ray; since it does not matter which point is picked, suppose we pick $P(-1,1)$ in the illustrative example of 135°. Dropping a perpendicular from P to the x-axis gives us what is termed a reference triangle, one of 45° in this case. The sine and cosine of 135° are defined in terms of the reference triangle formed as follows:

$$\sin 135° = \sin 45° = \frac{1}{\sqrt{2}} = 0.7071$$

$$\cos 135° = \cos 45° = \frac{-1}{\sqrt{2}} = -0.7071$$

In general, we proceed in the same fashion. For example, $\sin 105° = \sin 75° = 0.9659$ (Table 8); $\cos 105° = -\cos 75° = -0.2588$. Sine values are positive for angles that put the terminal ray in the second quadrant and cosine values are negative.

Case 3. Suppose angle A puts the terminal ray in the third quadrant, such as 225°, shown in Figure 10.6.

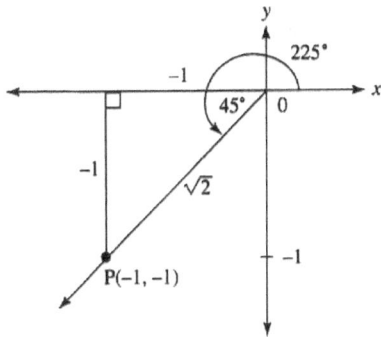

Figure 10.6

Pick a point P on the terminal ray, $P(-1,-1)$, for example, for the 225° angle. Extend a line from $P(-1,-1)$ perpendicular to the x-axis. The sine and cosine of 225° are defined in terms of the reference triangle formed as follows:

$$\sin 225° = \sin 45° = \frac{-1}{\sqrt{2}} = -0.7071$$

$$\cos 225° = \cos 45° = \frac{-1}{\sqrt{2}} = -0.7071$$

In general, we proceed in the same fashion. For example,
$$\sin 200° = -\sin 20° = -0.3420; \cos 200° = -\cos 20° = -0.9397.$$
Sine and cosine values are negative for angles that put the terminal ray in the third quadrant.

Case 4. Suppose angle A puts the terminal ray in the fourth quadrant, such as 315°, shown in Figure 10.7.

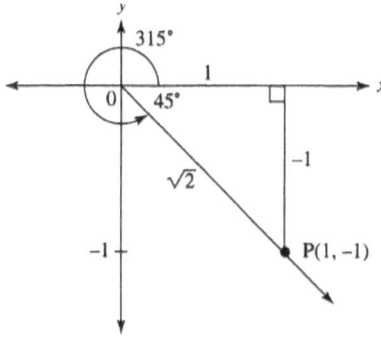

Figure 10.7

Pick a point $P(x, y)$ on the terminal ray, $P(1, -1)$, for example, for the 315° angle. Extend a line from $P(1, -1)$ perpendicular to the x-axis. The sine and cosine of 315° are defined in terms of the reference triangle formed as follows:

$$\sin 315° = \sin 45° = \frac{-1}{\sqrt{2}} = -0.7071$$

$$\cos 315° = \cos 45° = \frac{1}{\sqrt{2}} = 0.7071$$

In general, we proceed in the same way. For example,
$\sin 330° = -\sin 30° = -0.5000$; $\cos 330° = \cos 30° = 0.8660$.

Sine values are negative for angles that put the terminal ray in the fourth quadrant and cosine values are positive.

To help remember where sine and cosine are positive and negative it is useful to label the four quadrants according to the CAST system as shown in Figure 10.8.

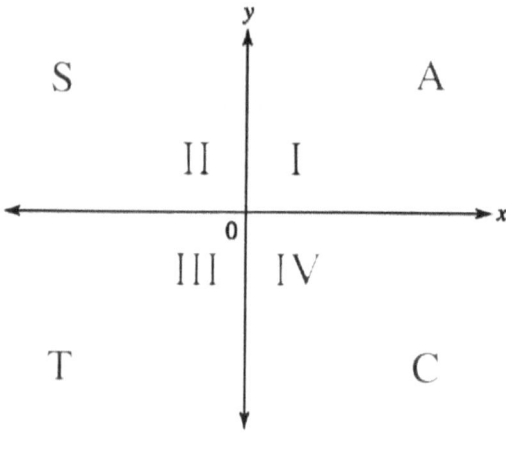

Figure 10.8

The A in quadrant I means that all trigonometric functions (sine, cosine and others) are positive there; S in II means that sine is positive there as well; C in IV means that cosine is positive there as well; elsewhere they are negative.

In summary:

> The sine and cosine of any angle which has a reference angle are the
> same as those of the reference angle except for sign. The sign may be
> determined by the CAST device.

If the terminal ray of angle A lies on one of the coordinate axes, we do not
have a reference triangle. We proceed as follows in such a case.

To find sin 180° and cos 180° pick a point on the terminal ray (the negative
x-axis), $P(-1,0)$, for example (see Figure 10.9).

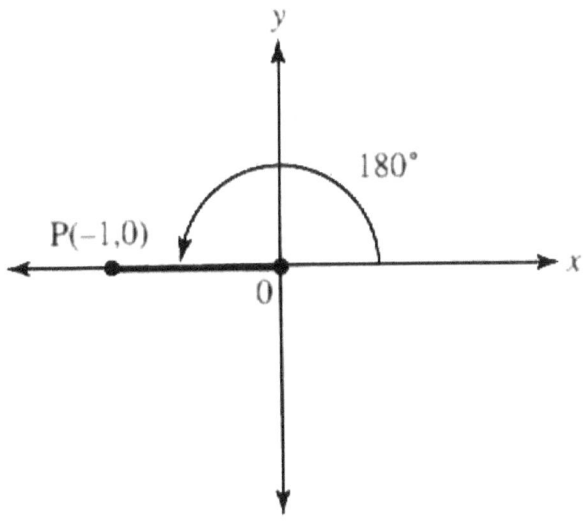

Figure 10.9

Then $r=1$ and $\sin 180° = \dfrac{0}{1} = 0$ and $\cos 180° = -\dfrac{1}{1} = -1$.

To find sin 270° and cos 270° pick a point on the terminal ray (the negative
y-axis), $P(0,-1)$, for example (see Figure 10.10).

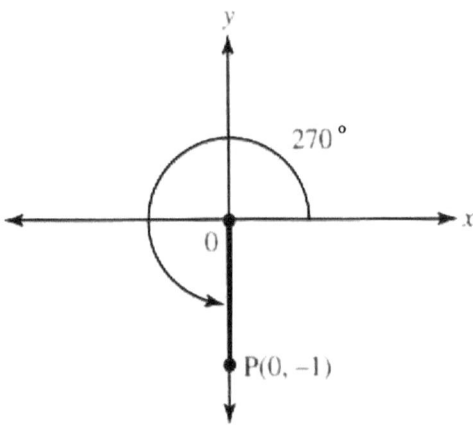

Figure 10.10

Then $r = 1$ and $\sin 270° = \dfrac{-1}{1} = -1$ and $\cos 270° = \dfrac{0}{1} = 0$.

Consider the angle 405°, shown in Figure 10.11.

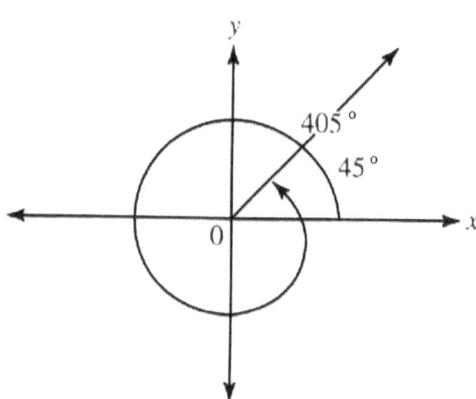

Figure 10.11

An angle represents a rotation and 405° is equivalent to 45° in the sense that the terminal ray for a rotation of 405° is in the same position as the terminal ray for a rotation of 45°. Thus, $\sin 405° = \sin 45°$ and $\cos 405° = \cos 45°$.

More generally, we have that the sine and cosine of $\cos 45° + n(360°)$ are equal to the sine and cosine of 45°, respectively.

Still more generally:

> For any angle of $A°$, the sine and cosine of $A° + n(360°)$ are equal
> to the sine and cosine of $A°$, respectively.

To determine cosine 840°, for example, we first note that $840° = 120° + 2(360°)$.
Thus, $\cos 840° = \cos 120° = -\cos 60° = -0.9397$.

EXERCISES

Find the values of sine and cosine of the following angles:

1. 25°
2. 64°
3. 130°
4. 250°
5. 210°

6. 290°
7. 650°
8. 720°
9. 270°
10. 90°

11. 950°
12. 230°
13. 150°
14. 580°
15. 650°

10.6 POLAR FORM OF A COMPLEX-NUMBER

A point P may be pinpointed by specifying its x and y coordinates, as noted in
Figure 10.12,

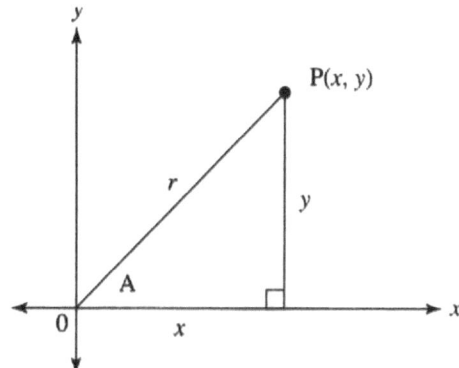

Figure 10.12

but this is not the only way to pinpoint P. We can also do this by stating the
distance r of the point from the origin O and the angle A that the terminal ray

OP makes with the positive part of the x-axis. The ordered pair (r, A) is called **polar coordinates** of P. Thus, for example, $(\sqrt{2}, 45°)$ are polar coordinates of $P(1, 1)$ (see Figure 10.13).

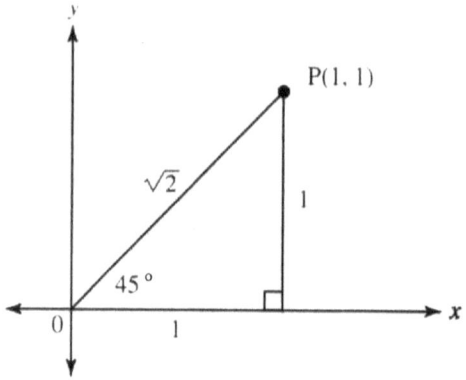

Figure 10.13

Polar coordinates of a point are not unique since we may rotate the terminal ray OP any number of times we desire and still end up at P; thus $(\sqrt{2}, 45° + 1(360°))$, $(\sqrt{2}, 45° + 2(360°))$, and so on, are also polar coordinates of $P(1, 1)$.

From Figure 10.12 we have the following relations:

$$\sin A = \frac{y}{r}, \qquad \cos A = \frac{x}{r}, \qquad r = \sqrt{x^2 + y^2} \qquad (10.6)$$

By solving the first two equations of (9.6) for x and y we obtain:

$$x = r \cos A, \qquad y = r \sin A \qquad (10.7)$$

(10.6) allows us to move from rectangular coordinates (x, y) to polar coordinates of a point; (10.7) allows us to go the other way.

Consider the complex-number $z = (x, y)$, whose standard form is $x + iy$. From (10.7) we obtain the **polar form** of z:

$$z = r \cos A + ir \sin A$$
$$= r(\cos A + i \sin A)$$

(10.8)

r, as we have noted, is called the **modulus**, or **length**, or **absolute value** of z. The angle A is called its **amplitude** or **argument**.

EXAMPLE 1

Write $z = -1 - i$ in polar form, using the smallest value of the amplitude.

$z = -1$ and $y = -1$, which puts z in the third quadrant (see Figure 10.14).

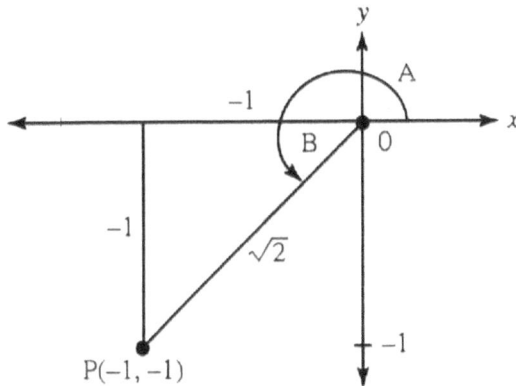

Figure 10.14

$r = \sqrt{(-1)^2 + (-1)^2} = \sqrt{2}$ (whose approximate value is 1.414);

$\sin A = -\sin B = \dfrac{1}{1.414} = 0.7072$. To find B, or as close an approximation to that Appendix Table 8 (p. 437) will allow, we look down the column of sine values and come as close to 0.7072 as possible. We can come as close as 0.7071, which yields $B = 45°$. Thus $A = 180° + 45° = 225°$.

The polar form of $z = -1 - i$ is $\sqrt{2}(\cos 225° + i \sin 225°)$.

EXAMPLE 2

Write −8 in polar form, using the smallest value of the amplitude.

In standard form $-8 = -8 + 0i$. Thus $r = 8$ and $A = 180°$ (see Figure 10.15).

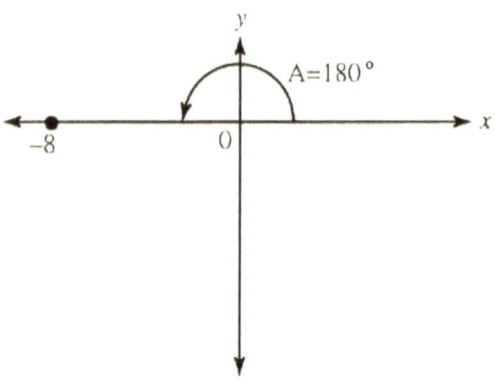

Figure 10.15

We have $-8 = 8(\cos 180° + i \sin 180°)$.

EXAMPLE 3

Write $2(\cos 30° + i \sin 30°)$ in standard form.

From Appendix Table 8 we have: $\cos 30° = 0.8660$ (to four places) and $\sin 30° = 0.5000$.
Thus, $2(\cos 30° + i \sin 30°) = 2(0.8660 + 0.5i) = 1.7320 + i$, approximately.

EXERCISES

Write the following in standard form:

1. $3(\cos 50° + i \sin 50°)$
2. $2(\cos 60° + i \sin 60°)$
3. $8(\cos 405° + i \sin 405°)$
4. $5(\cos 170° + i \sin 170°)$

5. $3(\cos 135° + i \sin 135°)$
6. $2(\cos 270° + i \sin 270°)$
7. $3(\cos 90° + i \sin 90°)$
8. $5(\cos 300° + i \sin 300°)$

Write the following in polar form, using the smallest amplitude value.

9. $1+i$

10. $-1+i$

11. $1-i$

12. -8

13. $1+i\sqrt{3}$

14. $-3+4i$

15. 16

16. $8i$

10.7 de MOIVRE'S THEOREM

Return to Multiplication and Division

It is interesting and useful to examine multiplication and division of complex-numbers stated in polar form. Consider z_1 and z_2 in polar form:

$$z_1 = r(\cos A + i \sin A)$$
$$z_2 = R(\cos B + i \sin B)$$

Multiplying them yields:

$$rR[\underbrace{(\cos A \cos B - \sin A \sin B)}_{\cos(A+B)} + i\underbrace{(\sin A \cos B + \cos A \sin B)}_{\sin(A+B)}] \quad (10.9)$$

In trigonometry it is shown that

$$\cos(A+B) = \cos A \cos B - \sin A \sin B$$

and

$$\sin(A+B) = \sin A \cos B + \cos A \sin B,$$

so that (10.9) takes the form:

$$z_1 z_2 = rR[\cos(A+B) + i \sin(A+B)] \qquad (10.10)$$

By means of a similar analysis it is shown that:

$$\frac{z_1}{z_2} = \frac{r}{R}[\cos(A-B) + i \sin(A-B)] \qquad (10.11)$$

Thus, for example, the product of $z_1 = 2(\cos 10° + i \sin 10°)$ and $z_2 = 5(\cos 80° + i \sin 80°) = 10(\cos 90° + i \sin 90°) = 10(0 + i \cdot 1) = 10i$.

The quotient z_1 / z_2 with $z_1 = 4(\cos 45° + i \sin 45°)$ with $z_2 = 2(\cos 15° + i \sin 15°) = 2(\cos 30° + i \sin 30°) = 2(0.8660 + 0.5i) = 1.7320 + i$, approximately.

de Moivre's Theorem

As a special case of (10, 10) we have that z^2 of $z = r(\cos A + i \sin A)$ is given by:

$$z^2 = r^2(\cos 2A + i \sin 2A).$$

We also have:

$$z^3 = r^3(\cos 3A + i \sin 3A)$$
$$z^4 = r^4(\cos 4A + i \sin 4A)$$

The general result suggested, called de Moivre's theorem after Abraham de Moivre (1667-1754), is the following:

If $z = r(\cos A + i \sin A)$, then $z^n = r^n(\cos nA + i \sin nA)$ (10.12)

Thus, for example, if $z = 2(\cos 15° + i \sin 15°)$, $z^2 = 4(\cos 30° + i \sin 30°) = 4(0.8660 + 0.5i) = 3.4640 + 2i$, approximately.

$z^3 = 8(\cos 45° + i \sin 45°) = 8(0.7071 + 0.7071i) = 5.6568 + 5.6568i$, approximately.

EXERCISES

Find the following products, quotients, and powers.

1. $z_1 z_2$, where $z_1 = 3(\cos 15° + i \sin 15°)$, $z_2 = 4(\cos 30° + i \sin 30°)$.

2. $z_1 z_2$, where $z_1 = 6(\cos 50° + i \sin 50°)$, $z_2 = 2(\cos 30° + i \sin 30°)$.

3. z_1/z_2 for z_1 and z_2 given in 2.

4. $z_1 z_2$ for $z_1 = 8(\cos 120° + i \sin 120°)$, $z_2 = 4(\cos 30° + i \sin 30°)$.

5. z_1/z_2 for z_1 and z_2 given in 4.

6. $(1+i)^2$ 7. $(1+i)^4$

8. $(-2+3i)^4$ 9. $(3-i)^4$

10. 2^5 11. $(3i)^5$

12. Give a geometric interpretation to the product of two complex-numbers in polar form.

10.8 THE nth ROOTS OF A NUMBER

Through use of de Moivre's theorem we can find the nth roots of any complex-number.

EXAMPLE 1

Find the cube roots of -8.

Let us recall from section 1.6 that the principal cube root of -8 is -2. It has two other cube roots in the complex-number system which we now proceed to determine.

In polar form $-8 = 8(\cos 180° + i \sin 180°)$. But $180°$ may be replaced by $180° + n(360°)$, where n can take any integer value. Thus we may write:

$$-8 = 8[\cos(180° + n(360°)) + i \sin(180° + n(360°))] \qquad (10.13)$$

Let $z = r(\cos A + i \sin A)$ denote a cube root of -8. Then $z^3 = -8$, given by (10.13). From de Moivre's theorem (10.12) we have:

$$r^3(\cos 3A + i \sin 3A) = 8[\cos(180° + n(360°)) + i \sin(180° + n(360°))]$$

Thus:

$$r^3 = 8 \tag{10.14}$$
$$3A = 180° + n(360°) \tag{10.15}$$

From (10.14) we have $r = 2$. From (10.15) we have:

$$A = 60° + n(120°)$$

Thus:

$$z = 2[\cos(60° + n(120°)) + i \ \sin(60° + n(120°))]$$

describes the cube roots of -8

For $n = 0$, we have:

$$z = 2(\cos 60° + i \sin 60°)$$
$$= 1 + 1.7320i, \ \text{approximately}$$

For $n = 1$, we have:

$$z = 2(\cos 180° + i \sin 180°)$$
$$= 2(-1 + 0) = -2, \ \text{the principal cube root of } -8.$$

For $n = 2$, we have:

$$z = 2(\cos 300° + i \sin 300°)$$
$$= 2(\cos 60° - i \sin 60°)$$
$$= 2(0.5 - 0.8660i)$$
$$= 1 - 1.7320i, \ \text{approximately}$$

Taking $n = 3, 4, \ldots$ yields repetitions of these results. -8 has three distinct cube roots: $1 + 1.7320i$, -2, and $1 - 1.7320i$.

Each has modulus 2 and graphically they divide the circle centered at 0 with radius 2 into three arcs of equal length (Figure 10.16).

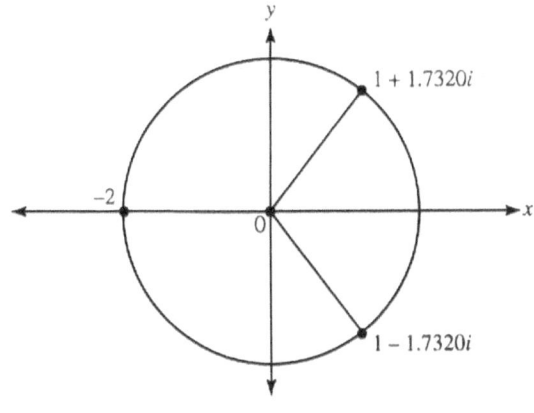

Figure 10.16

More generally, to obtain the nth roots of a non-zero complex-number we begin by expressing it in polar form:

$$R[\cos(B+k(360°))+i\ \sin(B+k(360°))] \qquad (10.16)$$

Let $z = r(\cos A + i\ \sin A)$ denote an nth root of (10.16). Follow, with suitable refinements, in the tracks laid down in Example 1.

The k integral values 1, 2, . . . , n yield n distinct roots. Taking further integral values for k only yields repetitions. Each of the n roots has modulus r and geometrically speaking they break the circle centered at O with radius r into n subarcs of equal length.

EXERCISES

Find the indicated roots, which should be stated in standard form.

1. Cube roots of 1.

2. Cube roots of 8.

3. Cube roots of $8i$

4. Square roots of $4i$

5. Fifth roots of 32.

6. Cube roots of -1.

7. Sixth roots of 1.

8. Fourth roots of $-8+i8\sqrt{3}$

10.9 GENERAL OBSERVATIONS

In sections 1.2 and 10.2 we observed that both the real-numbers and complex-numbers satisfy a number of conditions: closure for addition and multiplication, commutativity for addition and multiplication, associativity for addition and multiplication, the existence of additive and multiplicative identity numbers called zero and one, the existence of an additive inverse for each real-number and complex-number, the existence of a multiplicative inverse for each non-zero real-number and complex-number, and the distributive property of multiplication over addition.

More generally, any collection of objects for which addition and multiplication operations are defined satisfying these conditions is called a **field.**

The real-number and complex-number systems are fields. Since the real-number system is contained within the complex-number field and is a field with respect to its addition and multiplication operations, it is said to be a **subfield** of the complex-number field.

Whatever results can be established for the real-number system that depend only on the field properties are applicable to other fields as well, particularly the complex-number field. Thus, much of the algebra that was developed in the context of the real-number system—factoring, machinery for solving equations, and so on—carries over intact to the complex-numbers.

> The real-number field is said to be an **ordered field** because it satisfies the order conditions 01 through 04 described in section 1.3. As we saw in section 10.3, the complex-number system cannot be ordered and is thus **not** an ordered field. Properties of the real-number system that depend on its ordering do not carry over to the complex-number field.

> In working with complex-numbers we must be very careful not to slip into procedures and arguments based on complex-number inequality since such is meaningless.

In enlarging our number system from the real-numbers to complex-numbers we have gained mathematical strength in the scope of internal mathematical problems and real world applications. Two internal mathematical developments that we saw open up are concerned with solving the general quadratic equation and obtaining the nth roots of real-numbers and complex-numbers.

> Real world applications of complex-numbers are beyond the scope of this book, but include the development of the theory of flight and modern electrical engineering; indeed, a very rich harvest of applications.

EXERCISES

1. Is $G = \{-1, 0, 1\}$ with respect to real-number addition and multiplication a sub-field of the real-number system? Explain.

2. Consider the collection Q of all rational numbers. (r is rational if it can be expressed as a ratio of two integers, a and b; $r = a/b$. 0, 1, -2, 1/2, 5/8, $-2/3$ are examples of rational numbers. $\sqrt{2}$, $1+\sqrt{3}$, and ϖ are examples of real-numbers that are not rational.)

   (a) Is Q with respect to real-number addition and multiplication a subfield of the real-number system? Explain.

   (b) Is Q an ordered subfield of the real-number system? Explain.

   (c) If the real-number system R is an ordered field and the rational-number system Q is an ordered subfield of R, then how can we tell R and Q apart in terms of its algebraic properties? That is, what additional condition distinguishes the real-number system from the rational-number system? Hint: Look up the least upper bound property.

10.10 SELF-TESTS FOR CHAPTER 10

Allow 70 or so minutes for each self-test. Review the first one before going on to the second.

Self-Test 1

1. For $z_1 = -2 + 3i$ and $z_2 = 3 - i$ find:

   (a) $z_1 + z_2$ (d) z_1 / z_2

   (b) $z_1 - z_2$ (e) The additive inverse of z_1.

   (c) $z_1 z_2$ (f) The multiplicative inverse of z_2.

2. For what values of z, if any, is $\dfrac{z+3}{z^2+4}$ not defined? Explain.

3. Would it be correct to say that $-2i < 2i$ Explain.

4. Solve: (a) $x^2+4=0$, (b) $2x^2-x+2=0$

5. Write the following numbers in polar form, using the smallest value of the amplitude.

   (a) -16 (b) $-2+i$

6. By using de Moivre's theorem, find z^8 for $z=1+i$. Express z^8 in standard form.

7. Find the fourth roots of 16.

8. Is the collection of imaginary-numbers of the form $0+yi$ a subfield of the complex-number system with respect to complex-number addition and multiplication? Explain.

Self-Test 2

1. For $z_1=1-2i$ and $z_2=-2+5i$ find:

   (a) z_1+z_2 (d) z_1/z_2

   (b) z_1-z_2 (e) The additive inverse of z_1.

   (c) z_1z_2 (f) The multiplicative inverse of z_2.

2. Does $0=(0,0)$ have a multiplicative inverse? Explain.

3. Consider $z_1=1+i$ and $z_2=-2+3i$.

   (a) Determine $|z_1|$ and $|z_2|$.

   (b) Would it be correct to say that $z_1<z_2$ because $|z_1|<|z_2|$? Explain.

4. Solve: $3x^2+x+1=0$

5. Write the following numbers in polar form, using the smallest value of the amplitude.

   (a) $3i$ (b) $-2 + 2i\sqrt{3}$

6. Write the following numbers in standard form:

   (a) $2(\cos 300° + i \sin 300°)$ (b) $5(\cos 135° + i \sin 135°)$

7. By using de Moivre's theorem, find z^4 for $z = 1 - i$. Express z^4 in standard form.

8. Find the sixth roots of $-64i$.

9. Is the collection of complex-numbers G of the form $n + mi$, where n and m are integers, a subfield of the complex-number field with respect to complex-number addition and multiplication? Explain.

10. Consider the following argument which is based on a product and quotient property of radicals first introduced for real-numbers in section 1.6 and carried over to the complex-numbers.

$$1 = \sqrt{1} = \sqrt{(-1)(-1)} = \sqrt{-1}\sqrt{-1} = i \cdot i = i^2 = -1.$$

Thus, $1 = -1$? How can you explain this?

CHAPTER 11

Life in the
Land of Matrices

11.1 ALGEBRAIC OPERATIONS ON MATRICES

A matrix of numbers is a rectangular array of numbers, called entries, in which the locations of the entries are distinguished. Figure 11.1(a) illustrates a 2 by 1 matrix (2 rows and 1 column); Figure 11.1(b) a 1 by 3 matrix (1 row and 3 columns), and Figure 11.1(c) a 3 by 3 matrix.

More generally, a matrix with m rows and n columns is called an **m by n matrix.** Two matrices with the same number of rows and same number of columns are said to be the **same size.** No two matrices in Figure 11.1 are the same size.

$$A = \begin{bmatrix} 1 \\ -1 \end{bmatrix} \qquad B = \begin{bmatrix} 4 & -\frac{1}{2} & 8 \end{bmatrix} \qquad C = \begin{bmatrix} 2 & -1 & 1 \\ 1 & 4 & 0 \\ 3 & 0 & -1 \end{bmatrix}$$

$$(a) \qquad\qquad (b) \qquad\qquad (c)$$

Figure 11.1

Matrices C and D shown in Figure 11.2 are the same size; both are 2 by 2 matrices.

$$C = \begin{bmatrix} 2 & -1 \\ 3 & -2 \end{bmatrix} \qquad\qquad D = \begin{bmatrix} 2 & -1 \\ 3 & 4 \end{bmatrix}$$

Figure 11.2

Two matrices are said to be **equal** if they are the same size and corresponding entries are equal. Matrices C and D in Figure 11.2 are not equal since the entry in row 2, column 2 of C is not equal to the entry in row 2, column 2 of D.

If a matrix has the same number of rows as columns, it is said to be a **square matrix.** Matrices C and D in Figure 11.2 are square 2 by 2 matrices, and matrix C in Figure 10.1(c) is a square 3 by 3 matrix. To simplify language, we shall term the entry in the ith row and jth column of a matrix the i-j entry. Thus the 1-2 entry of a matrix is the term in row 1 and column 2; the 2-1 entry is the term in row 2 and column 1. The 1-2 entry of matrix C in Figure 10.2 is -1; the 2-1 entry of matrix C is 3.

Matrices acquire mathematical life by virtue of the algebraic operations that are defined on them. It is to these algebraic operations that we now turn our attention.

Matrix Addition and Subtraction

If A and B are matrices of the same size, then

$$C = A + B$$

is the matrix obtained by adding corresponding entries of A and B.

$$D = A - B$$

is the matrix obtained by subtracting the entries of B from the corresponding entries of A. Thus if

$$A = \begin{bmatrix} 4 & 3 & 1 \\ 2 & -3 & -2 \end{bmatrix} \quad \text{and} \quad B = \begin{bmatrix} 2 & -1 & 4 \\ -3 & 2 & 1 \end{bmatrix}$$

then

$$C = A + B = \begin{bmatrix} 4+2 & 3-1 & 1+4 \\ 2-3 & -3+2 & -2+1 \end{bmatrix} = \begin{bmatrix} 6 & 2 & 5 \\ -1 & -1 & -1 \end{bmatrix}$$

and:

$$D = A - B = \begin{bmatrix} 4-2 & 3-(-1) & 1-4 \\ 2-(-3) & -3-2 & -2-1 \end{bmatrix} = \begin{bmatrix} 2 & 4 & -3 \\ 5 & -5 & -3 \end{bmatrix}$$

Inner Product of a Row and Column

In preparation for the definition of matrix multiplication, we introduce the following definition of inner product of a row and column. Consider the following two matrices A and B:

$$A = \begin{bmatrix} 2 & -1 \\ 3 & 2 \\ 0 & 4 \end{bmatrix}, \quad B = \begin{bmatrix} 2 & 0 & -1 \\ 4 & 3 & -2 \end{bmatrix}$$

The **inner product** of row 1 of matrix A, $[2 \ -1]$, and column 1 of matrix B, $\begin{bmatrix} 2 \\ 4 \end{bmatrix}$, is the number $2(2)+(-1)4 = 0$ obtained by multiplying corresponding entries (first entry in row 1 of A and first entry in column 1 of B, second entry in row 1 of A and second entry in column 1 of B) and adding.

The **inner product** of row 1 of B, $[2 \ \ 0 \ -1]$, and column 1 of A, $\begin{bmatrix} 2 \\ 3 \\ 0 \end{bmatrix}$, is $2(2)+0(3)+(-1)0 = 4$.

The inner product of row 1 of A and column 1 of A is not defined, since this row and column do not have the same number of entries.

More generally, if A and B are matrices and the number of entries in a row of A equals the number of entries in a column of B, the **inner product of row i of matrix A and column j of matrix B** is the number obtained by multiplying corresponding entries in row i and column j and adding.

EXERCISES

1. For $A = \begin{bmatrix} 2 & -1 & 3 \\ 1 & 2 & 0 \end{bmatrix}$ and $B = \begin{bmatrix} 4 & 2 & -1 \\ 5 & 2 & 6 \end{bmatrix}$, find $A+B$, $A-B$, and $B-A$.

2. For $A = \begin{bmatrix} 2 & 1 \\ 5 & -2 \\ -4 & 3 \end{bmatrix}$ and $B = \begin{bmatrix} -1 & 2 \\ 3 & 1 \\ 2 & -5 \end{bmatrix}$, find $A + B$ and $A - B$.

3. For $A = \begin{bmatrix} 2 & -1 \\ 6 & -2 \end{bmatrix}$ and $B = \begin{bmatrix} 8 & -7 \\ 3 & -4 \end{bmatrix}$, find $A + B$ and $A - B$.

4. For $A = \begin{bmatrix} 2 & 3 \\ 4 & 5 \end{bmatrix}$ and $I_2 = \begin{bmatrix} 1 & 0 \\ 0 & 1 \end{bmatrix}$, find $I_2 - A$.

5. For $A = \begin{bmatrix} 3 & 2 & -1 \\ 4 & 6 & 2 \\ 3 & 0 & 1 \end{bmatrix}$ and $I_3 = \begin{bmatrix} 1 & 0 & 0 \\ 0 & 1 & 0 \\ 0 & 0 & 1 \end{bmatrix}$, find $I_3 - A$.

6. $A = \begin{bmatrix} 1 \\ 2 \\ 4 \end{bmatrix}$ and $B = \begin{bmatrix} -1 \\ 3 \\ -2 \end{bmatrix}$, find $A + B$ and $B - A$.

7. For $A = \begin{bmatrix} 3 & 2 & 4 \end{bmatrix}$ and $B = \begin{bmatrix} 2 & -1 & -5 \end{bmatrix}$, find $A + B$ and $A - B$.

8. For the matrices $A = \begin{bmatrix} 2 & -1 \\ 3 & 2 \\ 0 & 4 \end{bmatrix}$ and $B = \begin{bmatrix} 2 & 0 & -1 \\ 4 & 3 & -2 \end{bmatrix}$, find the inner

   product of (a) row 1 of A and column 2 of B, (b) row 1 of A and column 3 of B, (c) row 2 of A and column 1 of B, (d) row 2 of A and column 2 of B, (e) row 2 of A and column 3 of B, (f) row 3 of A and column 1 of B, (g) row 3 of A and column 2 of B, and (h) row 3 of A and column 3 of B.

9. For the matrices A and B defined in the preceding exercise, find the inner product of (a) row 1 of B and column 1 of A, (b) row 1 of B and column 2 of A, (c) row 2 of B and column 1 of A, (d) row 2 of B and column 2 of A.

Matrix Multiplication

Consider the 3 by 2 matrix A and 2 by 3 matrix B defined as follows:

$$A = \begin{bmatrix} 2 & -1 \\ 3 & 2 \\ 0 & 4 \end{bmatrix}, \qquad B = \begin{bmatrix} 2 & 0 & -1 \\ 4 & 3 & -2 \end{bmatrix}$$

The matrix AB has as many rows as A and as many columns as B, and is thus a 3 by 3 matrix whose entries are determined in the following way:

$$AB = \begin{bmatrix} 0 & - & - \\ - & - & - \\ - & - & - \end{bmatrix}$$

The 1-1 entry of AB is the inner product of row 1 of A and column 1 of B; $2(2)+(-1)4 = 0$.

$$AB = \begin{bmatrix} 0 & -3 & - \\ - & - & - \\ - & - & - \end{bmatrix}$$

The 1-2 entry of AB is the inner product of row 1 of A and column 2 of B; $2(0)+(-1)3 = -3$.

$$AB = \begin{bmatrix} 0 & -3 & 0 \\ - & - & - \\ - & - & - \end{bmatrix}$$

The 1-3 entry of AB is the inner product of row 1 of A and column 3 of B; $2(-1)+(-1)(-2) = 0$.

$$AB = \begin{bmatrix} 0 & -3 & 0 \\ 14 & - & - \\ - & - & - \end{bmatrix}$$

The 2-1 entry of AB is the inner product of row 2 of A and column 1 of B; $3(2)+2(4) = 14$.

$$AB = \begin{bmatrix} 0 & -3 & 0 \\ 14 & 6 & \text{---} \\ \text{---} & \text{---} & \text{---} \end{bmatrix}$$

The 2-2 entry of AB is the inner product of row 2 of A and column 2 of B; $3(0) + 2(3) = 6$.

$$AB = \begin{bmatrix} 0 & -3 & 0 \\ 14 & 6 & -7 \\ \text{---} & \text{---} & \text{---} \end{bmatrix}$$

The 2-3 entry of AB is the inner product of row 2 of A and column 3 of B; $3(-1) + 2(-2) = -7$.

$$AB = \begin{bmatrix} 0 & -3 & 0 \\ 14 & 6 & -7 \\ 16 & \text{---} & \text{---} \end{bmatrix}$$

The 3-1 entry of AB is the inner product of row 3 of A and column 1 of B; $0(2) + 4(4) = 16$.

$$AB = \begin{bmatrix} 0 & -3 & 0 \\ 14 & 6 & -7 \\ 16 & 12 & \text{---} \end{bmatrix}$$

The 3-2 entry of AB is the inner product of row 3 of A and column 2 of B; $0(0) + 4(3) = 12$.

$$AB = \begin{bmatrix} 0 & -3 & 0 \\ 14 & 6 & -7 \\ 16 & 12 & -8 \end{bmatrix}$$

The 3-3 entry of AB is the inner product of row 3 of A and column 3 of B; $0(-1) + 4(-2) = -8$.

More generally, if A is an n by m matrix and B is an m by k matrix (that is, the number of columns of A equals the number of rows of B), then the **matrix product AB** is defined and is an n by k matrix (same number of rows as A and same number of columns as B) whose i-j entry is the inner product of row i of A and column j of B.

EXAMPLE 1

For the following matrices A and B find, provided that it is defined, the matrix product BA.

$$B = \begin{bmatrix} 2 & 0 & -1 \\ 4 & -3 & -2 \end{bmatrix}, \qquad A = \begin{bmatrix} 2 & -1 \\ 3 & 2 \\ 0 & 4 \end{bmatrix}$$

Since B is a 2 by 3 matrix and A is a 3 by 2 matrix, the matrix product BA is defined and is a 2 by 2 matrix. We have:

$$BA = \begin{bmatrix} 2(2)+0(3)+(-1)0 & 2(-1)+0(2)+(-1)4 \\ 4(2)+3(3)+(-2)0 & 4(-1)+3(2)+(-2)4 \end{bmatrix} = \begin{bmatrix} 4 & -6 \\ 17 & -6 \end{bmatrix}$$

Let us note that the matrix products AB and BA are not the same:

$$AB = \begin{bmatrix} 0 & -3 & 0 \\ 14 & 6 & -7 \\ 16 & 12 & -8 \end{bmatrix}, \qquad BA = \begin{bmatrix} 4 & -6 \\ 17 & -6 \end{bmatrix}$$

In multiplying matrices one must pay **careful attention** to the **order in which they are multiplied.**

EXAMPLE 2

Find, if they exist, the matrix products AB and BA, where:

$$A = \begin{bmatrix} -2 \\ 4 \\ 3 \end{bmatrix}, \qquad B = \begin{bmatrix} -1 & 3 & 5 \end{bmatrix}$$

Since A is a 3 by 1 matrix and B is a 1 by 3 matrix, the product AB is defined and is a 3 by 3 matrix with the following entries:

$$AB = \begin{bmatrix} -2(-1) & -2(3) & -2(5) \\ 4(-1) & 4(3) & 4(5) \\ 3(-1) & 3(3) & 3(5) \end{bmatrix} = \begin{bmatrix} 2 & -6 & -10 \\ -4 & 12 & 20 \\ -3 & 9 & 15 \end{bmatrix}$$

Since B is a 1 by 3 matrix and A is a 3 by 1 matrix, the product BA is a 1 by 1 matrix:

$$BA = [-1(-2) + 3(4) + 5(3)] = [29]$$

EXAMPLE 3

Find the matrix product PC, where P and C are defined as follows:

Service Departments

	Acc.	Main.	Mar.	Pur.	
$P =$	0.24	0.20	0.30	0.20	Production dept. P_1
	0.26	0.20	0.30	0.20	Production dept. P_2
	0.28	0.20	0.40	0.30	Production dept. P_3

$$C = \begin{bmatrix} 23,423 \\ 19,902 \\ 87,756 \\ 14,333 \end{bmatrix}$$

Total cost: accounting
Total cost: maintenance
Total cost: marketing
Total cost: purchasing

Referring back to the service-charge-allocation problem considered in Example 1 of section 4.6 (pp. 196-199), we see that the entries in the first row of matrix P express the fraction of the total costs of the service departments—accounting, maintenance, marketing, and purchasing, respectively—that are assigned to production department P_1 of the Arkin Company. Similarly, the entries in the second row relate to production department P_2, and the entries in the third row relate to P_3. Matrix C states the total costs of the service departments. The matrix product PC is given by:

$$PC = \begin{bmatrix} 0.24(23,423)+0.20(19,902)+0.30(87,756)+0.20(14,333) \\ 0.26(23,423)+0.20(19,902)+0.30(87,756)+0.20(14,333) \\ 0.28(23,423)+0.20(19,902)+0.40(87,756)+0.30(14,333) \end{bmatrix}$$

$$= \begin{bmatrix} 38,795 \\ 39,264 \\ 49,941 \end{bmatrix} \begin{matrix} \text{Service departments' costs allocated to } P_1 \\ \text{Service departments' costs allocated to } P_2 \\ \text{Service departments' costs allocated to } P_3 \end{matrix}$$

Matrix PC describes the allocation of service departments' costs to the production departments of the firm.

EXERCISES

10. For the matrices $G = \begin{bmatrix} 1 & 2 \\ 3 & -1 \end{bmatrix}$ and $H = \begin{bmatrix} -2 & 4 \\ 1 & 3 \end{bmatrix}$, find the matrix products $GH,\ HG,\ G^2,\ H^2, G^2H^2$.

11. For $A = \begin{bmatrix} 1 & 3 & 2 \\ -1 & 0 & 1 \\ -2 & 1 & 3 \end{bmatrix}$ and $B = \begin{bmatrix} 2 & 1 & -1 \\ 1 & 3 & 0 \\ 0 & 1 & -2 \end{bmatrix}$, find AB, BA, A^2, and B^2.

12. For $A = \begin{bmatrix} 1 \\ 2 \\ -1 \end{bmatrix}$ and $B = \begin{bmatrix} 2 & 4 & -7 \end{bmatrix}$, find, if possible, AB and BA.

13. For $A = \begin{bmatrix} 1 & 1 \\ 1 & 1 \end{bmatrix}$ and $B = \begin{bmatrix} -1 & -1 \\ 1 & 1 \end{bmatrix}$, find AB.

14. For $A = \begin{bmatrix} 1 & 2 \\ 2 & 4 \end{bmatrix}$ and $B = \begin{bmatrix} 2 & 4 \\ -1 & -2 \end{bmatrix}$, find AB.

15. For $A = \begin{bmatrix} 2 & 4 & -1 \\ -4 & -8 & 2 \\ -2 & -4 & 1 \end{bmatrix}$ and $B = \begin{bmatrix} 4 & 1 & -1 \\ 1 & 1 & 1 \\ 12 & 6 & 2 \end{bmatrix}$, find AB.

16. For $A = \begin{bmatrix} a & b \\ c & d \end{bmatrix}$ and $I_2 = \begin{bmatrix} 1 & 0 \\ 0 & 1 \end{bmatrix}$, find AI_2 and $I_2 A$.

17. For $A = \begin{bmatrix} a & b & c \\ d & e & f \\ g & h & i \end{bmatrix}$ and $I_3 = \begin{bmatrix} 1 & 0 & 0 \\ 0 & 1 & 0 \\ 0 & 0 & 1 \end{bmatrix}$, find AI_3 and $I_3 A$.

18. $A = \begin{bmatrix} 2 & -1 \\ 1 & -2 \end{bmatrix}$, $X = \begin{bmatrix} 0 & 1 \\ 3 & 2 \end{bmatrix}$, and $Y = \begin{bmatrix} -1 & -1 \\ 1 & -2 \end{bmatrix}$, find AX and AY and

    compare.

19. For $A = \begin{bmatrix} 1 & 3 \\ -1 & 2 \end{bmatrix}$, $B = \begin{bmatrix} -1 & 3 \\ 2 & -1 \end{bmatrix}$, and $C = \begin{bmatrix} 2 & -1 \\ 3 & -2 \end{bmatrix}$,

    (a) Find $A(BC)$ and $(AB)C$ and compare.
    (b) Find $A(B+C)$ and $AB + AC$ and compare.
    (c) Find $(B+C)A$ and $BA + CA$ and compare.

20. For $A = \begin{bmatrix} 1 & 2 & -1 \\ 2 & -1 & 3 \\ 1 & 0 & 2 \end{bmatrix}$, $B = \begin{bmatrix} 2 & 0 & 1 \\ 1 & 3 & 2 \\ 0 & 1 & -1 \end{bmatrix}$, and $C = \begin{bmatrix} -2 & 2 & -1 \\ 3 & -1 & 0 \\ 2 & 0 & 1 \end{bmatrix}$,

    (a) Find BC, AB, and AC.
    (b) Find $A(BC)$ and $(AB)C$ and compare.
    (c) Find $A(B+C)$ and $AB + AC$ and compare.
    (d) Find $(B+C)A$ and $BA + CA$ and compare.

21. For $A = \begin{bmatrix} a & b \\ c & d \end{bmatrix}$, $B = \begin{bmatrix} e & f \\ g & h \end{bmatrix}$, and $C = \begin{bmatrix} j & k \\ m & n \end{bmatrix}$,

    (a) Find BC, AB, and AC.
    (b) Find $A(BC)$ and $(AB)C$ and show that they are equal.

(c) Find $A(B+C)$ and $AB+AC$ and show that they are equal.
(d) Find $(B+C)A$ and $BA+CA$ and show that they are equal.

22. This concerns the Sonin Company's situation described in section 4.6, Exercise 1, p. 196.

$$
\begin{array}{ccc}
S_1 & S_2 & S_3
\end{array}
$$

$$
P = \begin{bmatrix} 0.22 & 0.40 & 0.30 \\ 0.40 & 0.35 & 0.42 \end{bmatrix} \begin{matrix} P_1 \\ P_2 \end{matrix} \quad \text{and } C = \begin{bmatrix} 22,000 \\ 30,000 \\ 44,000 \end{bmatrix}
$$

Matrix P describes the fractions of the total costs of the service departments S_1, S_2, and S_3 of a firm that are assigned to its production departments P_1 and P_2. Matrix C describes the total costs of the service departments.

Find the matrix that describes the allocation of service departments' costs to the production departments.

23. The Karkin Steel Works operates out of two plants, P_1 and P_2, and produces two types of steel, a chromium-manganese-molybdenum alloy (CMM steel) and a chromium-silicon-tungsten alloy (CST steel). Of concern are three main types of pollutants—particulate matter, sulfur oxides, and hydrocarbons—which are by-products of the operation of the plants. Air-quality standards require that specified minimal amounts of pollutants be removed. These minimal amounts (in tons) for the two types of steel are shown in matrix B. The cost (in dollars) of removing 1 ton of each kind of pollutant at each plant is shown in matrix E.

Determine matrix BE and explain what it describes.

Parti-culates	Sulfur oxides	Hydro-carbons	
20	30	25	CMM steel
30	25	40	CST steel

$$
B = \begin{bmatrix} 20 & 30 & 25 \\ 30 & 25 & 40 \end{bmatrix}
$$

Plant P_1	Plant P_2	
200	220	Particulates
160	150	Sulfur oxides
180	200	Hydrocarbons

$$
E = \begin{bmatrix} 200 & 220 \\ 160 & 150 \\ 180 & 200 \end{bmatrix}
$$

11.2 PROPERTIES OF MATRICES

It is interesting to compare the properties of matrix addition and multiplication with those of real-number addition and multiplication. There are many similarities, but also some striking differences. First, from section 1.2, we recall some properties of real-number addition. All letters in the following list represent any real-numbers.

1. *Closure property of real-number addition:* If a and b are any numbers in the real-number system R, then the sum of a and b, denoted by $a+b$, is in R.

2. *Commutative property of real-number addition:* $a+b=b+a$.

3. *Associative property of real-number addition:* $(a+b)+c=a+(b+c)$.

4. *Zero element:* There is an unique real-number, called *zero* and denoted by 0, with the property $a+0=0+a$, where a is any real-number.

5. *Negative of a real-number:* If a is any real-number, there is an unique real-number, called the negative of a and denoted by $-a$, with the property $a+(-a)=(-a)+a=0$.

From the way in which matrix addition is defined, it is not difficult to see that it also has these properties. All letters in the following list denote any matrices of the same size.

Closure property of matrix addition. If A and B are any matrices of the same size, then there is an unique matrix of that size, denoted by $A+B$, called the sum of A and B.
   As we have seen, $A+B$ is obtained by adding corresponding entries of A and B.

Commutative property of matrix addition. $A+B=B+A$.
   This follows from the definition of matrix addition and the commutativity of real-number addition.

Associative property of matrix addition. $(A+B)+C=A+(B+C)$.
   This follows from the definition of matrix addition and the associativity of real-number addition.

Zero matrix. There is an unique matrix, called the **zero matrix** and denoted by $\overline{0}$, with the property $A + \overline{0} = \overline{0} + A = A$.

$\overline{0}$ is the matrix that is the same size as A and whose entries are all zeros. Thus for the class of 2 by 2 matrices:

$$\overline{0} = \begin{bmatrix} 0 & 0 \\ 0 & 0 \end{bmatrix}$$

For the class of 3 by 3 matrices:

$$\overline{0} = \begin{bmatrix} 0 & 0 & 0 \\ 0 & 0 & 0 \\ 0 & 0 & 0 \end{bmatrix}$$

Negative of a matrix, also called the **additive inverse of** a matrix. If A is any matrix, then there is an unique matrix, called the **negative of A** or **additive inverse of A** and denoted by $-A$, with the property $A + (-A) = (-A) + A = \overline{0}$. For example, if

$$A = \begin{bmatrix} a & b \\ c & d \end{bmatrix}$$

then the **negative of** A is

$$-A = \begin{bmatrix} -a & -b \\ -c & -d \end{bmatrix}.$$

Thus the **negative of** $A = [4 \quad -2]$ is $-A = [-4 \quad 2]$.

From section 1.2 we have the following properties of real-number multiplication.

1. *Closure property of real-number multiplication:* If a and b are any real-numbers in the real-number system R, then the product of a and b, denoted by $a \cdot b$ or ab, is in R.

2. *Commutative property of real-number multiplication:* $ab = ba$.

3. *Associative property of real-number multiplication:* $(ab)c = a(bc)$.

4. *Identity element:* There is an unique real-number, called one and denoted by 1, with the property $1a = a1 = a$, where a is any real-number.

5. *Inverse property:* If a is any nonzero real-number, then there is an unique real-number, called the multiplicative inverse of a, or reciprocal of a, and denoted by a^{-1}, such that $aa^{-1} = a^{-1}a = 1$.

6. *Nonzero products of nonzero factors property:* The product of any two nonzero real-numbers is nonzero; that is, if $a \neq 0$ and $b \neq 0$, then $ab \neq 0$.

7. *Cancellation property:* If $ax = ay$, where $a \neq 0$, then $x = y$.

In considering the behavior of matrix multiplication, we restrict ourselves to square matrices of the same size to ensure that all matrix products are at least defined. In the following discussion all letters denote any square matrices of the same size.

Closure property of matrix multiplication. If A and B are square matrices of the same size, then there is an unique square matrix of that size, denoted by AB, called the product of A and B. This follows from the way in which matrix multiplication is defined.

Commutativity of matrix multiplication? NO. Matrix multiplication is not commutative since $AB = BA$ does *not* hold for all square matrices of the same size.

For example, if

$$A = \begin{bmatrix} 1 & 3 \\ 2 & -1 \end{bmatrix} \quad \text{and} \quad B = \begin{bmatrix} 1 & 4 \\ -2 & -3 \end{bmatrix}$$

then $AB \neq BA$ since

$$AB = \begin{bmatrix} -5 & -5 \\ -2 & -3 \end{bmatrix} \quad \text{whereas} \quad BA = \begin{bmatrix} 9 & -1 \\ -8 & -3 \end{bmatrix}.$$

Associative property of matrix multiplication. $(AB)C = A(BC)$. Associativity for the class of 2 by 2 matrices is established in Exercise 21(b) of section 11.1.

Identity matrices. The analog for the class of 2 by 2 matrices, for example, of the identity element 1 of the real-number system is a matrix, call it I_2, with the property that $AI_2 = I_2 A = A$, where A is any 2 by 2 matrix. For the class of 2 by 2 matrices, the **identity matrix** I_2 is defined by:

$$I_2 = \begin{bmatrix} 1 & 0 \\ 0 & 1 \end{bmatrix}$$

For any 2 by 2 matrix $A = \begin{bmatrix} a & b \\ c & d \end{bmatrix}$, $AI_2 = I_2 A = A$ (see Exercise 16 of Section 11.1).

For the class of 3 by 3 matrices, the **identity matrix** is

$$I_3 = \begin{bmatrix} 1 & 0 & 0 \\ 0 & 1 & 0 \\ 0 & 0 & 1 \end{bmatrix}$$

since for any 3 by 3 matrix $AI_3 = I_3 A = A$ (see Exercise 17 of Section 11.1).

More generally, for the class of n by n matrices, **the identity matrix is the n by n matrix**

$$I_n = \begin{bmatrix} 1 & 0 & 0 & \cdots & 0 \\ 0 & 1 & 0 & \cdots & 0 \\ 0 & 0 & 1 & \cdots & 1 \\ & \cdot & \cdot & & \cdot \\ & \cdot & \cdot & & \cdot \\ & \cdot & \cdot & & \cdot \\ 0 & 0 & 0 & \cdots & 1 \end{bmatrix}$$

whose entries are 0's except for 1's that run down the main diagonal.

Multiplicative Inverse of a matrix. If $A \neq \overline{0}$ is an n by n matrix, and there is an n by n matrix B with the property $AB = BA = I_n$ then B is denoted by A^{-1}, is called the **multiplicative inverse of A,** and A is said to be **nonsingular.**

Thus, for example, the **multiplicative inverse of**

$$A = \begin{bmatrix} 1 & 3 \\ -1 & 2 \end{bmatrix} \quad \text{is} \quad B = \begin{bmatrix} \dfrac{2}{5} & -\dfrac{3}{5} \\ \dfrac{1}{5} & \dfrac{1}{5} \end{bmatrix}$$

since $AB = BA = I_2$, the identity matrix for the class of 2 by 2 matrices.

If the multiplicative inverse of a matrix A does not exist, then A is said to be a **singular matrix.**

Many nonzero square matrices are singular. Consider, for example,

$$A = \begin{bmatrix} 1 & 0 \\ 0 & 0 \end{bmatrix}.$$

There is no 2 by 2 matrix B such that $AB = BA = I_2$. To verify this, let $B = \begin{bmatrix} a & b \\ c & d \end{bmatrix}$ and consider the product AB:

$$AB = \begin{bmatrix} 1 & 0 \\ 0 & 0 \end{bmatrix} \begin{bmatrix} a & b \\ c & d \end{bmatrix} = \begin{bmatrix} a & b \\ 0 & 0 \end{bmatrix}$$

Clearly, $AB = \begin{bmatrix} a & b \\ 0 & 0 \end{bmatrix}$ cannot be made equal to $I_2 = \begin{bmatrix} 1 & 0 \\ 0 & 1 \end{bmatrix}$, no matter how a and b are chosen.

Distributive property. A link between addition and multiplication that holds for real-numbers as well as matrices is the distributive property. If a, b, and c are any real-numbers, then $a(b+c) = ab + ac$. If A, B and C are any square matrices of the same size, then $A(B+C) = AB + AC$ and $(B+C) = BA + CA$.

The special case of the distributive property for the class of 2 by 2 matrices is established in Exercises 21(c) and 21(d) of section 11.1.

Some Matrix Surprises

> **Products of nonzero matrices.** If A and B are nonzero matrices, it does not follow that their product is a nonzero matrix.

For example, for:

$$A = \begin{bmatrix} 1 & 2 \\ 2 & 4 \end{bmatrix} \quad \text{and} \quad B = \begin{bmatrix} 2 & 4 \\ -1 & -2 \end{bmatrix}, \quad AB = \begin{bmatrix} 0 & 0 \\ 0 & 0 \end{bmatrix}$$

> **Matrix Cancellation? NO.** The cancellation property does not hold for matrices. If $AX = AY$, where $A \neq \overline{0}$, it does not follow that $X = Y$.

For example, if A, X, and Y are defined by

$$A = \begin{bmatrix} 2 & -1 \\ 4 & -2 \end{bmatrix}, \quad X = \begin{bmatrix} 0 & 1 \\ 3 & 2 \end{bmatrix}, \quad Y = \begin{bmatrix} -1 & -1 \\ 1 & -2 \end{bmatrix}$$

then $AX = AY$, but $X \neq Y$.

EXERCISES

1. What is the zero matrix for the class of 4 by 2 matrices? Explain.

2. What is the negative of $\begin{bmatrix} 3 & 1 \\ -2 & 2 \\ 1 & -4 \\ -3 & 6 \end{bmatrix}$? Explain.

3. What is the identity matrix for the class of 4 by 4 matrices? Explain.

4. Is $\begin{bmatrix} -\dfrac{1}{4} & \dfrac{3}{8} \\ \dfrac{1}{2} & -\dfrac{1}{4} \end{bmatrix}$ the multiplicative inverse of $\begin{bmatrix} 2 & 3 \\ 4 & 2 \end{bmatrix}$? Explain.

5. Is $\begin{bmatrix} 2 & \dfrac{1}{2} \\ \dfrac{1}{3} & -1 \end{bmatrix}$ the multiplicative inverse of $\begin{bmatrix} 4 & 1 \\ 2 & 3 \end{bmatrix}$? Explain.

6. Is $\begin{bmatrix} \dfrac{1}{2} & \dfrac{1}{2} & -\dfrac{1}{2} \\ -\dfrac{1}{2} & \dfrac{1}{6} & \dfrac{1}{6} \\ \dfrac{1}{2} & -\dfrac{1}{2} & \dfrac{1}{2} \end{bmatrix}$ the multiplicative inverse of $\begin{bmatrix} 1 & 0 & 1 \\ 2 & 3 & 1 \\ 1 & 3 & 2 \end{bmatrix}$? Explain.

11.3 MATRIX INVERSION

A matrix A is nonsingular if it has a multiplicative inverse A^{-1}. As we have noted,

$A = \begin{bmatrix} 1 & 3 \\ -1 & 2 \end{bmatrix}$ is nonsingular with multiplicative inverse $A^{-1} = \begin{bmatrix} \dfrac{2}{5} & -\dfrac{3}{5} \\ \dfrac{1}{5} & \dfrac{1}{5} \end{bmatrix}$. But

how is A^{-1} determined? This is the problem we now take up. To begin, we again turn to the 2 by 2 matrix:

$$A = \begin{bmatrix} 1 & 3 \\ -1 & 2 \end{bmatrix}$$

By placing the identity matrix $I_2 = \begin{bmatrix} 1 & 0 \\ 0 & 1 \end{bmatrix}$ next to A, we obtain the tableau of numbers:

$$[A\,|\,I_2] = \begin{bmatrix} 1 & 3 & \vdots & 1 & 0 \\ -1 & 2 & \vdots & 0 & 1 \end{bmatrix}$$

To obtain A^{-1} we employ the following row operations, familiar from the tableau method for solving systems of linear equations (section 4.4), on the tableau $[A \mid I_2]$ to convert the A part of $[A \mid I_2]$ to I_2; in the course of doing so, the I_2 part of $[A \mid I_2]$ will be converted to A^{-1}.

> **Pivoting.** To convert a chosen number c to 1, multiply the row containing c by $1/c$.
>
> **Conversion to Zero.** To convert a number n to 0, multiply the one row (that is, the row containing 1 from the pivoting operation) by n and add the resulting row to the row containing n.
>
> **Interchanging Rows.** Any two rows in a tableau may be interchanged.

The sequence of inversion tableaus that leads to the multiplicative inverse of

$$A = \begin{bmatrix} 1 & 3 \\ -1 & 2 \end{bmatrix}$$

is shown in Figure 11.3.

$[A \mid I_2]$	(1)	(1)	3	1	0	
	(2)	-1	2	0	1	
T_2	(3)	1	3	1	0	row (1)
	(4)	0	(5)	1	1	row (1) + row (2)
T_3	(5)	1	0	$\frac{2}{5}$	$-\frac{3}{5}$	(-3)row (6) + row (3)
	(6)	0	1	$\frac{1}{5}$	$\frac{1}{5}$	$(\frac{1}{5})$row (4)

Figure 11.3

From tableau T_3 we see that the conversion of A to I_2 has been completed, and that I_2 in turn has been converted to:

$$A^{-1} = \begin{bmatrix} \dfrac{2}{5} & -\dfrac{3}{5} \\ \dfrac{1}{5} & \dfrac{1}{5} \end{bmatrix}$$

EXAMPLE 1

Find, provided that it exists, the multiplicative inverse of:

$$A = \begin{bmatrix} 2 & 4 \\ 1 & -1 \end{bmatrix}$$

By placing I_2 next to A we obtain the initial inversion tableau $[A \mid I_2]$.

$$[A \mid I_2] = \begin{bmatrix} 2 & 4 & \vdots & 1 & 0 \\ 1 & -1 & \vdots & 0 & 1 \end{bmatrix}$$

The sequence of inversion tableaus in which the A part of $[A \mid I_2]$ is converted to I_2 and the I_2 part of $[A \mid I_2]$ is converted to A^{-1} is displayed in Figure 11.4.

Figure 11.4

The simplest way to begin the conversion of the A part of $[A \mid I_2]$ to I_2 is to interchange rows ① and ② in tableau $[A \mid I_2]$; this interchange yields tableau

T_2 in Figure 11.4. From tableau T_4 we see that the conversion of the A part of $[A \mid I_2]$ to I_2 has been completed, and that I_2 has been converted to

$$A^{-1} = \begin{bmatrix} \dfrac{1}{6} & \dfrac{4}{6} \\ \dfrac{1}{6} & -\dfrac{2}{6} \end{bmatrix}.$$

As a check against error, we verify that $AA^{-1} = A^{-1}A = I_2$.

$$\begin{bmatrix} 2 & 4 \\ 1 & -1 \end{bmatrix}\begin{bmatrix} \dfrac{1}{6} & \dfrac{4}{6} \\ \dfrac{1}{6} & -\dfrac{2}{6} \end{bmatrix} = \begin{bmatrix} 1 & 0 \\ 0 & 1 \end{bmatrix}$$

$$\begin{bmatrix} \dfrac{1}{6} & \dfrac{4}{6} \\ \dfrac{1}{6} & -\dfrac{2}{6} \end{bmatrix}\begin{bmatrix} 2 & 4 \\ 1 & -1 \end{bmatrix} = \begin{bmatrix} 1 & 0 \\ 0 & 1 \end{bmatrix}$$

EXAMPLE 2

Find, provided that it exists, the multiplicative inverse of

$$E = \begin{bmatrix} 1 & 0 & 1 \\ 2 & 3 & 1 \\ 1 & 3 & 2 \end{bmatrix}.$$

By placing I_3, the identity matrix for the class of 3 by 3 matrices, next to E, we obtain the initial inversion tableau $[E \mid I_3]$ shown in Figure 11.5.

$[E\,|\,I_3]$	①	$\begin{matrix}①\end{matrix}$	0	1	1	0	0	
	②	2	3	1	0	1	0	
	③	1	3	2	0	0	1	
T_2	④	1	0	1	1	0	0	row ①
	⑤	0	3	−1	−2	1	0	(−2)row ④ + row ②
	⑥	0	3	①	−1	0	1	(−1)row ④ + row ③
T_3	⑦	1	−3	0	2	0	−1	(−1)row ⑥ + row ④
	⑧	0	⑥	0	−3	1	1	row ⑥ + row ⑤
	⑨	0	3	1	−1	0	1	row ⑥
T_4	⑩	1	0	0	$\frac{1}{2}$	$\frac{1}{2}$	$-\frac{1}{2}$	(3)row ⑪ + row ⑦
	⑪	0	1	0	$-\frac{1}{2}$	$\frac{1}{6}$	$\frac{1}{6}$	$(\frac{1}{6})$row ⑧
	⑫	0	0	1	$\frac{1}{2}$	$-\frac{1}{2}$	$\frac{1}{2}$	(−3)row ⑪ + row ⑨

Figure 11.5

In going from tableau T_2 to T_3, we pivot on 1 in row ⑥ and column (3), rather than on 3 in row ⑤ and column (2). This is for the sake of convenience; to obtain I_3, we need a 1 in that position, and it makes sense to take advantage of good fortune when it comes our way. Although 1's must be placed in the main diagonal, they need not be placed there in order. From tableau T_4 in Figure 11.5, we see that the conversion of the E part of $[E\,|\,I_3]$ has been completed, and that I_3 has been converted to E^{-1} given by:

$$
E^{-1} = \begin{bmatrix} \dfrac{1}{2} & \dfrac{1}{2} & -\dfrac{1}{2} \\[2mm] -\dfrac{1}{2} & \dfrac{1}{6} & \dfrac{1}{6} \\[2mm] \dfrac{1}{2} & -\dfrac{1}{2} & \dfrac{1}{2} \end{bmatrix}
$$

As a check against error, this result can be verified by showing that $EE^{-1} = E^{-1}E = I_3$.

EXAMPLE 3

Find, provided that it exists, the multiplicative inverse of

$$A = \begin{bmatrix} 1 & 2 \\ 2 & 4 \end{bmatrix}.$$

The sequence of inversion tableaus is displayed in Figure 11.6. The entry in row ④, column (2) is 0, and thus cannot be converted to 1 by pivoting. Thus the A part of $[A \,|\, I_2]$ cannot be converted to I_2 which means that **matrix A is singular and does not have a multiplicative inverse.**

Such is the case in general; if a matrix A cannot be reduced to the identity matrix by means of the row operations described, then A does not have a multiplicative inverse.

$$[A\,|\,I_2] \quad \begin{matrix} ① \\ ② \end{matrix} \begin{bmatrix} ① & 2 & \vdots & 1 & 0 \\ 2 & 4 & \vdots & 0 & 1 \end{bmatrix}$$

$$T_2 \quad \begin{matrix} ③ \\ ④ \end{matrix} \begin{bmatrix} 1 & 2 & \vdots & 1 & 0 \\ 0 & 0 & \vdots & -2 & 1 \end{bmatrix} \quad \begin{matrix} \text{row ①} \\ (-2)\text{row ③} + \text{row ②} \end{matrix}$$

Figure 11.6

EXERCISES

Find, provided that it exists, the multiplicative inverse of each of the following matrices.

1. $\begin{bmatrix} 1 & 3 \\ 4 & -1 \end{bmatrix}$ 2. $\begin{bmatrix} 1 & -2 \\ 2 & 3 \end{bmatrix}$ 3. $\begin{bmatrix} 2 & 3 \\ 4 & 1 \end{bmatrix}$

4. $\begin{bmatrix} 2 & 4 \\ 1 & 2 \end{bmatrix}$ 5. $\begin{bmatrix} 1 & 0 & 0 \\ 2 & 2 & 0 \\ 5 & 5 & 5 \end{bmatrix}$ 6. $\begin{bmatrix} 2 & -1 & 0 \\ 1 & 0 & 1 \\ 1 & -2 & 0 \end{bmatrix}$

7. $\begin{bmatrix} 2 & 0 & 1 \\ 1 & 1 & 3 \\ 1 & -1 & -2 \end{bmatrix}$
8. $\begin{bmatrix} 1 & 1 & 2 \\ 2 & 0 & 2 \\ 3 & 0 & 1 \end{bmatrix}$
9. $\begin{bmatrix} 1 & 2 & 3 \\ 3 & 2 & 1 \\ 5 & 4 & 5 \end{bmatrix}$

10. $\begin{bmatrix} 2 & -1 & 3 \\ 1 & -2 & 1 \\ -1 & 1 & 2 \end{bmatrix}$
11. $\begin{bmatrix} 2 & 3 & -1 \\ 1 & -2 & 3 \\ 3 & -1 & -2 \end{bmatrix}$
12. $\begin{bmatrix} 1 & -3 & 2 \\ 3 & -4 & 1 \\ 2 & -1 & -1 \end{bmatrix}$

11.4 MATRIX SOLUTIONS TO SYSTEMS OF LINEAR EQUATIONS

There are many methods for solving systems of linear equations, with each one possessing its own advantages and disadvantages. One of these methods is based on matrix inversion. To illustrate, consider the problem of determining x_1, x_2, and x_3 so as to satisfy the following system:

$$\begin{aligned} x_1 \qquad + x_3 &= 4 \qquad\qquad (11.1) \\ 2x_1 + 3x_2 + x_3 &= 6 \qquad\qquad (11.2) \\ x_1 + 3x_2 + 2x_3 &= 12 \qquad\qquad (11.3) \end{aligned}$$

If we define matrices E, X, and B by

$$E = \begin{bmatrix} 1 & 0 & 1 \\ 2 & 3 & 1 \\ 1 & 3 & 2 \end{bmatrix}, \qquad X = \begin{bmatrix} x_1 \\ x_2 \\ x_3 \end{bmatrix}, \qquad B = \begin{bmatrix} 4 \\ 6 \\ 12 \end{bmatrix}$$

then

$$EX = \begin{bmatrix} x_1 \qquad + x_3 \\ 2x_1 + 3x_2 + x_3 \\ x_1 + 3x_2 + 2x_3 \end{bmatrix}$$

and the problem, in matrix terms, is to find X such that:

$$EX = B$$

If E^{-1} exists, then by multiplying both sides of $EX = B$ on the left by E^{-1}, we obtain:

$$E^{-1}EX = E^{-1}B$$
$$I_3X = E^{-1}B$$

$$X = E^{-1}B$$

Thus, if E, the matrix of coefficients of the system, is nonsingular, then the system has exactly one solution and that solution is given by $E^{-1}B$.

From Example 2 of section 11.3, we have

$$E^{-1} = \begin{bmatrix} \dfrac{1}{2} & \dfrac{1}{2} & -\dfrac{1}{2} \\ -\dfrac{1}{2} & \dfrac{1}{6} & \dfrac{1}{6} \\ \dfrac{1}{2} & \dfrac{1}{2} & -\dfrac{1}{2} \end{bmatrix}.$$

Thus

$$X = E^{-1}B = \begin{bmatrix} \dfrac{1}{2} & \dfrac{1}{2} & -\dfrac{1}{2} \\ -\dfrac{1}{2} & \dfrac{1}{6} & \dfrac{1}{6} \\ \dfrac{1}{2} & \dfrac{1}{2} & -\dfrac{1}{2} \end{bmatrix} \begin{bmatrix} 4 \\ 6 \\ 12 \end{bmatrix} = \begin{bmatrix} -1 \\ 1 \\ 5 \end{bmatrix}$$

and we obtain $x_1 = -1$, $x_2 = 1$, $x_3 = 5$. It is easily verified, as a check against error, that $(-1, 1, 5)$ is the solution of system (11.1) through (11.3).

More generally, consider the n by n system (n equations and n unknowns) of linear equations.

$$c_{11}x_1 + c_{12}x_2 + \cdots + c_{1n}x_n = b_1$$
$$c_{21}x_1 + c_{22}x_2 + \cdots + c_{2n}x_n = b_2$$

$$\cdot \qquad \cdot$$
$$\cdot \qquad \cdot$$
$$\cdot \qquad \cdot$$

$$c_{n1}x_1 + c_{n2}x_2 + \cdots + c_{nn}x_n = b_n$$

and define matrices:

$$E = \begin{bmatrix} c_{11} & c_{12} & \cdots & c_{1n} \\ c_{21} & c_{22} & \cdots & c_{2n} \\ \cdot & \cdot & & \cdot \\ \cdot & \cdot & & \cdot \\ \cdot & \cdot & & \cdot \\ c_{n1} & c_{n2} & \cdots & c_{nn} \end{bmatrix}, \quad X = \begin{bmatrix} x_1 \\ x_2 \\ \cdot \\ \cdot \\ \cdot \\ x_n \end{bmatrix}, \quad B = \begin{bmatrix} b_1 \\ b_2 \\ \cdot \\ \cdot \\ \cdot \\ b_n \end{bmatrix}$$

E is the coefficient matrix of the system, X is the matrix of unknowns, and B is the matrix of constants. In terms of matrices, the problem is to find X such that

$$EX = B.$$

If E^{-1} exists, then the system has an unique solution, which is given by:

$$\boxed{X = E^{-1}B}$$

If E^{-1} does not exist, then the system either has no solution or infinitely many solutions. Let us observe that the existence of an unique solution depends only on the existence of E^{-1} and thus on the nature of the coefficient matrix E, and not on the nature of the matrix of constants B.

The matrix-inversion approach to solving n by n systems with an unique solution is advantageous in situations in which the coefficient matrix E does not change, but the matrix of constants B is to be varied. As long as E does not change, only one matrix inversion is required, and through the solution equation $X = E^{-1}B$ solutions corresponding to a number of possible B matrices can easily be obtained.

We examine such situations in the next two sections.

EXERCISES

Solve each of the following systems of linear equations by means of matrix inversion.

1. $\begin{aligned} -x_1 + 2x_2 &= 4 \\ 3x_1 + x_2 &= 10 \end{aligned}$ 2. $\begin{aligned} x_1 - 2x_2 &= 6 \\ 2x_1 + 3x_2 &= 9 \end{aligned}$ 3. $\begin{aligned} 3x + 5y &= 8 \\ x + 3y &= 6 \end{aligned}$

4. $\begin{aligned} x_1 + x_2 + 2x_3 &= 3 \\ 2x_1 + 2x_3 &= 4 \\ 3x_1 + x_3 &= 6 \end{aligned}$ 5. $\begin{aligned} 2x_1 + 3x_2 - x_3 &= 11 \\ x_1 - 2x_2 + 3x_3 &= 2 \\ 3x_1 - x_2 - 2x_3 &= 5 \end{aligned}$ 6. $\begin{aligned} x_1 - x_2 - x_3 &= -4 \\ 3x_1 - 5x_2 + 5x_3 &= 6 \\ 2x_1 + 3x_2 + 4x_3 &= 7 \end{aligned}$

11.5   RETURN TO SERVICE CHARGE ALLOCATION AND INCOME CONSOLIDATION PROBLEMS

Service Charge Allocation

In Example 1 of section 4.6 (p. 196), we saw that the problem of determining the total costs of the service departments of the Arkin Company leads to the system of equations

$$x - 0.1y \quad -0.1w = 20,000$$
$$-0.02x + \quad y \quad -0.1w = 18,000$$
$$-0.1x - 0.2y + z - 0.1w = 80,000$$
$$-0.1x - 0.1y \quad + \quad w = 10,000$$

where x, y, z, and w denote the total costs of the accounting, maintenance, marketing, and purchasing departments, respectively, and $20,000, $18,000, $80,000 and $10,000 are the respective overhead costs of these departments for the month of January.

In matrix terms we have

$$EX = B$$

where:

$$E = \begin{bmatrix} 1 & -0.1 & 0 & -0.1 \\ -0.02 & 1 & 0 & -0.1 \\ -0.1 & -0.2 & 1 & -0.1 \\ -0.1 & -0.1 & 0 & 1 \end{bmatrix}, \quad X = \begin{bmatrix} x \\ y \\ z \\ w \end{bmatrix}, \quad B = \begin{bmatrix} 20,000 \\ 18,000 \\ 80,000 \\ 10,000 \end{bmatrix}$$

For E^{-1} we obtain:

$$E^{-1} = \begin{bmatrix} 1.0135134 & 0.1126125 & 0 & 0.1126125 \\ 0.0307124 & 1.0135134 & 0 & 0.1044225 \\ 0.1179360 & 0.2252251 & 1 & 0.1343160 \\ 0.1044225 & 0.1126125 & 0 & 1.0217034 \end{bmatrix}$$

Thus

$$X = E^{-1}B = \begin{bmatrix} 23,423.43 \\ 19,901.72 \\ 87,755.93 \\ 14,332.51 \end{bmatrix}$$

describes the total costs of the service departments.

> If for each service department there is **no change** in the percentage of its total cost assigned to the other service departments, so that matrix E, which reflects these percentages, is unchanged, then with one matrix inversion the total costs of the service departments can be obtained for **each financial period** as the matrix B, expressing overhead costs, changes.

For further discussion of the cost-allocation problem and the use of matrix methods, consult the following literature:

[1] N. CHURCHILL. "Linear Algebra and Cost Allocations: Some Examples." *Accounting Review*, vol. 39, no. 4 (October 1964), pp. 894-904.

[2] R. S. KAPLAN. "Variable and Self-Service Costs in Reciprocal Allocation Models." *Accounting Review*, vol. 48, no. 4 (October 1973), pp. 738-748.

[3] R. P. MANES. "Comment on Matrix Theory and Cost Allocation." *Accounting Review*, vol. 40, no. 3 (July 1965), pp. 640-643.

[4] R. MINCH and E. PETRI. "Matrix Models of Reciprocal Service Cost Allocation." *Accounting Review*, vol. 47, no. 3 (July 1972), pp. 576-580.

[5] T. H. WILLIAMS and C. H. GRIFFIN. "Matrix Theory and Cost Allocation." *Accounting Review*, vol. 39, no. 3 (July 1964), pp. 671-678.

Income Consolidation

In Example 2 of section 4.6 (p. 199) we saw that the problem of determining the net incomes of the Russel, Ferrara, and Thomas Companies on a consolidated basis leads to the system of equations

$$x - 0.8y - 0.6z = 100,000$$
$$y - 0.2z = 80,000$$
$$-0.2x - 0.1y + z = 60,000$$

where x, y, and z denote the net incomes of the Russel, Ferrara, and Thomas Companies, respectively, on a consolidated basis, and $100,000, $80,000 and $60,000 are the respective net incomes of these companies from their own operations.

In matrix terms this system is expressed by

$$EX = B$$

where:

$$E = \begin{bmatrix} 1 & -0.8 & -0.1 \\ 0 & 1 & -0.1 \\ -0.1 & -0.1 & 1 \end{bmatrix}, \quad X = \begin{bmatrix} x \\ y \\ z \end{bmatrix}, \quad B = \begin{bmatrix} 100,000 \\ 80,000 \\ 60,000 \end{bmatrix}$$

Calculating E^{-1} yields

$$E^{-1} = \begin{bmatrix} 1.183575 & 1.038647 & 0.917874 \\ 0.048309 & 1.062802 & 0.241546 \\ 0.241546 & 0.314009 & 1.207729 \end{bmatrix}.$$

We thus obtain:

$$X = E^{-1}B = \begin{bmatrix} 1.183575 & 1.038647 & 0.917874 \\ 0.048309 & 1.062802 & 0.241546 \\ 0.241546 & 0.314009 & 1.207729 \end{bmatrix} \begin{bmatrix} 100,000 \\ 80,000 \\ 60,000 \end{bmatrix} = \begin{bmatrix} 256,521.7 \\ 104,347.8 \\ 121,739.1 \end{bmatrix}$$

As is not surprising, this result agrees with the one obtained in Example 2 of section 4.6.

The advantage of this solution procedure over the other is that if there are no changes in the intercorporate shareholdings, so that matrix E is not changed, then with one matrix inversion the consolidated basis net incomes can easily be obtained for each financial period (month, quarter, etc.) as the matrix B, expressing the net incomes of these affiliates from their own operations, changes.

Thus, for example, if in the next financial period the net incomes of these affiliates from their own operations were $110,000, $90,000 and $50,000, respectively, then the consolidated basis net incomes would simply be obtained as follows:

$$X = E^{-1}B = \begin{bmatrix} 1.183575 & 1.038647 & 0.917874 \\ 0.048309 & 1.062802 & 0.241546 \\ 0.241546 & 0.314009 & 1.207729 \end{bmatrix} \begin{bmatrix} 110,000 \\ 90,000 \\ 50,000 \end{bmatrix} = \begin{bmatrix} 269,565.2 \\ 113,043.5 \\ 115,217.3 \end{bmatrix}$$

EXERCISES

1. In connection with the service-charge-allocation problem, determine the costs of the accounting, maintenance, marketing and purchasing departments (a) if the overhead costs of these departments are $22,000, $20,000, $76,000, and $12,000, respectively; (b) if the overhead costs of these departments are $24,000, $22,000, $90,000, and $14,000, respectively.

2. In Exercise 1 of section 4.6 (p. 196), the fraction of the total cost of each service department of the Sonin Company that is assigned to the service and production departments of the firm is specified. The overhead of the service departments for the month of March is also given.

   (a) State, in matrix terms, the system of linear equations that determines the total costs of the service departments.

   (b) Solve this system by matrix inversion to determine these total costs.

   (c) If the overhead of S_1, S_2 and S_3 for April is $42,000, $36,000, and $24,000, respectively, and the assignment of costs to the service and production departments remains the same, determine the total costs of the service departments.

   (d) State, in terms of a suitable matrix product, the allocation of the total costs of the service departments to the production departments.

3. In Exercise 2 of section 4.6 (p. 199), the interdependency structure between the Ramunė, Algis, and Charles companies is shown.

   (a) State in matrix terms the system of linear equations that determines the net incomes of these affiliate companies on a consolidated basis.

(b) Solve this system by matrix inversion.

(c) If the net incomes of these companies are $90,000, $70,000, and $70,000, respectively, find the net incomes of these affiliate companies on a consolidated basis.

11.6 LEONTIEF INPUT-OUTPUT MODELS

Input-output models for economic systems were pioneered by the economist Wassily Leontief, a recipient of the 1973 Nobel Prize in Economics. In input-output analysis an economic system is viewed as a collection of interacting industries in which each industry produces an output that serves as raw materials, or input, for the industries of the system and requires input from the industries of the system. Let a_{ij} denote the amount of input (dollar's worth) of commodity i needed to produce $1 worth of commodity j; the first subscript refers to input, the second to output. Thus, for example, the equation $a_{21} = 0.20$ asserts that 20¢ worth of commodity 2 is needed to produce $1 worth of commodity 1. For an n-industry economy, the matrix

$$A = \begin{bmatrix} a_{11} & a_{12} & \cdots & a_{1n} \\ a_{21} & a_{22} & \cdots & a_{2n} \\ . & & & . \\ . & & & . \\ . & & & . \\ a_{n1} & a_{n2} & \cdots & a_{nn} \end{bmatrix}$$

called the **input-coefficient matrix** of the system, specifies the amount of each commodity that is needed to produce $1 worth of each commodity. The entries in the first column, for example, specify the inputs required from each of the n industries to produce $1 worth of the commodity produced by industry 1. The entries in the first row specify the amount of the commodity provided by industry 1 needed to produce $1 worth of the commodities produced by the n industries of the system.

We also **assume** that there is an **open sector** in the economy (consisting of households, for example) that absorbs a noninput demand for the product of each industry and supplies the primary input, labor. Let $d_1, d_2, \ldots, d_n$ denote the demand of the open sector for the commodities produced by industries $1, 2, \ldots,$

n and let $x_1, x_2, \ldots, x_n$ denote the **total output (dollar's worth) of industries 1, 2, \ldots, n**. The product $a_{ij}x_j$ is (the amount of commodity i needed to produce $1 worth of commodity j) x (total dollar's worth of commodity j produced), and thus expresses the input requirement of industry j for commodity i. For example, if $a_{ij} = 0.20$ (20¢ worth of commodity i is needed to produce $1 worth of commodity j) and $x_j = 5000$ ($5000 worth of commodity j is produced), then $0.20(5000) = \$1000$ worth of commodity i is needed to produce commodity j. Thus $a_{11}x_1$ is the input requirement of industry 1 for commodity 1, $a_{12}x_2$ is the input requirement of industry 2 for commodity 1, $a_{13}x_3$ is the input requirement of industry 3 for commodity 1, and so on. The sum

$$a_{11}x_1 + a_{12}x_2 + \cdots + a_{1n}x_n + d_1$$

is the sum of the input requirements of the n industries and the open sector for commodity 1. For x_1, the total output of industry 1, to satisfy this demand, we must have:

$$x_1 = a_{11}x_1 + a_{12}x_2 + \cdots + a_{1n}x_n + d_1$$

Similarly, for x_2, the total output of industry 2, to satisfy the demand for commodity 2, we must have:

$$x_2 = a_{21}x_1 + a_{22}x_2 + \cdots + a_{2n}x_n + d_2$$

More generally, for x_n, the total output of industry n, to satisfy the demand for commodity n, we must have:

$$x_n = a_{n1}x_1 + a_{n2}x_2 + \cdots + a_{nn}x_n + d_n$$

Thus the conditions that must be satisfied by output levels $x_1, x_2, \ldots, x_n$ of the n industries in the economy to satisfy the demands of the open sector and the industries themselves are expressed by the following system of n equations:

$$x_1 = a_{11}x_1 + a_{12}x_2 + \cdots + a_{1n}x_n + d_1$$
$$x_2 = a_{21}x_1 + a_{22}x_2 + \cdots + a_{2n}x_n + d_2$$
$$\vdots \qquad\qquad \vdots$$
$$x_n = a_{n1}x_1 + a_{n2}x_2 + \cdots + a_{nn}x_n + d_n$$

Rewriting this system so that terms involving x_1, $x_2, \ldots,$ x_n appear on one side and the constants d_1, $d_2, \ldots,$ d_n appear on the other side yields:

$$
\begin{aligned}
(1-a_{11})x_1 - \quad & a_{12}x_2 - \cdots - a_{1n}x_n = d_1 \\
-a_{21}x_1 + (1-a_{22})x_2 & - \cdots - a_{2n}x_n = d_2 \\
& \vdots \\
-a_{n1}x_1 - a_{n2}x_2 - \cdots & + (1-a_{nn}x_n) = d_n
\end{aligned}
\tag{11.4}
$$

It is advantageous, as we shall see, to express this system in terms of a matrix equation involving a matrix product. To do so we introduce matrices I_n, X, and D, as follows:

$$
I_n = \begin{bmatrix} 1 & 0 & 0 & \cdots & 0 \\ 0 & 1 & 0 & \cdots & 0 \\ 0 & 0 & 1 & \cdots & 0 \\ \cdot & \cdot & \cdot & & \cdot \\ \cdot & \cdot & \cdot & & \cdot \\ \cdot & \cdot & \cdot & & \cdot \\ 0 & 0 & 0 & \cdots & 1 \end{bmatrix}, \quad X = \begin{bmatrix} x_1 \\ x_2 \\ \cdot \\ \cdot \\ \cdot \\ x_n \end{bmatrix}, \quad D = \begin{bmatrix} d_1 \\ d_2 \\ \cdot \\ \cdot \\ \cdot \\ d_n \end{bmatrix}
$$

Matrix I_n is an n by n matrix with 1's in the main diagonal and 0's elsewhere; X is called the **output matrix** of the system and D is called the **final-demand matrix** of the system. Subtracting A, the input-coefficient matrix, from I_n yields:

$$
I_n - A = \begin{bmatrix} (1-a_{11}) & -a_{12} & \cdots & -a_{1n} \\ -a_{21} & (1-a_{22}) & \cdots & -a_{2n} \\ \cdot & & \cdot & \\ \cdot & & \cdot & \\ \cdot & & \cdot & \\ -a_{n1} & -a_{n2} & \cdots & (1-a_{nn}) \end{bmatrix}
$$

Taking the product of $(I_n - A)$ and X, $(I_n - A)X$, yields the left side of system (11.4); the right side of system (11.4) is expressed by matrix D. Thus, in matrix terms, system (11.4) is expressed by the following matrix equation:

$$(I_n - A)X = D$$

In summary, then, the problem of satisfying the needs of the n industries of the economy, expressed by the input-coefficient matrix A, and the needs of the open sector of the system, expressed by the final-demand matrix D, reduces to the matrix problem of determining output matrix X such that the product $(I_n - A)X$ equals D.

As long as the input-coefficient matrix A does not change, $(I_n - A)^{-1}$ does not change, and with one matrix inversion a variety of possible final-demand situations can be studied.

To illustrate, consider a two-industry economy governed by the input-coefficient matrix

$$A = \begin{bmatrix} 0.4 & 0.3 \\ 0.3 & 0.2 \end{bmatrix}$$

and having final-demand matrix $D = \begin{bmatrix} d_1 \\ d_2 \end{bmatrix}$. The problem is to find an output matrix $X = \begin{bmatrix} x_1 \\ x_2 \end{bmatrix}$ such that $(I_2 - A)X = D$. If $(I_2 - A)^{-1}$ exists, then $X = (I_2 - A)^{-1} D$.

$$I_2 - A = \begin{bmatrix} 0.6 & -0.3 \\ -0.3 & 0.8 \end{bmatrix}, \qquad (I_2 - A)^{-1} = \begin{bmatrix} \dfrac{80}{39} & \dfrac{10}{13} \\ \dfrac{10}{13} & \dfrac{20}{13} \end{bmatrix}$$

Thus

$$X = (I_2 - A)^{-1} D = \begin{bmatrix} \dfrac{80}{39} & \dfrac{10}{13} \\ \dfrac{10}{13} & \dfrac{20}{13} \end{bmatrix} \begin{bmatrix} d_1 \\ d_2 \end{bmatrix} = \begin{bmatrix} \dfrac{80}{39}d_1 + \dfrac{10}{13}d_2 \\ \dfrac{10}{13}d_1 + \dfrac{20}{13}d_2 \end{bmatrix}$$

and we have:

$$x_1 = \frac{80}{39}d_1 + \frac{10}{13}d_2$$

$$x_2 = \frac{10}{13}d_1 + \frac{20}{13}d_2$$

If $d_1 = \$39,000$ and $d_2 = \$61,000$ ($\$39,000$ worth of commodity 1 and $\$61,100$ worth of commodity 2 are required by the open sector), then $x_1 = \$127,000$ and $x_2 = \$124,000$. $\$127,000$ worth of commodity 1 and $\$124,000$ worth of commodity 2 must be produced to satisfy the input needs of industries 1 and 2 and the requirements of the open sector.

If the projected requirements of the open sector should change to $d_1 = \$46,800$ and $d_2 = \$65,000$, then $x_1 = \$146,000$ and $x_2 = \$136,000$. $\$146,000$ worth of commodity 1 and $\$136,000$ worth of commodity 2 must be produced to satisfy the input needs of industries 1 and 2 and the requirements of the open sector.

The successful application of the input-output model to an economy requires that a **realistic input-coefficient matrix A and final demand matrix D** be developed for the economy. **Assuming** that this could be done, the second major problem is to **suitably refine these matrices** so that they are **realistic over time**.

Leontief's pioneering studies of the American economy from an input-output analysis point of view are discussed in [5] and [6] in the references noted.

For further discussion of input-output analysis, see the following works:

References

[1] W. J. BAUMOL. *Economic Theory and Operations Analysis*, 4th ed. Englewood Cliffs, N.J.: Prentice-Hall, Inc., 1977, Chapter 22.

[2] H. B. CHENERY and P. G. CLARK. *Interindustry Economics*. New York: John Wiley & Sons, Inc., 1959.

[3] Conference on Research in Income and Wealth, National Bureau of Economic Research. *Input-Output Analysis: An Appraisal*. Princeton, N.J.: Princeton University Press, 1955.

[4] R. DORFMAN, P. A. SAMUELSON, and R. M. SOLOW. *Linear Programming and Economic Analysis*. New York: McGraw-Hill Book Company, 1958, Chapters 9-12.

[5] W. W. LEONTIEF. *The Structure of American Economy, 1919-1939*, 2d ed. New York: Oxford University Press, 1951.

[6] W. W. LEONTIEF, ed. *Studies in the Structure of the American Economy*. New York: Oxford University Press, 1953.

EXERCISES

1. For a two-industry economy, let us **assume** that $0.20 and $0.30 worth of the first industry's commodity is needed by the first and second industries to produce $1 worth of their respective commodities and that $0.30 and $0.10 worth of the second industry's commodity is needed by the first and second industries to produce $1 worth of their respective commodities.

   (a) Set up the input-coefficient matrix of the economy.

   (b) If the open sector of the economy requires $2520 worth of commodity 1 and $3150 worth of commodity 2, what output levels will satisfy the input needs of the industries and the requirements of the open sector? How much of these outputs will be consumed by industries 1 and 2?

   (c) If the open sector of the economy requires $3465 worth of commodity 1 and $3780 worth of commodity 2, what output levels will satisfy the input needs of the industries and the requirements of the open sector? How much of these outputs will be consumed by industries 1 and 2?

11.7 SELF-TESTS FOR CHAPTER 11

Allow 75 or so minutes for each self-test. Go over the first one before undertaking the second.

Self-Test 1

1. For $E = \begin{bmatrix} 4 & 2 & 1 \\ 6 & -4 & 3 \end{bmatrix}$ and $F = \begin{bmatrix} -2 & 1 & 3 \\ 6 & -2 & 1 \end{bmatrix}$, find $E + F$ and $E - F$.

2. For $A = \begin{bmatrix} 4 & -2 \\ 1 & 3 \end{bmatrix}$ and $B = \begin{bmatrix} -1 & 2 \\ 3 & 5 \end{bmatrix}$, find AB, BA, A^2, and B^2.

3. For $A = \begin{bmatrix} 4 & 3 & 1 \\ 1 & -1 & -2 \end{bmatrix}$ and $B = \begin{bmatrix} 2 & 1 \\ 4 & -2 \\ 3 & -1 \end{bmatrix}$, find, if possible, AB and BA.

4. What is the negative of $A = \begin{bmatrix} 5 & 3 \\ -4 & 1 \end{bmatrix}$? Explain.

5. Is $B = \begin{bmatrix} 1 & -2 \\ 2 & -3 \end{bmatrix}$ the multiplicative inverse of $A = \begin{bmatrix} -3 & 2 \\ -2 & 1 \end{bmatrix}$? Explain.

Find, provided that it exists, the multiplicative inverse of each of the following matrices.

6. $B = \begin{bmatrix} 1 & 4 \\ -2 & -8 \end{bmatrix}$

7. $F = \begin{bmatrix} 1 & -2 & 3 \\ 2 & 1 & -3 \\ -1 & 1 & 2 \end{bmatrix}$

Solve each of the following systems of equations by matrix inversion.

8. $4x_1 + 3x_2 = 10$
    $2x_1 + 5x_2 = 12$

9. $x_1 + x_2 + 3x_3 = 2$
    $x_1 + 2x_2 - x_3 = -4$
    $x_1 + x_2 + x_3 = 8$

Write True or False next to each of the following statements. If a statement is false, give an example which shows that it's false.

10. For square matrices A and B of the same size, $AB = BA$.

11. If the matrix product $AB = \overline{0}$, then $A = \overline{0}$ or $B = \overline{0}$.

12. If the matrix product $AX = AY$, where $A \neq \overline{0}$, then $X = Y$.

13. Multiplication of square matrices of the same size is associative.

14. Every nonzero square matrix has an additive inverse.

15. Every nonzero square matrix has a multiplicative inverse.

Self-Test 2

1. For $G = \begin{bmatrix} 3 & -1 \\ 4 & 2 \end{bmatrix}$ and $H = \begin{bmatrix} 4 & 1 \\ -2 & 0 \end{bmatrix}$, find $G + H$ and $G - H$.

2. For $A = \begin{bmatrix} 1 & 3 & -2 \end{bmatrix}$ and $B = \begin{bmatrix} 3 \\ -2 \\ 1 \end{bmatrix}$, find, if possible, AB and BA.

3. For $A = \begin{bmatrix} 2 & 4 \\ 1 & 2 \end{bmatrix}$, $X = \begin{bmatrix} 1 & 2 \\ 1 & -1 \end{bmatrix}$, and $Y = \begin{bmatrix} 5 & 6 \\ -1 & -3 \end{bmatrix}$, find AX and AY and

   compare.

4. Is $B = \begin{bmatrix} -\dfrac{2}{11} & -\dfrac{3}{11} \\ \dfrac{5}{11} & \dfrac{2}{11} \end{bmatrix}$ the multiplicative inverse of $A = \begin{bmatrix} 2 & 3 \\ -5 & -2 \end{bmatrix}$?

   Explain.

Find, provided that it exists, the multiplicative inverse of each of the following matrices.

5. $A = \begin{bmatrix} 3 & 5 \\ 1 & 2 \end{bmatrix}$

6. $E = \begin{bmatrix} 1 & 3 & 2 \\ 0 & 1 & 2 \\ 2 & -1 & -3 \end{bmatrix}$

Solve each of the following systems of linear equations by means of matrix inversion.

7. $\begin{aligned} x_1 + 2x_2 &= 8 \\ -2x_1 + x_2 &= 4 \end{aligned}$

8. $\begin{aligned} 2x_1 + x_2 - x_3 &= 1 \\ 3x_1 + 2x_2 + 5x_3 &= -2 \\ x_1 + x_2 + 5x_3 &= 3 \end{aligned}$

9. For a two-industry economy, let us suppose that 50¢ and 20¢ worth of the first industry's product is needed by the first and second industries to produce $1 worth of their respective commodities, and that 20¢ and 50¢ worth of the second industry's product is needed by the first and second industries to produce $1 worth of their respective commodities.

   (a) Set up the input-coefficient matrix of the economy.

   (b) If the open sector of the economy requires $1575 worth of commodity 1 and $1890 worth of commodity 2, what output levels will satisfy the input needs of the industries and the requirements of the open sector? How much of these outputs will be consumed by industries 1 and 2?

(c) If the open sector of the economy requires \$2940 worth of commodity 1 and \$3675 worth of commodity 2, what output levels will satisfy the input needs of the industries and the requirements of the open sector? How much of these outputs will be consumed by industries 1 and 2?

APPENDIX ON TABLES

Table 1 SQUARES AND SQUARE ROOTS

No.	Square	Square Root	No.	Square	Square Root	No.	Square	Square Root
1	1	1.000	51	2,601	7.141	101	10,201	10.050
2	4	1.414	52	2,704	7.211	102	10,404	10.100
3	9	1.732	53	2,809	7.280	103	10,609	10.149
4	16	2.000	54	2,916	7.348	104	10,816	10.198
5	25	2.236	55	3,025	7.416	105	11,025	10.247
6	36	2.449	56	3,136	7.483	106	11,236	10.296
7	49	2.646	57	3,249	7.550	107	11,449	10.344
8	64	2.828	58	3,364	7.616	108	11,664	10.392
9	81	3.000	59	3,481	7.681	109	11,881	10.440
10	100	3.162	60	3,600	7.746	110	12,100	10.488
11	121	3.317	61	3,721	7.810	111	12,321	10.536
12	144	3.464	62	3,844	7.874	112	12,544	10.583
13	169	3.606	63	3,969	7.937	113	12,769	10.630
14	196	3.742	64	4,096	8.000	114	12,996	10.677
15	225	3.873	65	4,225	8.062	115	13,225	10.724
16	256	4.000	66	4,356	8.124	116	13,456	10.770
17	289	4.123	67	4,489	8.185	117	13,689	10.817
18	324	4.243	68	4,624	8.246	118	13,924	10.863
19	361	4.359	69	4,761	8.307	119	14,161	10.909
20	400	4.472	70	4,900	8.367	120	14,400	10.954
21	441	4.583	71	5,041	8.426	121	14,641	11.000
22	484	4.690	72	5,184	8.485	122	14,884	11.045
23	529	4.796	73	5,329	8.544	123	15,129	11.091
24	576	4.899	74	5,476	8.602	124	15,376	11.136
25	625	5.000	75	5,625	8.660	125	15,625	11.180
26	676	5.099	76	5,776	8.718	126	15,876	11.225
27	729	5.196	77	5,929	8.775	127	16,129	11.269
28	784	5.292	78	6,084	8.832	128	16,384	11.314
29	841	5.385	79	6,241	8.888	129	16,641	11.358
30	900	5.477	80	6,400	8.944	130	16,900	11.402
31	961	5.568	81	6,561	9.000	131	17,161	11.446
32	1,024	5.657	82	6,724	9.055	132	17,424	11.489
33	1,089	5.745	83	6,889	9.110	133	17,689	11.533
34	1,156	5.831	84	7,056	9.165	134	17,956	11.576
35	1,225	5.916	85	7,225	9.220	135	18,225	11.619
36	1,296	6.000	86	7,396	9.274	136	18,496	11.662
37	1,369	6.083	87	7,569	9.327	137	18,769	11.705
38	1,444	6.164	88	7,744	9.381	138	19,044	11.747
39	1,521	6.245	89	7,921	9.434	139	19,321	11.790
40	1,600	6.325	90	8,100	9.487	140	19,600	11.832
41	1,681	6.403	91	8,281	9.539	141	19,881	11.874
42	1,764	6.481	92	8,464	9.592	142	20,164	11.916
43	1,849	6.557	93	8,649	9.644	143	20,449	11.958
44	1,936	6.633	94	8,836	9.695	144	20,736	12.000
45	2,025	6.708	95	9,025	9.747	145	21,025	12.042
46	2,116	6.782	96	9,216	9.798	146	21,316	12.083
47	2,209	6.856	97	9,409	9.849	147	21,609	12.124
48	2,304	6.928	98	9,604	9.899	148	21,904	12.166
49	2,401	7.000	99	9,801	9.950	149	22,201	12.207
50	2,500	7.071	100	10,000	10.000	150	22,500	12.247

Table 2 COMPOUND INTEREST: $(1+i)^n$

n	1%	2%	3%	4%	5%
1	1.01000	1.02000	1.03000	1.04000	1.05000
2	1.02010	1.04040	1.06090	1.08160	1.10250
3	1.03030	1.06121	1.09273	1.12486	1.15762
4	1.04060	1.08243	1.12551	1.16986	1.21551
5	1.05101	1.10408	1.15927	1.21665	1.27628
6	1.06152	1.12616	1.19405	1.26532	1.34010
7	1.07214	1.14869	1.22987	1.31593	1.40710
8	1.08286	1.17166	1.26677	1.36857	1.47746
9	1.09369	1.19509	1.30477	1.42331	1.55133
10	1.10462	1.21899	1.34392	1.48024	1.62889
11	1.11567	1.24337	1.38423	1.53945	1.71034
12	1.12683	1.26824	1.42576	1.60103	1.79586
13	1.13809	1.29361	1.46853	1.66507	1.88565
14	1.14947	1.31948	1.51259	1.73168	1.97993
15	1.16097	1.34587	1.55797	1.80094	2.07893
16	1.17258	1.37279	1.60471	1.87298	2.18287
17	1.18430	1.40024	1.65285	1.94790	2.29202
18	1.19615	1.42825	1.70243	2.02582	2.40662
19	1.20811	1.45681	1.75351	2.10685	2.52695
20	1.22019	1.48595	1.80611	2.19112	2.64330
21	1.23239	1.51567	1.86029	2.27877	2.78596
22	1.24472	1.54598	1.91610	2.36992	2.92526
23	1.25716	1.57690	1.97359	2.46472	3.07152
24	1.26973	1.60844	2.03279	2.56330	3.22510
25	1.28243	1.64061	2.09378	2.66584	3.38635
26	1.29526	1.67342	2.15659	2.77247	3.55567
27	1.30821	1.70689	2.22129	2.88337	3.73346
28	1.32129	1.74102	2.28793	2.99870	3.92013
29	1.33450	1.77584	2.35657	3.11865	4.11614
30	1.34785	1.81136	2.42726	3.24340	4.32194
31	1.36133	1.84759	2.50008	3.37313	4.53804
32	1.37494	1.88454	2.57508	3.50806	4.76494
33	1.38869	1.92223	2.65234	3.64838	5.00319
34	1.40258	1.96068	2.73191	3.79432	5.25335
35	1.41660	1.99989	2.81386	3.94609	5.51602
36	1.43077	2.03989	2.89828	4.10393	5.79182
37	1.44508	2.08069	2.98523	4.26809	6.08141
38	1.45953	2.12230	3.07478	4.43881	6.38548
39	1.47412	2.16474	3.16703	4.61637	6.70475
40	1.48886	2.20804	3.26204	4.80102	7.03999

Appendix on Tables

Table 3 PRESENT VALUE: $(1+i)^{-n}$

n	1%	2%	3%	4%	5%
1	.99010	.98039	.97087	.96154	.95238
2	.98030	.96117	.94260	.92456	.90703
3	.97059	.94232	.91514	.88900	.86384
4	.96098	.92385	.88849	.85480	.82270
5	.95147	.90573	.86261	.82193	.78353
6	.94205	.88797	.83748	.79031	.74622
7	.93272	.87056	.83109	.75992	.71068
8	.92348	.85349	.78941	.73069	.67684
9	.91434	.83676	.76642	.70259	.64461
10	.90529	.82035	.74409	.67556	.61391
11	.89632	.80426	.72242	.64958	.58468
12	.88745	.78849	.70138	.62460	.55684
13	.87866	.77303	.68095	.60057	.53032
14	.86996	.75788	.66112	.57748	.50507
15	.86135	.74301	.64186	.55526	.48102
16	.85282	.72845	.62317	.53391	.45811
17	.84438	.71416	.60502	.51337	.43630
18	.83602	.70016	.58739	.49363	.41552
19	.82774	.68643	.57029	.47464	.39573
20	.81954	.67297	.55368	.45639	.37689
21	.81143	.65978	.53755	.43883	.35894
22	.80340	.64684	.52189	.42196	.34185
23	.79544	.63416	.50669	.40573	.32557
24	.78757	.62172	.49193	.39012	.31007
25	.77977	.60953	.47761	.37512	.29530
26	.77205	.59758	.46369	.36069	.28124
27	.76440	.58586	.45019	.34682	.26785
28	.75684	.57437	.43708	.33348	.25509
29	.74934	.56311	.42435	.32065	.24295
30	.74192	.55207	.41199	.30832	.23138
31	.73458	.54125	.39999	.29646	.22036
32	.72730	.53063	.38834	.28506	.20987
33	.72010	.52023	.37703	.27409	.19987
34	.71297	.51003	.36604	.26355	.19035
35	.70591	.50003	.35538	.25342	.18129
36	.69892	.49022	.34503	.24367	.17266
37	.69200	.48061	.33498	.23430	.16444
38	.68515	.47119	.32523	.22529	.15661
39	.67837	.46195	.31575	.21662	.14915
40	.67165	.45289	.30656	.20829	.14205

Table 4 FUTURE VALUE OF AN ANNUITY: $s_{\overline{n}|i}$

n	1%	2%	3%	4%	5%
1	1.00000	1.00000	1.00000	1.00000	1.00000
2	2.01000	2.02000	2.03000	2.04000	2.05000
3	3.03010	3.06040	3.09090	3.12160	3.15250
4	4.06040	4.12161	4.18363	4.24646	4.31012
5	5.10101	5.20404	5.30914	5.41632	5.52563
6	6.15202	6.30812	6.46841	6.63298	6.80191
7	7.21354	7.43428	7.66246	7.89829	8.14201
8	8.28567	8.58297	8.89234	9.21423	9.54911
9	9.36853	9.75463	10.15911	10.58280	11.02656
10	10.46221	10.94972	11.46388	12.00611	12.57789
11	11.56683	12.16872	12.80780	13.48635	14.20679
12	12.68250	13.41209	14.19203	15.02581	15.91713
13	13.80933	14.68033	15.61779	16.62684	17.71298
14	14.94742	15.97394	17.08632	18.29191	19.59863
15	16.09690	17.29342	18.59891	20.02359	21.57856
16	17.25786	18.63929	20.15688	21.82453	23.65749
17	18.43044	20.01207	21.76159	23.69751	25.84037
18	19.61475	21.41231	23.41444	25.64541	28.13238
19	20.81090	22.84056	25.11687	27.67123	30.53900
20	22.01900	24.29737	26.87037	29.77808	33.06595
21	23.23919	25.78332	28.67649	31.96920	35.71925
22	24.47159	27.29898	30.53678	34.24797	38.50521
23	25.71630	28.84496	32.45288	36.61789	41.43048
24	26.97346	30.42186	34.42647	39.08260	44.50200
25	28.24320	32.03030	36.45926	41.64591	47.72710
26	29.52563	33.67091	38.55304	44.31174	51.11345
27	30.82089	35.34432	40.70963	47.08421	54.66913
28	32.12910	37.05121	42.93092	49.96758	58.40258
29	33.45039	38.79223	45.21885	52.96629	62.32271
30	34.78489	40.56808	47.57542	56.08494	66.43885
31	36.13274	42.37944	50.00268	59.32834	70.76079
32	37.49407	44.22703	52.50276	62.70147	75.29883
33	38.86901	46.11157	55.07784	66.20953	80.06377
34	40.25770	48.03380	57.73018	69.85791	85.06696
35	41.66028	49.99448	60.46208	73.65222	90.32031
36	43.07688	51.99437	63.27594	77.59831	95.83632
37	44.50765	54.03425	66.17422	81.70225	101.62814
38	45.95272	56.11494	69.15945	85.97034	107.70955
39	47.41225	58.23724	72.23423	90.40915	114.09502
40	48.88637	60.40198	75.40126	95.02552	120.79977

Appendix on Tables

Table 5 $\dfrac{1}{s_{\overline{n}|}}$

n	1%	2%	3%	4%	5%
1	1.00000	1.00000	1.00000	1.00000	1.00000
2	.49751	.49505	.49261	.49020	.48780
3	.33002	.32675	.32353	.32035	.31721
4	.24628	.24262	.23903	.23549	.23201
5	.19604	.19216	.18835	.18463	.18097
6	.16255	.15853	.15460	.15076	.14702
7	.13863	.13451	.13051	.12661	.12282
8	.12069	.11651	.11246	.10853	.10472
9	.10674	.10252	.09843	.09449	.09069
10	.09558	.09133	.08723	.08329	.07950
11	.08645	.08218	.07808	.07415	.07039
12	.07885	.07456	.07046	.06655	.06283
13	.07241	.06812	.06403	.06014	.05646
14	.06690	.06260	.05853	.05467	.05102
15	.06212	.05783	.05377	.04994	.04634
16	.05794	.05365	.04961	.04582	.04227
17	.05426	.04997	.04595	.04220	.03870
18	.05098	.04670	.04271	.03899	.03555
19	.04805	.04378	.03981	.03614	.03275
20	.04542	.04116	.03722	.03358	.03024
21	.04303	.03878	.03487	.03128	.02800
22	.04086	.03663	.03275	.02920	.02597
23	.03889	.03467	.03081	.02731	.02414
24	.03707	.03287	.02905	.02559	.02247
25	.03541	.03122	.02743	.02401	.02095
26	.03387	.02970	.02594	.02257	.01956
27	.03245	.02829	.02456	.02124	.01829
28	.03112	.02699	.02329	.02001	.01712
29	.02990	.02578	.02211	.01888	.01605
30	.02875	.02465	.02102	.01783	.01505
31	.02768	.02360	.02000	.01686	.01413
32	.02667	.02261	.01905	.01595	.01328
33	.02573	.02169	.01816	.01510	.01249
34	.02484	.02082	.01732	.01431	.01176
35	.02400	.02000	.01654	.01358	.01107
36	.02321	.01923	.01580	.01289	.01043
37	.02247	.01851	.01511	.01224	.00984
38	.02176	.01782	.01446	.01163	.00928
39	.02109	.01717	.01384	.01106	.00876
40	.02046	.01656	.01326	.01052	.00828

Table 6 PRESENT VALUE OF AN ANNUITY $a_{\overline{n}|i}$

n	1%	2%	3%	4%	5%
1	0.99010	0.98039	0.97087	0.96154	0.95238
2	1.97040	1.94156	1.91347	1.88609	1.85941
3	2.94099	2.88388	2.82861	2.77509	2.72325
4	3.90197	3.80773	3.71710	3.62990	3.54595
5	4.85343	4.71346	4.57971	4.45182	4.32948
6	5.79548	5.60143	5.41719	5.24214	5.07569
7	6.72819	6.47199	6.23028	6.00205	5.78637
8	7.65168	7.32548	7.01969	6.73274	6.46321
9	8.56602	8.16224	7.78611	7.43533	7.10782
10	9.47130	8.98259	8.53020	8.11090	7.72173
11	10.36763	9.78685	9.25262	8.76048	8.30641
12	11.25508	10.57534	9.95400	9.38507	8.86325
13	12.13374	11.34837	10.63496	9.98565	9.39357
14	13.00370	12.10625	11.29607	10.56312	9.89864
15	13.86505	12.84926	11.93794	11.11839	10.37966
16	14.71787	13.57771	12.56110	11.65230	10.83777
17	15.56225	14.29187	13.16612	12.16567	11.27407
18	16.39827	14.99203	13.75351	12.65930	11.68959
19	17.22601	15.67846	14.32380	13.13394	12.08532
20	18.04555	16.35143	14.87747	13.59033	12.46221
21	18.85698	17.01121	15.41502	14.02916	12.82115
22	19.66038	17.65805	15.93692	14.45112	13.16300
23	20.45582	18.29220	16.44361	14.85684	13.48857
24	21.24339	18.91393	16.93554	15.24696	13.79864
25	22.02316	19.52346	17.41315	15.62208	14.09394
26	22.79520	20.12104	17.87684	15.98277	14.37519
27	23.55961	20.70690	18.32703	16.32959	14.64303
28	24.31644	21.28127	18.76411	16.66306	14.89813
29	25.06579	21.84438	19.18845	16.98371	15.14107
30	25.80771	22.39646	19.60044	17.29203	15.37245
31	26.54229	22.93770	20.00043	17.58849	15.59281
32	27.26959	23.46833	20.38877	17.87355	15.80268
33	27.98969	23.98856	20.76579	18.14765	16.00255
34	28.70267	24.49859	21.13184	18.41120	16.19290
35	29.40858	24.99862	21.48722	18.66461	16.37419
36	30.10751	25.48884	21.83225	18.90828	16.54685
37	30.79951	25.96945	22.16724	19.14258	16.71129
38	31.48466	26.44064	22.49246	19.36786	16.86789
39	32.16303	26.90259	22.80822	19.58448	17.01704
40	32.83469	27.35548	23.11477	19.79277	17.15909

Appendix on Tables

Table 7 $\dfrac{1}{a_{\overline{n}|i}}$

n	1%	2%	3%	4%	5%
1	1.01000	1.02000	1.03000	1.04000	1.05000
2	0.50751	0.51505	0.52261	0.53020	0.53780
3	0.34002	0.34675	0.35353	0.36035	0.36721
4	0.25628	0.26262	0.26903	0.27549	0.28201
5	0.20604	0.21216	0.21835	0.22463	0.23097
6	0.17255	0.17853	0.18460	0.19076	0.19702
7	0.14863	0.15451	0.16051	0.16661	0.17282
8	0.13069	0.13651	0.14246	0.14853	0.15472
9	0.11674	0.12252	0.12843	0.13449	0.14069
10	0.10558	0.11133	0.11723	0.12329	0.12950
11	0.09645	0.10218	0.10808	0.11415	0.12039
12	0.08885	0.09456	0.10046	0.10655	0.11283
13	0.08241	0.08812	0.09403	0.10014	0.10646
14	0.07690	0.08260	0.08853	0.09467	0.10102
15	0.07212	0.07783	0.08377	0.08994	0.09634
16	0.06794	0.07365	0.07961	0.08582	0.09227
17	0.06426	0.06997	0.07595	0.08220	0.08870
18	0.06098	0.06670	0.07271	0.07899	0.08555
19	0.05805	0.06378	0.06881	0.07614	0.08275
20	0.05542	0.06116	0.06722	0.07358	0.08024
21	0.05303	0.05878	0.06487	0.07128	0.07800
22	0.05086	0.05663	0.06275	0.06920	0.07597
23	0.04889	0.05467	0.06081	0.06731	0.07414
24	0.04707	0.05287	0.05905	0.06559	0.07247
25	0.04541	0.05122	0.05743	0.06401	0.07095
26	0.04387	0.04970	0.05594	0.06257	0.06956
27	0.04245	0.04829	0.05456	0.06124	0.06829
28	0.04112	0.04699	0.05329	0.06001	0.06712
29	0.03990	0.04578	0.05211	0.05888	0.06605
30	0.03875	0.04465	0.05102	0.05783	0.06505
31	0.03768	0.04360	0.05000	0.05686	0.06413
32	0.03667	0.04261	0.04905	0.05595	0.06328
33	0.03573	0.04169	0.04816	0.05510	0.06249
34	0.03484	0.04082	0.04732	0.05431	0.06176
35	0.03400	0.04000	0.04654	0.05358	0.06107
36	0.03321	0.03923	0.04580	0.05289	0.06043
37	0.03247	0.03851	0.04511	0.05224	0.05984
38	0.03176	0.03782	0.04446	0.05163	0.05928
39	0.03109	0.03717	0.04384	0.05106	0.05876
40	0.03046	0.03656	0.04326	0.05052	0.05828

Table 8 *SINE AND COSINE VALUES*

Deg.	Sin	Cos	Deg.	Sin	Cos
0	0.0000	1.0000	45	0.7071	0.7071
1	0.0175	0.9998	46	0.7193	0.6947
2	0.0349	0.9994	47	0.7314	0.6820
3	0.0523	0.9986	48	0.7431	0.6691
4	0.0698	0.9976	49	0.7547	0.6561
5	0.0872	0.9962	50	0.7660	0.6428
6	0.1045	0.9945	51	0.7771	0.6293
7	0.1219	0.9925	52	0.7880	0.6157
8	0.1392	0.9903	53	0.7986	0.6018
9	0.1564	0.9877	54	0.8090	0.5878
10	0.1736	0.9848	55	0.8192	0.5736
11	0.1908	0.9816	56	0.8290	0.5592
12	0.2079	0.9781	57	0.8387	0.5446
13	0.2250	0.9744	58	0.8480	0.5299
14	0.2419	0.9703	59	0.8572	0.5150
15	0.2588	0.9659	60	0.8660	0.5000
16	0.2756	0.9613	61	0.8746	0.4848
17	0.2924	0.9563	62	0.8829	0.4695
18	0.3090	0.9511	63	0.8910	0.4540
19	0.3256	0.9455	64	0.8988	0.4384
20	0.3420	0.9397	65	0.9063	0.4226
21	0.3584	0.9336	66	0.9135	0.4067
22	0.3746	0.9272	67	0.9205	0.3907
23	0.3907	0.9205	68	0.9272	0.3746
24	0.4067	0.9135	69	0.9336	0.3584
25	0.4226	0.9063	70	0.9397	0.3420
26	0.4384	0.8988	71	0.9455	0.3256
27	0.4540	0.8910	72	0.9511	0.3090
28	0.4695	0.8829	73	0.9563	0.2924
29	0.4848	0.8746	74	0.9613	0.2756
30	0.5000	0.8660	75	0.9659	0.2588
31	0.5150	0.8572	76	0.9703	0.2419
32	0.5299	0.8480	77	0.9744	0.2250
33	0.5446	0.8387	78	0.9781	0.2079
34	0.5592	0.8290	79	0.9816	0.1908
35	0.5736	0.8192	80	0.9848	0.1736
36	0.5878	0.8090	81	0.9877	0.1564
37	0.6018	0.7986	82	0.9903	0.1392
38	0.6157	0.7880	83	0.9925	0.1219
39	0.6293	0.7771	84	0.9945	0.1045
40	0.6428	0.7660	85	0.9962	0.0872
41	0.6561	0.7547	86	0.9976	0.0698
42	0.6691	0.7431	87	0.9986	0.0523
43	0.6820	0.7314	88	0.9994	0.0349
44	0.6947	0.7193	89	0.9998	0.0175
45	0.7071	0.7071	90	1.0000	0.0000

ANSWERS TO ODD-NUMBERED EXERCISES AND SELF-TESTS

Section 1.2 (p. 30)

1. -6

3. 6

5. -1

7. 10

9. 1

11. 1

13. $15a$

15. $5wz$

17. $2ab$

19. $10xyz$

21. $-4x - 6y - 8z$

23. $-2x - 9$

25. $2n - 8m$

27. $3xy + a$

29. $[2(x+2y)$

31. $x(y+z)$

33. $4xy(z+2w)$

35. $2x(1+2x)$

37. $a^2(y+z+w)$

39. $k^2(4w-3+a)$

41. $3a(bc+2b+c)$

43. $x(2y+3z+1)$

45. yes, since $(-x+4)+(x-4)=0$

47. no, since $-2(2)\neq1$

49. 1 is the multiplicative inverse of 1 since $1\cdot1=1$.

51. $\dfrac{1}{6}$, $-1, 4, 1/\sqrt{2}$, and $1/\pi$, respectively

Section 1.4 (p. 38)

1. -2

3. -7

5. -7

7.   8

9.   3

11.   4

13.   2

15. -9

17. $-\dfrac{1}{2}$

19. -8

21. $-\dfrac{1}{8}$

23. not defined

25. $-\dfrac{1}{16}$

27. $-9a+6b+9c$

29. $a-2b-4c$

31. $-x+2y+3z$

33. ab

35. 0

37. $-xy+6y$

39. $b(4a^2-3x-1)$

41. $2y^2(x-2+3z)$

43. $gh(1-3h)$

45. $xy(3xy^2-2y+1)$

47. No; division by 0 is not defined and, thus, so is cancellation by 0 in the denominator.

49. No; Burt's conclusion is based on incorrect results.

Section 1.5 (p. 44)

1. incorrect

3. incorrect

21. $\dfrac{2a+b}{a+b}$

23. -1

5.   incorrect

7.   incorrect

9. $\dfrac{1}{3}$

11. $\dfrac{3}{11}$

13. $\dfrac{3ac}{4}$, $b \neq 0$

15. 4, $a \neq -b$

17. $\dfrac{2+(x+y)}{x+y}$

19. $\dfrac{3(x+y)}{2}$, $x \neq y$

21. $\dfrac{2a+b}{a+b}$

23. -1, $x \neq 3$

25. $\dfrac{2x+1}{2x}$

27. 2, where $x \neq 2$

29. 1, where $x \neq -1$

31. 4, where $x - 4y \neq 0$

33. $-3/2$, where $x \neq 5$

35. $1/x^2$, where $4m + 3n \neq 0$

37. c/e, where $4d - 1 \neq 0$

39. $ab/3$, where $3 - c \neq 0$

40.   For 23, $y = 3$; for 25, $x = 0$; for 27, $x = 0$.

41.   not equal

43.   not equal

45.   equal

47. $\dfrac{4}{7}$

49. $\dfrac{4}{5}$

51. $\dfrac{3(8y-1)}{4y}$, where $x \neq 0$

53. $4/b$, where $a \neq 0$

55. $\dfrac{3yx}{(x+1)(6y-1)}$

57. $12/x$, where $y \neq 0$

63. $-8/n$, where $n \neq 3$

65. $\dfrac{2}{3}$

67. $-\dfrac{12}{5}$

69. $-\dfrac{1}{8}$

71. $2a^2/3$, where $a \neq 0, b \neq 0$

73. $2/b$, where $a \neq 0$

75. $2/x$, where $x \neq 2$

77. $-\dfrac{1}{2}$, where $x \neq 2$,

79. $-1/a$, where $a \neq 3$

59. 1/2x, where $3x - 2 \neq 0$

61. $-\dfrac{3}{4}$, where $x \neq 2$

81. $-\dfrac{1}{2}$, where $2x - y \neq 0$

Section 1.5 (p. 54)

83. 3

85. $\dfrac{11}{13}$

87. $\dfrac{21}{40}$

89. $\dfrac{29}{28}$

91. $\dfrac{19}{12}$

93. $\dfrac{2x - 1}{2y}$

95. $-x/3$

97. $\dfrac{2x}{3y}$

99. 1, where $x \neq -2$

101. $\dfrac{8x + 8}{2x + 1}$

103. $\dfrac{22}{15m}$

105. $\dfrac{11y - 1}{15}$

107. $\dfrac{11x - y}{12}$

109. $\dfrac{2x + 1}{x(x + 1)}$

111. $\dfrac{5x - 6}{x(x - 2)}$

113. $\dfrac{3x+10}{x+2}$

115. $\dfrac{4t+15}{t+3}$

117. $\dfrac{-x^2+2x+3}{x-2}$

119. $\dfrac{x^2-x+y^2}{y(x-1)}$

121. $\dfrac{x^2+7x-12}{(x-1)(2x-3)}$

Section 1.5 (p. 57)

123. $-\dfrac{6}{35}$

125. $\dfrac{1}{12}$

127. $-\dfrac{15}{4}$

129. $\dfrac{x+10}{3x}$

131. $\dfrac{-2x+1}{3(x+1)}$

133. $\dfrac{-2x+3}{2x}$

135. $\dfrac{-6x--7}{3x+4}$

137. $7/3x$

139. $\dfrac{a+5b}{6}$

141. $\dfrac{13b-7}{10}$

143. 5. x and y not 0; 6. $b \neq 0$, $a \neq 1$; 8. $b \neq 0, 1$, $a \neq 0$; 9. $a \neq 0, 1$; 10. $x \neq 0$; 11. $a \neq 0, 1$; 13. none; 14. $x \neq 2$, -1. Division by 0 is not defined.

Section 1.6 (p. 63)

1. -32

3. $-\dfrac{1}{8}$

5. $\dfrac{3}{16}$

7. $-\dfrac{9}{8}$

9. 1000

11. $\dfrac{1}{324}$

13. 1

15. $-\dfrac{25}{64}$

17. $4/xy^2$

19. x^6

21. 1

23. 1

25. $6x^5y^2$

27. $25/3x^8$

29. $2+(1/x)+(1/x^3)$

31. $\dfrac{1}{8a^3b^6}$

33. $\dfrac{a^3}{8b^6}$

35. $\dfrac{4x^4}{3y^5}$

37. $8a^5 + \dfrac{4}{a}$

39. $108x^8 y$

41. $x^2 + x$

43. $\dfrac{1}{125m^3 n^3 (1 + 2m^2 n^3)}$

45. $3 + \dfrac{2}{v^5}$

47. $2 + \dfrac{2}{y} + \dfrac{2}{y^3}$

Section 1.6 (p. 65)

49. 8

51. −4

53. −2

55. 3

57. 5

59. 4

61. 6

63. $\dfrac{25}{2}$

Section 1.6 (p. 68)

65. $4\sqrt{3}$

67. $3\sqrt{6}$

69. $3\sqrt{14}$

71. $15\sqrt{7}$

73. $-\sqrt{7}$

75. $10/\sqrt{3}$

77. $2\sqrt{15}$

79. $\sqrt{5}$

81. $2\sqrt[4]{3}$

83. $2a\sqrt{b}$

85. $2x\sqrt{xy}$

87. $2\sqrt{x^2 + 4y^2}$

Section 1.6 (p. 71)

89. 5

91. 5

93. $\dfrac{1}{6}$

95. $\dfrac{1}{2}$

97. 9

99. 27

101. $\dfrac{1}{16}$

103. 36

105. 32

107. $3x^2$

109. x

111. $1/x^5$

113. x^3

115. $x^{47/30}$

Section 2.1 (p. 78)

1. $x = 6$

3. $x = 4$

5. $x = 14$

7. $x = \dfrac{4}{3}$

9. $x = 22$

11. $x = -1$

13. $x = 3$

15. $x = 2$

17. $(0,-6)$, $(2,0)$, $(1,-3)$

19. $(0,-2)$, $(4,0)$, $(2,-1)$

21. $(0,-7)$, $(2,0)$, $(4,7)$

23. $(0,0,12)$, $(0,-6,0)$, $(4,0,0)$

25. $y = \dfrac{10-5x}{2}$

27. $y = \pm\sqrt{25-x^2}$

29. $z = 10 + 3y - \dfrac{1}{x}$

31. $z = \dfrac{12-4xy}{-3}$

33. $z = \dfrac{4}{10-xy}$

35. $z = 3A - x - y$

37. $r = C / 2\pi$

Section 2.2 (p. 88)

3.

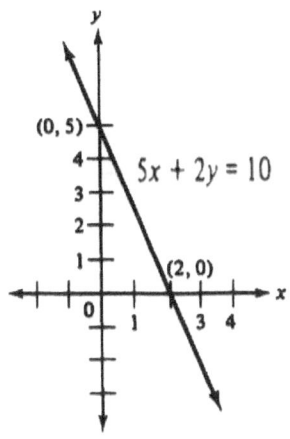

7.

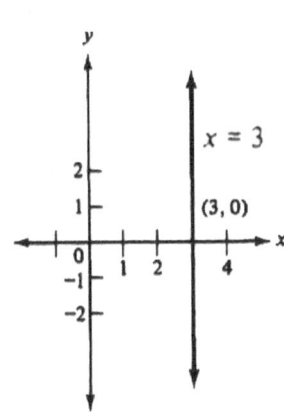

5.

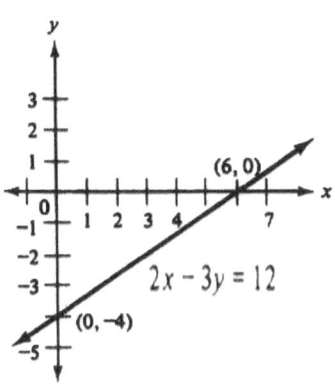

9.

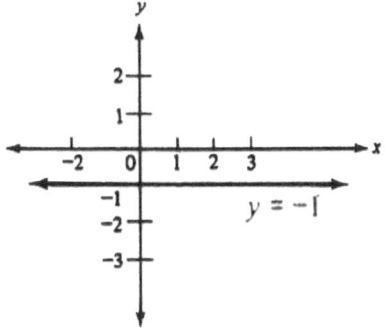

11.

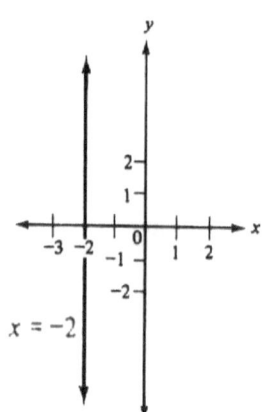

15.

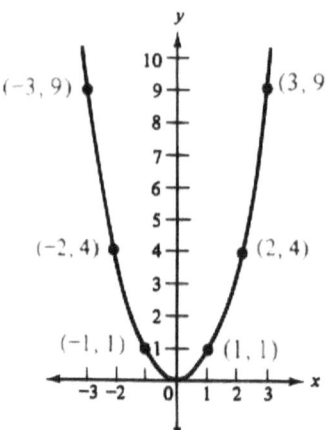

13.

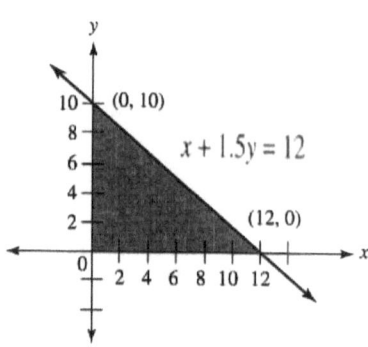

17.

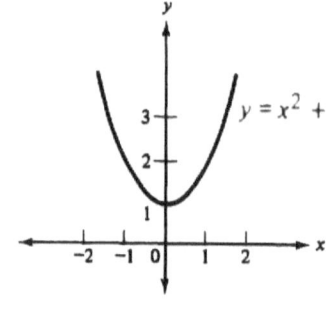

19.

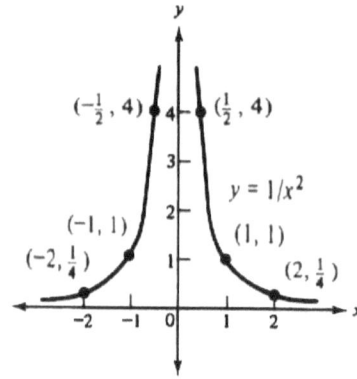

Section 2.3 (p. 91)

1. $240   3. $1890   5. $I = \$144$; $A = \$1944$

Section 2.3 (p. 97)

7. (a) $4105.71,
   (b) $4118.37

9. $4973.76

11. (a) $940,
    (b) 6.38%

13. $1276.60

15. 8.286%; 8.286% compounded annually yields the same interest at the end of the year as 8% per annum compounded 8 times a year.

17. For 12% per annum compounded 3 times a year, $v = 12.486\%$; for 11% per annum compounded 11 times a year, $v = 11.567\%$. $124.86; $115.67

Section 2.4 (p. 100)

1. $(0,2)$, $(3,-2)$, $(-4,0)$

3. $(2,3)$, $(6,2)$

5. $(3,-2)$, $(-4,0)$, $(2,3)$, $(6,2)$

7. $(3,-1,4)$, $(\frac{1}{2},\frac{1}{3},-3)$, $(-2,3,1)$

9. $(2,3,1)$, $(1,3,-1)$, $(4,2,1)$, $(\frac{1}{2},\frac{1}{3},-3)$

11. $(2,3,1)$, $(3,-1,4)$, $(\frac{1}{2},\frac{1}{3},-3)$, $(-2,3,1)$

Section 2.4 (p. 104)

21. $x > -\dfrac{2}{3}$

23. $x \le 2$

25. $y < 2$

27. $x < -15$

29. $x > \dfrac{5}{2}$

31. $y \ge 2$

Section 2.5 (p. 110)

1.

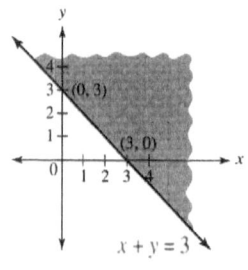

3.

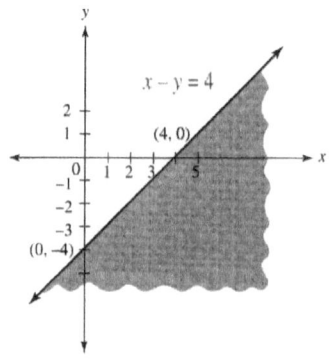

5.

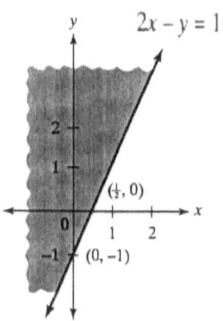

11.

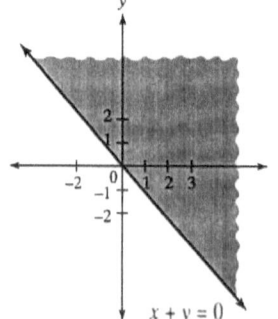

7.

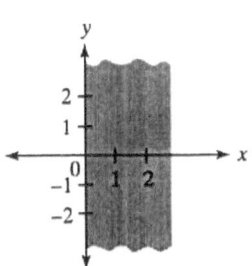

13.

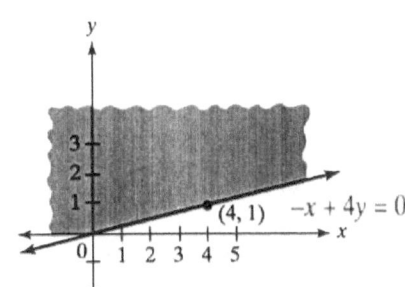

9.

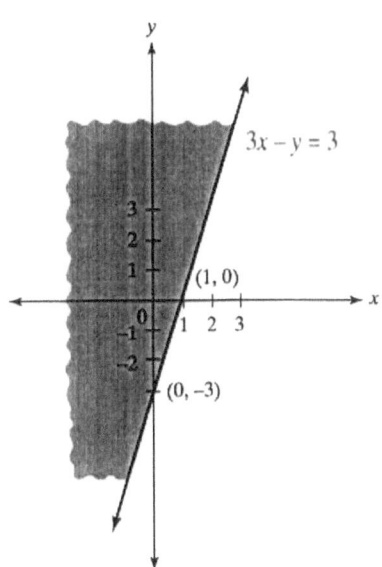

15.

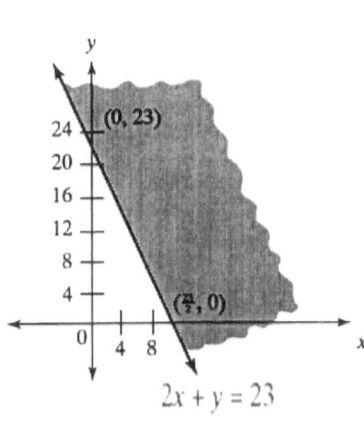

17.

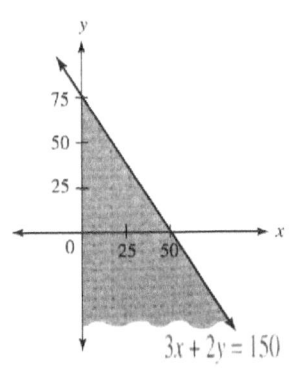

21.

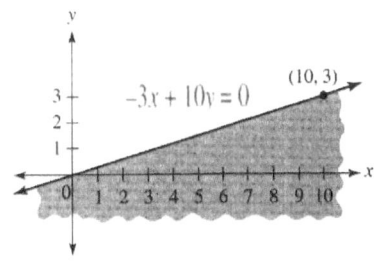

19.

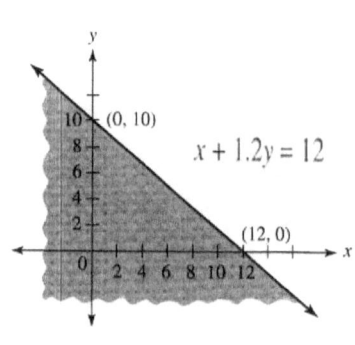

23.

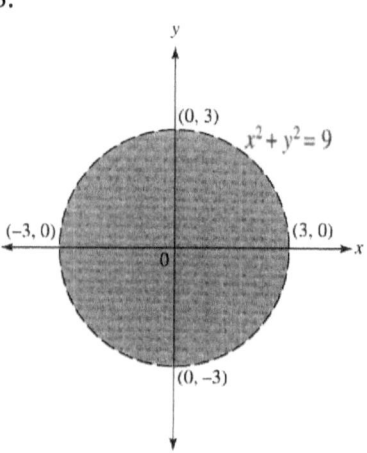

Section 3.1 (p. 114)

1. $f(1) = -4, f(3) = 2, f(-1) = -2, f(-2) = 2, f(0) = -4$

3. $f(-3) = 29, f(\frac{1}{2}) = \frac{17}{4}, f(3) = 17, f(10) = 276$

5. $f(0) = f(1) = f(-1) = f(10) = f(-2) = -1$

7. $f(1) = 1, f(2) = 4, f(3) = 7, f(-1) = -4$

Section 3.2 (p. 121)

1.

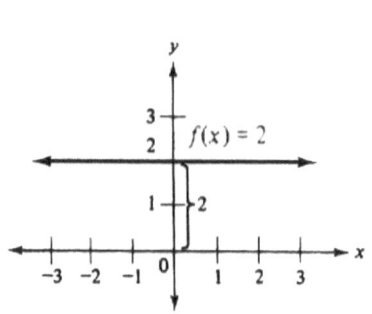

7.

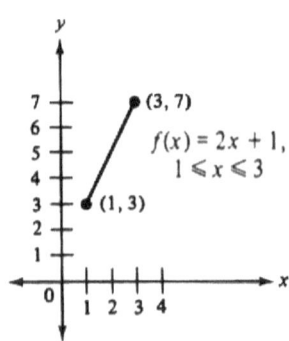

3.

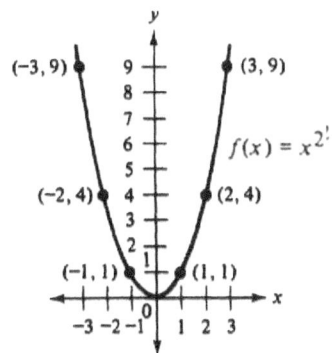

9.

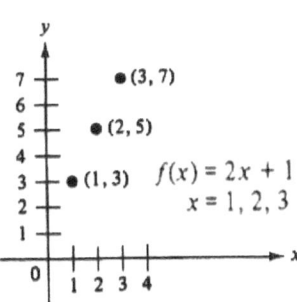

5.

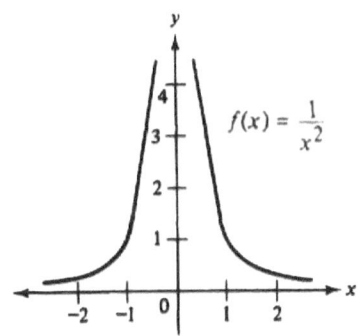

11.

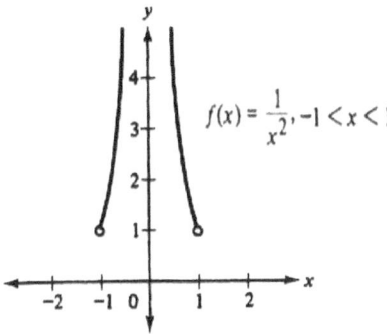

13.

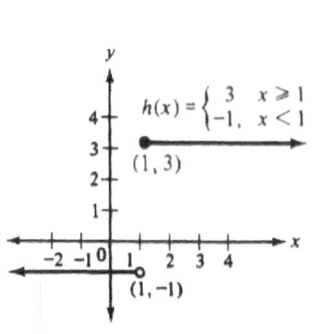

19.

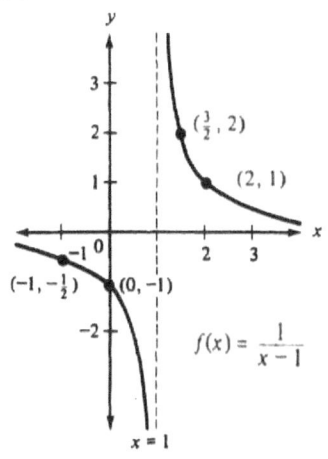

15.

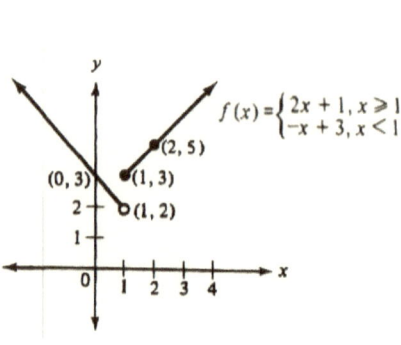

$$f(x) = \begin{cases} 2x + 1, x \geqslant 1 \\ -x + 3, x < 1 \end{cases}$$

21.

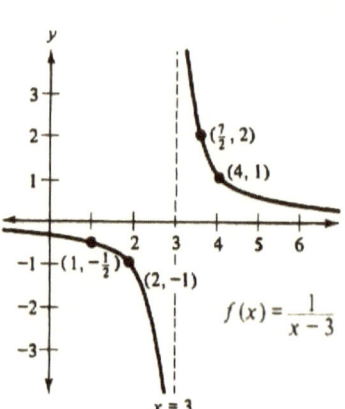

$$f(x) = \frac{1}{x - 3}$$

17.

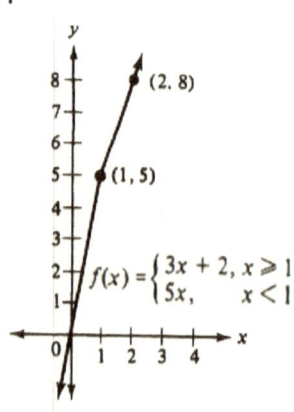

$$f(x) = \begin{cases} 3x + 2, x \geqslant 1 \\ 5x, \quad x < 1 \end{cases}$$

23.

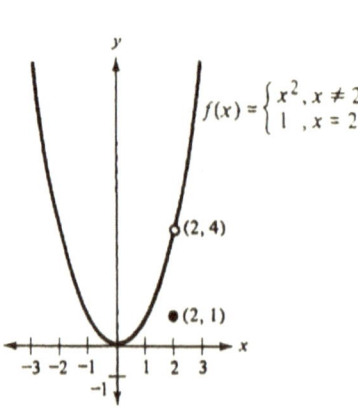

$$f(x) = \begin{cases} x^2, x \neq 2 \\ 1, x = 2 \end{cases}$$

25.

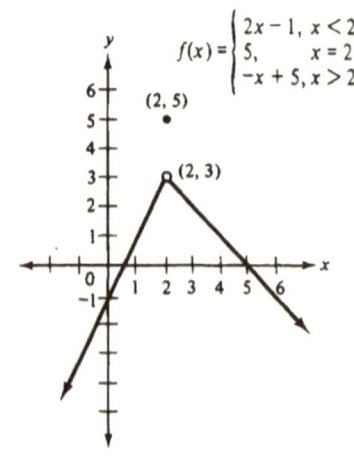

$$f(x) = \begin{cases} 2x - 1, x < 2 \\ 5, \quad x = 2 \\ -x + 5, x > 2 \end{cases}$$

Section 3.3 (p. 129)

1. $C(x) = \begin{cases} 8, & \text{where } 0 \le x \le 50 \\ 8 + 0.09(x - 50), & \text{for } x > 50, \text{ where } x \text{ is the number of message units} \end{cases}$

    accumulated.

3. $I(x) = (200 - 10x)(100 + x)$, where x is the number of taxis added to the fleet. $x = 0, 1, \ldots, 19$.

5. $y = 2(2^t) = 2^{t+1}$, for $t = 12$, $y = 2^{13} = 8192$

7. $C(x) = \begin{cases} 29, & \text{where } 0 < x \le 1 \\ 58, & \text{where } 1 < x \le 2, \text{ subject to the next increase in} \\ 87, & \text{where } 2 < x \le 3 \end{cases}$

    rates, where x is the weight of the letter in ounces (1994)

Section 3.5 (p. 133)

1. $\log_6 36 = 2$

3. $\log_{25} 5 = \dfrac{1}{2}$

5. $\log_{27} 3 = \dfrac{1}{3}$

7. $\log_5 \dfrac{1}{125} = -3$

9. $\log_9 27 = \dfrac{3}{2}$

11. $\log_{81} \dfrac{1}{27} = -\dfrac{3}{4}$

13. $2^4 = 16$

15. $5^3 = 125$

17. $10^{-1} = 0.1$

19. $7^3 = 343$

21. $10^{-2} = 0.01$

23. 3

25. $-\dfrac{1}{3}$

27. 2

29. 2

Section 3.6 (p. 144)

1.

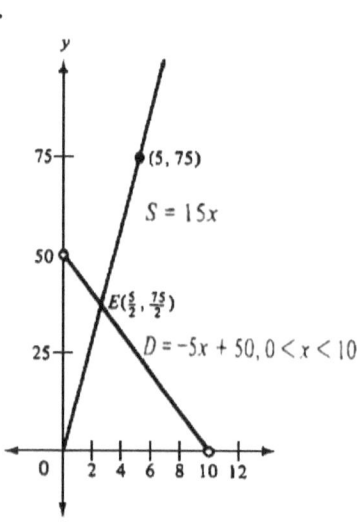

3.

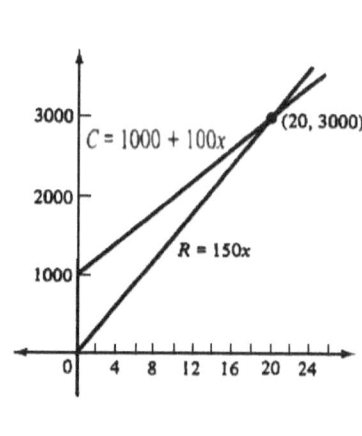

(c) No; math validity is established, but real world accuracy hinges on the real world accuracy of the raw materials we were given to work with

Section 3.7 (p. 148)

1. $f(7,3) = 84$, $f(-1,5) = 28$, $f(2,6) = -60$

3. $f(1,-1,3) = -12$, $f(4,2,1) = 23$, $f(-5,4,2)$ is not defined, $f(-2,4,3) = -33$

SELF-TESTS FOR CHAPTERS 1-3

Self-Test 1 (p. 148)

1. -11

2. $6xyz$

3. $-2x + xy + 3$

4. $x(3z + 4y + 1)$

5. $2y^2(2x - 1 + 3z)$

6. yes: -1 itself since $(-1)(-1) = 1$

7. $-12a + 5b$

8. (a) $z = \dfrac{4x + 2}{x + 2}$

   (b) -1 (c)x

9. (a) $x/2$, $x \neq 2$

   (b) $-1/x, x \neq 2$

10. yes; $0 \cdot \dfrac{1}{4} = 0$

11. (a) 3,   (b) 125

12. (a) $c(20) = 700$, $c(49) = 2991$, $R(20) = 8400$, $R(49) = 14{,}896$

    (b) $P(x) = -5x^2 + 490x - 100$

13. $6548.61

14. $x \geq -2$

15.

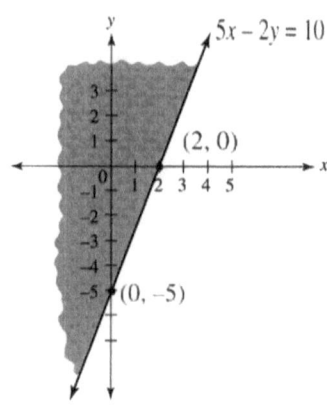

Self-Test 2 (p. 149)

1. $3rs(t+2+4k)$

2. No; there is no number x such that $0 \cdot x = 1$.

3. 11.1%

4. $f(-3)=-13,\ f(-1)=-1,\ f(-2)=-9,\ f(1)=8,\ f(3)=11$

5. No, since 1 is not in the domain of $f(x)$, but is in the domain of $g(x)$.

6. $x \geq -1$

7. $c(x) = \begin{cases} 1.00, & \text{for } 0 < x \leq 1 \\[2mm] 1.20, & \text{for } 1 < x \leq \dfrac{5}{4} \\[2mm] 1.40, & \text{for } \dfrac{5}{4} < x \leq \dfrac{6}{4} \\[2mm] 1.60, & \text{for } \dfrac{6}{4} < x \leq \dfrac{7}{4} \\[2mm] 1.80, & \text{for } \dfrac{7}{4} < x \leq 2 \end{cases}$

8. (a) y^2, $x \neq y$, $y \neq 0$

   (b) $\dfrac{2b}{a}$, $a \neq -3, 0$; $b \neq 0$

9. (a) $\dfrac{5x+2}{x(x+1)}$, $x \neq -1, 0$

   (b) $\dfrac{16x-5}{4x}$, $x \neq 0$

10. (a) 2/9,   (b) 1/16,   (c) 27/8

11. (a) x^3, $x \neq 0$

    (b) $\dfrac{2x^4}{y}$, $x \neq 0$, $y \neq 0$

    (c) $x^3 y^5$, $x \neq 0$, $y \neq 0$

12. (a) 1/81,   (b) −6

    (c) 1/4,   (d) 25

    (e) 1/4,   (f) 1/4

13. $x = 3$

14. $2060

Self-Test 3 (p. 151)

1. $2\sqrt{x^2 + 4y^2}$

2. 1/3

3. 1 has the property $x \cdot 1 = 1 \cdot x = x$ for any real-number x.

4. $\dfrac{-x^2 + 4x + 2}{(x-3)(x+2)}$, $x \neq -2, 3$.

5. −2 and 3 must be excluded because they lead to division by 0, which is undefined.

6. No, since $-\dfrac{1}{\pi}+\pi \neq 0$.

7. 12.6% compounded annually yields the same interest at the end of the year as 12% per annum compounded 6 times per year.

8. $5000(1.05)^{-12} = 2784.20$ or $2784.20

9. $I = (10,000 - 200x)(x + 30)$, where x is the number of new stores opened. $x = 0, 1, \cdots, 49$. The corresponding math model function is defined for $0 \leq x \leq 49$.

10.

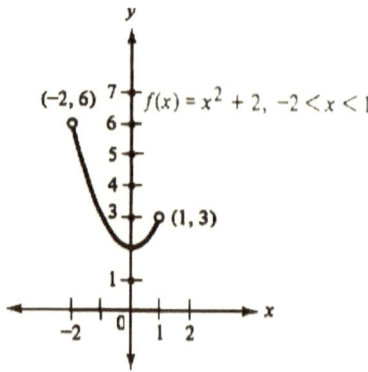

11.

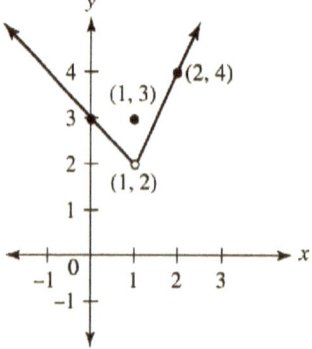

12.

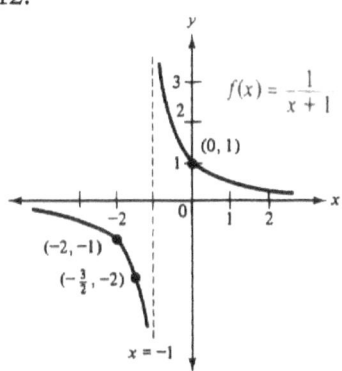

13. $f(1,2,-5) = -\dfrac{3}{5}$, $f(-4,1,3)$ is not defined,

$f(3,1,6) = \dfrac{5}{3}$, $f(3,2,-3) = -\dfrac{11}{3}$

Self-Test 4 (p. 152)

1. (a) $-3/5$,   (b) 64/9

2. (a) no; the denominator is not zero for $x = 3$.

   (b) $x = -1/2$, 2 since the denominator is zero for these values.

3. $z = \dfrac{7 - xm}{2m^2}$

4. $1700

5.

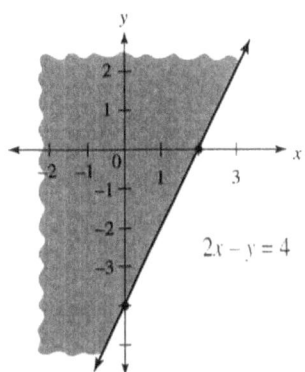

6. (a) $(1.05)^3 - 1 = 15.76$ or 15.76%

 (b) 15.76% compounded annually yields the same interest at the end of the year as 15% per annum compounded 3 times per year.

7. $375

8. $3\sqrt{m^2 + 9n^2}$

9. $y = 0.2(T_1 + T_2 + T_3) + 0.4T_4$

10. no; $f(2)$ is undefined: $g(2) = -1$

11.

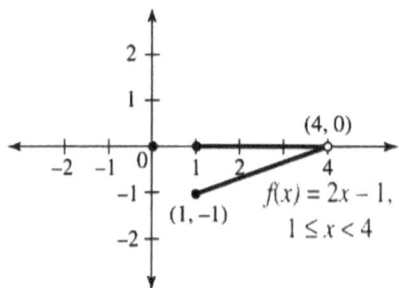

12. (a) 256,   (b) $-\dfrac{135}{16}$

13. $x = \dfrac{3}{12 - 2pq}$, $pq \neq 6$

14. $T(x) = \begin{cases} 0, & \text{for } 0 \leq x \leq 15{,}000 \\ 0.08(x - 15{,}000), & x > 15{,}000, \end{cases}$ where x denotes taxable

 income in dollars.

15.

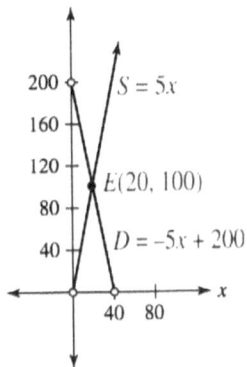

(c) No; math accuracy is established, but real world accuracy hinges on the real-world accuracy of the demand and supply functions we are given

Section 4.1 (p. 160)

1. $y = 2x + 2$

3. $y = x + 3$

5. $y = -2x - 1$

7. $3y = x + 2$

9. $x = 1$

Section 4.2 (p. 167)

1. $(5, 4)$

3. $(1, 4)$

5. $(2, 3)$

7. $(\dfrac{20}{3}, \dfrac{29}{3})$

9. $(50, 40)$

11. $(45, 10)$

13. $(\dfrac{211}{16}, \dfrac{155}{16})$

Section 4.3 (p. 177)

1. (a) $\hat{y} = 4.5x - 1.7$ (b) For $x = 5$, $y = 20.8$; if the president of the company is complimented five times per week for his intellectual keenness, then the predicted **average** annual increment is 20.8 hundred dollars.

3. (a) $\hat{y} = 13.946x - 32.704$ (b) For $x = 30$, $y = 385.68$; for sales personnel who obtain a score of 30 on the aptitude test, the predicted **average** first-year sales volume is $385.68 thousand.

5. (a) $\hat{y} = 0.0534x + 1.145$ (b) For $x = 35$, $y = 3.0$; for students who are 35 years old at the time of graduation, the predicted **average** grade-point average is 3.0.

Section 4.4 (p. 190)

1. $(5, 4)$

3. $(1, 4)$

5. $(\dfrac{20}{3}, \dfrac{29}{3})$

7. $(2, 3)$

9. $y = 2x + 5$, x may be chosen at will (arbitrarily)

11. $(150{,}000; 100{,}000)$

Section 4.5 (p. 195)

1. $(2, 1, -3)$

3. $(2, 1, 3)$

5. no solution

7. $(4, 0, 0)$

Section 4.6 (p. 201)

1. (a) Let x, y, and z denote the total costs of service departments S_1, S_2, and S_3, respectively. The condition that the total cost of each service department must equal the overhead of the department (for March) plus the charges for services provided by the other service departments leads to the following relations:

$$x = 40,000 + 0.1y + 0.2z$$
$$y = 30,000 + 0.1x + 0.05z$$
$$z = 20,000 + 0.05x + 0.1y$$

By transposing terms we obtain the following system of equations:

$$x - 0.1y - 0.2z = 40,000$$
$$-0.1x + y - 0.05z = 30,000$$
$$-0.05x - 0.1y + z = 20,000$$

(b) To the nearest dollar, $x = \$48,831$, $y = \$36,186$, $z = \$26,060$

(c) $P_1 = \$43,128$, $P_2 = \$46,872$

3.
$$x + 0.095y = 2,589.415$$
$$0.01x + 1.010y + 0.01z = 1,529.7$$
$$0.48x + 0.480y + z = 12,253.24$$

$$x = 2457.63, \ y = 1387.18, \ z = 10407.73$$

Section 4.7 (p. 210)

1. $x = 5 - z$, $y = z$, z arbitrary

3. $y = -6 + 2z$, $x = 10 - z$, z arbitrary

5. $x = -8 + w$, $y = -2$, $z = 2$, w arbitrary

7.   We introduce variables $s, t, u, v, w,$ and x as shown in the network diagram.

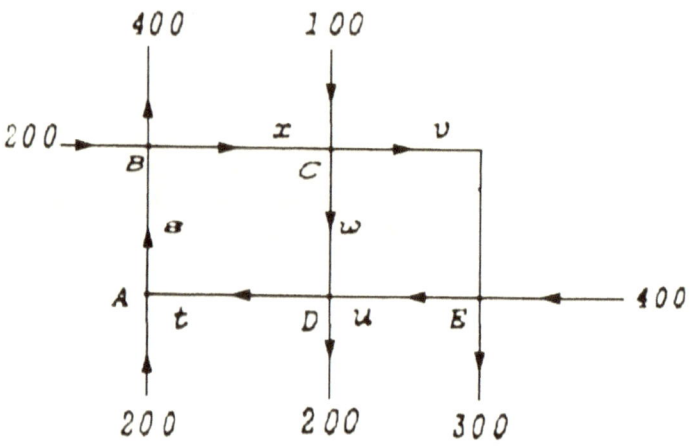

Variable x denotes the number of cars passing between points B and C per hour, v the number of cars passing between points C and E per hour, and so on. For equilibrium at the points A, B, C, D and E, the number of cars entering each of these points per hour must equal the number of cars leaving each of these points per hour. This leads to the following conditions:

$$A : t + 200 = s$$

$$B : s + 200 = x + 400$$

$$C : x + 100 = v + w$$

$$D : u + w = t + 200$$

$$E : v + 400 = u + 300$$

Transposing and rewriting terms yields the following system of equations:

$$s - t = 200$$
$$s - \qquad\qquad x = 200$$
$$v + w - x = 100$$
$$-t + u + \qquad w = 200$$
$$u - v = 100$$

Solving this system yields:

$$s = x + 200$$
$$t = x$$
$$u = -w + x + 200$$
$$v = -w + x + 100$$

w and x are arbitrary

For this solution to make sense in terms of the traffic background in question, the values given to the underlying variables must, at a minimum, be restricted to non-negative integers. From $s = x + 200$, we obtain

$$x = s - 200$$

$x \geq 0$ yields

$$s - 200 \geq 0$$
$$s \geq 200$$

Thus traffic equilibrium cannot be maintained if fewer than 200 cars per hour pass between points A and B, which is clear from the network structure.

The condition $t = x$ tells us that traffic equilibrium cannot be maintained if the number of cars passing between D and A per hour is not equal to the number of cars passing between B and C.

From $u \geq 0$ and $u = -w + x + 200$, we have

$$-w + x + 200 \geq 0$$
$$x + 200 \geq w$$
$$w \leq x + 200$$

This tells us that traffic equilibrium cannot be maintained if the number of cars passing between C and D per hour exceeds 200 plus the number of cars passing between B and C per hour.

Replacing x in $w \leq x + 200$ by $s - 200$ yields

$$w \leq s$$

which tells us that traffic equilibrium cannot be maintained if the number of cars passing between C and D per hour exceeds the number of cars passing between A and B per hour.

Section 4.8 (p. 216)

1. $(-1,3)$, $(\frac{1}{2}, 2)$, $(3,1)$

3. $(1, 4, 1), (1, 2, 1)$

5.

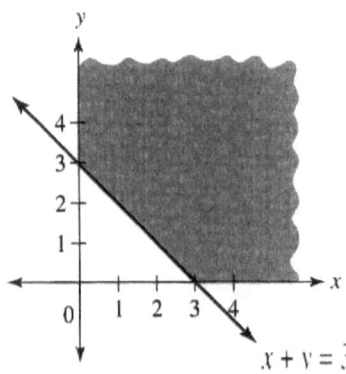

7.

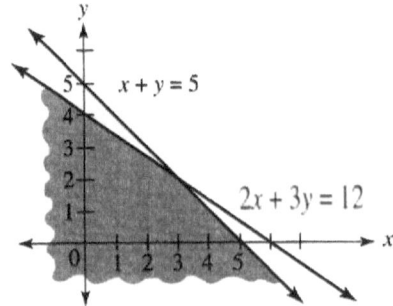

Section 4.9 (p. 219)

1. not linear

3. linear

5. $C(0, 15)$ is not defined, $C(0,23) = 92$,

   $C(\dfrac{23}{2},0)$ is not defined, $C(\dfrac{20}{3},\dfrac{29}{3}) = 72$,

   $C(\dfrac{75}{4},0) = \dfrac{375}{4}$

7. $I(8,2) = 0.84$, $I(25,0)$ is not defined, $I(8,17) = 1.74$, $I(20,5) = 2.1$.

Section 5.1 (p. 226)

1. Right in the sense of validity; wrong in the sense of realism. Jules and Janet are both right in the sense of validity. One or both might be wrong in the sense of realism.

Section 5.2 (p. 231)

1. $\dfrac{350}{50} = 7$; No; it's irrelevant. The "proof" of Henry's theorem establishes its validity with respect to Henry's postulates and has nothing to do with questions involving Andy's actual trip time.

3. Yes. To address the question of the realism of the valid conclusions obtained from Andy and Henry's models the trip would have to be taken along the routes for which these models were developed and a comparison between the deduced times of 6 hours (from Andy's model) and 7 hours (from Henry's model) made with the actual trip times. If the actual trip time for Andy's route is "close" to 6 hours, this would be evidence supporting the realism of Andy's model. If the actual trip time for Henry's route is close to 7 hours, this would be evidence supporting the realism of Henry's model.

5. (a) No. As noted in answer to the first question, the mathematical operation division, yielding $350 \div 50 = 7$, establishes the validity of Henry's conclusion from his postulates. Andy's trip time is irrelevant to the issue of validity.

(b) Andy's actual trip time of 5 hours and 55 minutes is "close" to the deduced trip time of 6 hours obtained from his model, which is evidence in support of the realism of his model. Review question 3.

7. (a) (i) Trip time: 5 hours, (ii) Trip time: 2 hours (iii) Trip time: 9 hours

(b) No; this information is irrelevant to the issue of validity.

(c) The actual trip time of 5 hours and 35 minutes is 10% off the deduced trip time of 5 hours, so that the difference between them is large enough that the times cannot be considered "close." This establishes that Ann's model is not realistic for the first leg of the trip. The actual trip times for the second leg of the trip and the overall trip are "close" enough to the validly deduced trip times that they provide evidence supporting the realism of Ann's model for the second leg of the trip and the overall trip.

(d) No. The validity of the theorems is irrelevant to the issue of their realism. The validity of the theorems is established by mathematical arguments. Their realism has to do with how accurate they are as descriptions of the real-world or institutional world that we have created.

Section 5.3 (p. 242)

1. (a) and (b) Mathematics is precise in the sense that its methods yield valid conclusions with respect to the assumptions made. Sometimes different assumptions are made about a situation under study, and these assumptions lead to different conclusions (solutions), each of which is valid (inescapable consequence) with respect to its assumptions. Different valid conclusions are not in conflict insofar as validity is concerned because they come from different assumptions. They are in conflict insofar as truth/realism is concerned.

(c) No; although both values are mathematically correct (that is, valid with respect to their respective assumptions), we are not assured that the larger value is more realistic than the smaller one, and that the true maximum profit is the predicted 230,000 value.

(d) The assumptions of the M1 model were highly unrealistic.

Section 6.2 (p. 251)

1. The basic data are summarized in Table 1.

Table 1

Product	No. made	Profit per unit ($)	Construction time per unit (hours)	Finishing time per unit (hours)
DT-1	x	140	8	3
DT-2	y	150	5	2
			≤ 2210	≤ 860

The assumptions underlying these data and the conditions that at least 50 DT-1 units and 50 DT-2 units must be made per week lead to LP-2.

Section 6.3 (p. 257)

1. $(6, 0)$; 30

3. $(1, 2)$; 3.7

5. $(45, 15)$; 5.7

7. $(\frac{50}{3}, \frac{65}{3})$; 3350

9. $(50, 40)$; 13.800

Section 6.4 (p. 262)

1. (a) The corner-point method cannot by itself guarantee that profit will be maximized when 300 ZKB-47 and 250 ZKB-82 units are made daily and sold. Whether these output levels or other ones will maximize profit is a question of truth, and the corner-point method, as a mathematical technique, can only ensure the validity of the predicted output levels with respect to the linear programming model that was set up for the profit maximization problem in question. If the assumptions that the model reflects are realistic, then the aforementioned output levels, obtained as a valid conclusion of these assumptions, will maximize profit; if these assumptions are not

realistic, then it might well happen that other output levels would yield a higher profit than that projected for 300 ZKB-47 and 250tt ZKB-82 units.

(b) The basis for implementing the conclusion that 300 ZKB-47 and 250 ZKB-82 units be made and sold daily is the belief, based on an analysis of the company's operations and the market, that the assumptions made are realistic.

Section 6.6 (pp. 268, 275)

1. No; obtaining the solution by means of the latest computer technology is irrelevant to the issue of whether the postulates underlying the Veronika Company's linear program model are realistic.

3. In their analysis of the Austin Company's production operations the Austin Company's operations research department and the Aleksa Company were led to impose the condition that a minimum number of DT-1 and DT-2 units be produced to cover production startup costs, which tend to be high, before assuming that the profit levels per unit for further production are constant. (25 DT-1 and 40 DT-2 units for the operations research department and 50 DT-1 and 50 DT-2 units for the Aleksa Company.)

     No such condition was imposed to cover initial startup costs for the RA5 and RA9 stereo systems which, at the very least, prompts a question about why it was not necessary to impose such a condition.

5. Maximize $P = 180x + 120y$ x and y are the number of K15 and
     subject to   K31 units, respectively, to be made

$$x \geq 0, y \geq 0$$ per week.

$$4x + 3y \leq 320$$

$$5x + 2y \leq 330$$

Sol. (50, 40); max. value 13,800 (see Exercise 9 of section 6.3.

7. Minimize $C = 45x + 50y$ x and y are the number of pounds
    subject to   of pork and beef, respectively, to be

    $x \geq 0, y \geq 0$ used in putting together a can of
                                    meat.
    $3x + 5y \geq 12$

    $x + y \geq 3$

    Sol. $(\frac{3}{2}, \frac{3}{2})$; minimum value 142.5.

9. Maximize $I = 0.09x + 0.06y$ x and y are the amounts (in millions
    subject to   of dollars) allocated for loans and
                                    securities, respectively.
    $x \geq 0, y \geq 0$

    $x + y \leq 25$

    $-x + 4y \geq 0$

    $x \geq 8$

    Sol. (20, 5); maximum value 2.1.

11. Two linear programs emerge, one for the faculty council and the other for
    the administration. The constraints are the same for both linear programs
    since the underlying conditions are the same.

    Max. $F = x + y$ subject to   Min. $C = 500x + 1000y$ subject
    $x \geq 0, y \geq 0$ to (same constraints); x and y
                                    denote the number of associate and
    $x \leq 22, y \geq 3$ full-professor slots, respectively, to
    $x + 2y \leq 30$ be established.
    $-x + 4y \leq 0$

    Sol. (22, 4); maximum value 26.   Sol. (12, 3); minimum value 9000.

15. Minimize $C = 1.2x + 1.8y$
    subject to

    $x \geq 0, y \geq 0$

    $x + y = 2,000,000$

    $30x + 32y \geq 62,400,000$

x and y denote the number of tons of steel produced subject to the F14 and F24 filter systems, respectively.

Sol. (800,000; 1,200,000); minimum value 3,120,000.

In all of these situations we are translating assumptions made into a linear program, omitting nothing that was explicitly stated and adding nothing of our own. Our concern is that the **assumptions made are realistic** and that the factors that were left out because they were viewed as negligible may be **realistically treated as such.**

Chapter 7 (p. 279)

Section 7.1 (p. 284)

1. If the marketing department reduced its requirement on afternoon time to at least 6 minutes of afternoon time be purchased from at least 9 minutes of afternoon time be purchased the constraints would be compatible.

   Alternatively, if the marketing department increased its advertising budget to $25,000, the constraints would be compatile. The region of feasible points would consist of one point, (8, 9).

Section 7.2 (p. 292)

1. Min. $C = x + 1.5y$ subject to
   $x \geq 0, y \geq 0$
   $x + 2y \geq 9$
   $3x + 2y \geq 12$

   x and y denote the number of pounds of bonemeal and processed vegetable matter, respectively, to be used to make up a can of Martins Miracle.

   Sol. (3/12, 15/4); min. value 7.125.

   (b) $x + y \geq 6$
   (c) $(3, 3)$

3. $2x + y \leq 5$; sol. $(1, 3)$, max value 34.

SELF-TESTS FOR CHAPTERS 4-7

Self-Test 1 (p. 293)

1. $y = \frac{1}{2}x + 4$

2. $x + 4y = 7$

3. $(16, 12)$

4. $(-2, 4)$

5. (a) $y = 0.4384x - 0.7371$; (b) for $x = 45$, $y = 19$; the predicted **average income** for employees of age 45 in this corporation is $19,000.

6. $(0, -8, -4)$

7. $x = -\frac{19}{7} + 2z$, $y = \frac{2}{7} - z$, z arbitrary

8. no; the 2^x term destroys linearity

9.

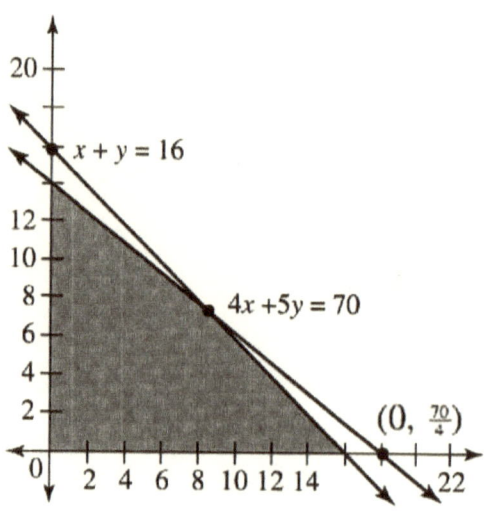

$x + y = 16$

$4x + 5y = 70$

$(0, \frac{70}{4})$

Self-Test 2 (p. 294)

1. (a) No; $F = 3x + y^2$ is not linear. (b) Yes. (c) No; we seek constraints that reflect reality, but there is no guarantee of success. (d) No; the validity of a solution means that it follows from the linear program, which may or may not be realistic. (e) No; see (d).

2. $(-2, 1, 3)$

3. $(1, 1); 5$

4. Max. $P = 2x + 3y$ subject to

   $x \geq 0, y \geq 0$

   $50x + 60y \leq 12{,}000$

   $3x + 4y \leq 760$

   where x and y are the number of K17 and K24 units, respectively, to be stocked.

   Solution: (0, 190); value 570.

5. (a) Different solutions were obtained on the basis of different assumptions. The mathematics is precise in the sense that its methods yield conclusions that are valid with respect to the underlying

assumptions. The fact that the simplex method was used in one situation whereas the corner-point method was used in the other instance is irrelevant as far as the validity of the conclusions obtained is concerned. **Both procedures, as mathematical methods, yield valid conclusions.**

(b) Both solutions are correct in the sense of being valid with respect to their underlying hypotheses. Clearly both cannot be true since the maximum profit cannot be $3000 per week and at the same time $2500 per week.

(c) The basic question is, which solution actually maximizes profit, or to allow a shade of gray, which solution comes closest to approximating the maximum profit? This is the solution that we wish to implement. The assumptions from which these solutions follow as valid consequences should be carefully reviewed as to their realism. If the assumptions that lead to the solution (450, 300), with a predicted maximum profit of $2500 per week, are more realistic than the assumptions that lead to the solution (500, 280), with a predicted maximum profit of $3000 per week, then the (450, 300) solution should be implemented since the predicted $2500 profit value will be closer to reality than the $3000 value. Without question, the predicted $3000 profit is more appealing, but what is more appealing is not necessarily what is realistic.

6. **What Math Can and Cannot Do for Business**
(Comments on "Math Will Rock Your World," *Business Week*, Jan. 23, 2006)

A KEY statement in the closing of the article notes that midcareer managers "still must understand enough about math to question the assumptions behind the numbers." Not so. Knowledge of math technique is useless here. It is intimate knowledge of the enterprise the math is being applied to that will prepare midcareer managers to challenge the realism of the assumptions.

Math methods applied to unrealistic assumptions cannot be expected to yield golden truths. Garbage in, garbage out—as the saying goes.

—William J. Adams
Professor of Mathematics
Pace University
New York

Self-Test 3 (p. 297)

1. (a) $x = 2y - 4$, y arbitrary   (b) no solution

2. (a)   (b)

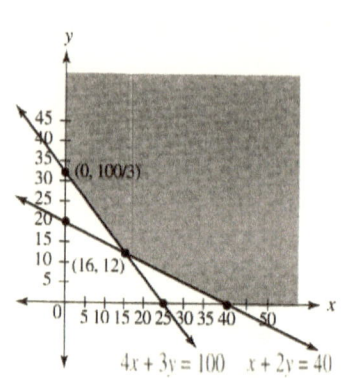

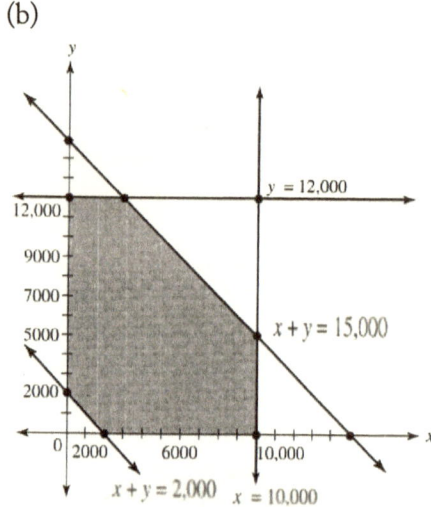

3. (a) Max. $F(x, y) = 1.1x + 0.8y$
   subject to
   $$x \geq 0, y \geq 0$$
   $$x \leq 16$$
   $$x \geq 8$$
   $$y \geq 6$$
   $$2000x + 1000y \leq 48,000$$
   or equivalently,
   $$2x + y \leq 48$$

   x is the number of minutes of morning time purchased, and y is ... of afternoon time ....

   (b) Solution: (8,32); max. value: 34.4

   (c) The reservations expressed by Bass, Lonsdale, and Kotler are applicable to the afore linear program model but, crude a model as it is, it might be sufficient for the purpose that the management of the Daniel Company has in mind. This is a matter for them to consider.

(d) (i)   Replace the constraints $x \geq 8$ and $y \geq 6$ by $x \geq 12$ and $y \geq 26$.

    (ii)   No; the advertising budget constraint $2000x + 1000y \leq 48,000$ is not satisfied. The modified constraints would put the Daniel Company $2000 over budget.

    (iii) Increase the advertising budget by $2000 or, if x (no. of minutes of morning time) and y (no. of minutes of afternoon time) are to be increased, make sure that $2x + y$ not exceed 48.

4. $0.99x - 0.02y - 0.02z = 30,000$ x, y, and z denote the total costs
     $-0.08x + y - 0.01z = 20,000$ of the accounting, shipping,
                                     and marketing departments,
     $-0.02x - 0.15y + z = 95,000$ respectively, for October.

5.   (a) $x + y \leq 33$ x and y denote the number of desks and bookcases, respectively, to be made. Sol. (15, 18); max. value 1392.

   (b)   Max. $P = 40x + 44y$
       subject to
        $x \geq 0, y \geq 0$
        $4x + 5y \leq 150$
        $6x + 7y \leq 228$
        $5x + 4y \leq 150$
        $x + y \leq 33$

Self-Test 4 (p. 300)

1. No solution

2.

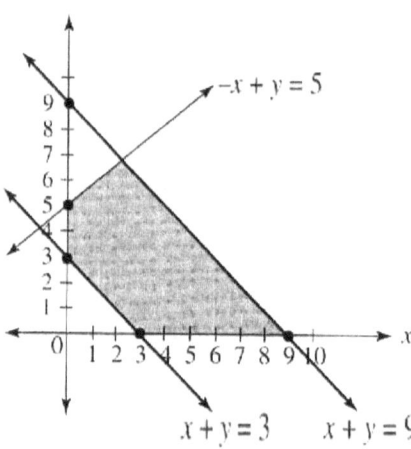

3. (a) Max. $P = 12x + 10y$
   subject to
   $$x \geq 0, y \geq 0$$
   $$x + 2y \leq 220$$
   $$3x + 2y \leq 540,$$

   where x and y are the number of B7 and B9 units, respectively, to be made. Sol. $(160, 30)$; maximum value 2220.

4. Disagree. Mathematical methods have the advantage of certitude in the sense of validity, not truth. The statement confuses truth with validity. Replace truth where it appears with validity and we have a correct assertion.

5. (a) because it's the only function that is linear.

6. (a) and (c) because they are linear inequalities.

7. No; we can make $F = 3x + 2y$ as large as we wish by taking suitable feasible points sufficiently far out along the x-axis. For example, to make F larger than 1000, take $P(1000, 0)$, to make F larger than 1 million, take

P(1 million, 0), and so on. We can do this because the feasible points of this linear program are unbounded.

Section 8.1 (p. 304)

1. $x^3 + 8x^2 + 4x - 4$

3. $-8x^2h + 5xh + h^2 - 1$

5. $4x^2y - 3xy - 3y + 7$

7. $2x^2 - x - 5$

9. $-5x^2h - 3xh + 2x + 1$

Section 8.1 (p. 306)

11. $3x^2 + 12x$

13. $-x^3 + x^2 - x$

15. $-3x^4 + 12x^3 - 21x$

17. $6x^3h^2 + 8xh^2 - 2xh$

19. $x^2 + x - 2$

21. $x^2 - h^2$

23. $2x^2 - x - 3$

25. $6x^2 + x - 2$

27. $x^3 - 3hx^2 + 3h^2x - h^3$

29. $x^4 + x^3 - 6x^2 + 4x - 8$

31. $x^5 + 2x^4 - 3x^3 - 4x^2 + 5x + 2$

33. $9x^4 + 6x^3 - 8x^2 + 21x - 14$

35. $16x^4 + 12x^3 + 12x^2 + 37x + 21$

37. $10x^5 - 30x^4 + 5x^3 - 15x^2$

39. $x^3 - 5x^2 - 18x + 72$

41. $x^5 + 5x^4h + 10x^3h^2 + 10x^2h^3 + 5xh^4 + h^5$

Section 8.1 (p. 311)

43. $x(5 + 3y + z)$

45. $2a(3x - 2ay)$

47. $xy(2 + x - y)$

49. $(x + 6)(x - 1)$

51. $(x + 11)(x + 1)$

53. $(t - 5)(t - 1)$

55. $(x - 7)(x - 3)$

57. does not factor

59. $3x(x - 7)(x - 2)$

61. $t(t - 10)(t - 6)$

63. $4x(x - 4)(x + 2)$

65. $(-x - 5)(x + 5)$

67. $(x - a)(x + a)$

69. does not factor

71. $(2x - 1)(x - 1)$

73. $(3x - 2)(2x + 3)$

Section 8.2 (p. 322)

1. -6 and 1

3. -3 and -1

5. -11 and -1

7. 6 and 12

9. $\dfrac{1}{2}$ and $\dfrac{5}{2}$

11. $\dfrac{-7 \pm \sqrt{41}}{2}$

13. $\dfrac{8 \pm \sqrt{56}}{4}$

15. $\dfrac{2 \pm \sqrt{32}}{2}$

17. no solution in the real-number system

19. 0 and $\dfrac{-5 \pm \sqrt{137}}{8}$

21. minimum point: $(-\dfrac{5}{2}, -\dfrac{57}{4})$

23. minimum point: $(\dfrac{3}{2}, -4)$

25. minimum point: $(-6, -25)$

27. (a) 3.36 and 48.64, (d) $P(x) = -\dfrac{11}{2}x^2 + 286x - 900$, $x = 26$; (b) and

(e) no; it establishes, mathematically speaking, the break-even levels of the firm and the output level for which the profit function is maximized. Their real world accuracy hinges on the real world accuracy of these functions.

29. (a) 47.56 (b) see the afore discussion of 27 (b) and (e).

Section 8.3 (p. 326)

1. (a) $y = 4x$, (b) 48

3. (a) $s = 12uv/t$ (b) 36

5. (a) $D = 4000/p$, (b) 800

7. $A = \pi r^2$, (b) 16ϖ

9. $I = 40/d^2$

Section 9.1 (p. 334)

1. 79; 820

3. 45,150

5. $19,200; $128,800

7. $r = \dfrac{2}{3}$, $l = \dfrac{64}{243}$, $S_6 = \dfrac{1330}{243}$

9. $40,725

13. The total declining balance depreciation of an asset that costs A dollars and has a useful life of n years is given by:

$$S_n = \frac{a(1 - r^n)}{1 - r}$$

where $a = \dfrac{2A}{n}$ and $r = 1 - \dfrac{2}{n}$. Substituting these values into S_n yields:

$$S_n = \frac{2A\left[1-\left(1-\dfrac{2}{n}\right)^n\right]}{n\left[1-\left(1-\dfrac{2}{n}\right)\right]}$$

$$= \frac{2A\left[1-\left(1-\dfrac{2}{n}\right)^n\right]}{n\left[\dfrac{2}{n}\right]}$$

$$= A\left[1-\left(1-\dfrac{2}{n}\right)^n\right]$$

Section 9.2 (p. 341)

1. (a) $394.91,   (b) $798.70,   (c) $1606.46

3.

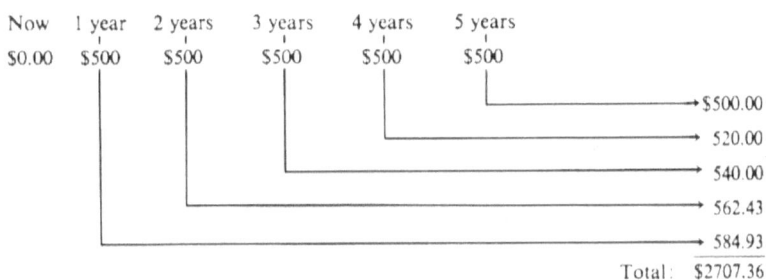

Section 9.2 (p. 347)

7. $11,652.30

9. Set A costs $167.55; set B costs $163.65. Set B is cheaper by $3.90.

11. $2256.08

13. $3951

15. $3211

17. $296

SELF-TESTS FOR CHAPTERS 8-9

Self-Test 1 (p. 349)

1. $3x^2 - 2x - 2$

2. $x^2 y^2 + 10x^2 y + xy - 1$

3. $x^2 h + 8xh + h^2 - 2$

4. $4x^3 + 20x^2 - 4x$

5. $15x^2 + x - 2$

6. $6x^4 + 13x^2 + 5$

7. $36x^2 + 15x - 6$

8. $3x^4 + 9x^3 y + 9x^2 y^2 + 3xy^3$

9. $(t + 7)(t - 3)$

10. $(2x - 5)(2x + 5)$

11. $4(x^3 + 2x^2 - 35)$

12. $(2x + 1)(x + 5)$

13. $(2y + 3)(y + 2)$

14. 1 and 5

15. 0, -3, and 9

16. $\dfrac{3 \pm \sqrt{69}}{6}$

17. Maximum point: $(1, 11)$

18. (a) $z = 6t$,   (b) 24

19. 9900

20. (a) $2667,   (b) $2311,   (c) $17,960

21. $223.32

22. $1139.68

Self-Test 2 (p. 350)

1. $5x^2 + 16x - 9$

2. $5x^2 + 5x - 3$

3. $-2x^2t^2 + 6xt^2 - 2xt - 23$

4. $6x^4h + 8x^3h^2 - 16x^2h$

5. $12x^3 + 3x^2 + 8x + 2$

6. $3x^5 - 2x^4 + 6x^3 - 13x^2 + 9x - 2$

7. $x^3 - 4x^2 - 11x + 30$

8. $4x^3y + 16x^2y^2 + 16xy^3$

9. $2x(x-4)(x+1)$

10. $(x-13)(x+1)$

11. $3x(x-7)(x+1)$

12. $(3t+4)(2t-1)$

13. $(3x+4)(x+2)$

14. $-5/3, 3/2$

15. $1, 3/2$

16. $\dfrac{-4 \pm \sqrt{88}}{-4}$

17. (a) 5.28, 94.72 (d) $P(x) = -2x^2 + 204x - 1000$; $x = 51$ yields the maximum value of $P(x)$; (b) and (d) no; see discussion of 27 (b) and (e) of sec. 8.2.

18. (a) $v = \dfrac{2wx}{5}$,   (b) $v = 12$

19. $r = -1/3$, $l = 1/1458$, $S_7 \approx 3/8$

20. $200(77.59831) = \$15,519.66$

21. $200(16.35143) = \$3270.29$

Section 10.2 (p. 355)

1. $(1, 6)$; $(-11, 3)$

3. $(4, 5)$; $(-9, 19)$

5. $(5, 4)$; $(0, 0)$

7. $(0, 0)$; $(5, -12)$

9. $(x + 1, y)$; (x, y)

11. $(-\dfrac{78}{25}, \dfrac{96}{25})$, $(1, 0)$

13. $(0, 2)$; $(-1, 0)$

15. (x, y); $(0, xy)$

17. $z_1 z_2 = z_2 z_1 = (-9, 19)$

19. $(z_1 z_2)z_3 = z_1(z_2 z_3) = (75, -11)$

21. $z_1(z_2 + z_3) = z_1 z_2 + z_1 z_3 = (-16, 2)$

Section 10.2 (p. 362)

23. No; their product is not $1 + 0i$.

25. (a) $\dfrac{7}{53} + \dfrac{2}{53}i$ (b) $\dfrac{-3}{25} - \dfrac{4}{25}i$

    (c) $\dfrac{2}{13} - \dfrac{3}{13}i$ (d) $\dfrac{-1}{17} + \dfrac{4}{17}i$

27. $6 - i$

29. $14 - 5i$

31. $\dfrac{11}{10} - \dfrac{7}{20}i$

33. $20 - 4i\sqrt{2}$

35. $-1 - 3i\sqrt{2}$

37. $\dfrac{4}{11} + \dfrac{5}{11}i\sqrt{2}$

39. No; the analysis given for real-numbers in section 1.4 applies to complex-numbers as well.

Section 10.3 (p. 365)

1. (a) $\sqrt{13}$ (b) $\sqrt{20}$ (c) $\sqrt{17}$

   (d) $\sqrt{40}$ (e) $\sqrt{34}$ (f) $\sqrt{58}$

3. No; an order relation cannot be defined on the non-real complex-numbers.

Section 10.4 (p. 367)

1. $\dfrac{-1 \pm i\sqrt{19}}{2}$

3. $\dfrac{-2 \pm i\sqrt{12}}{2} = -1 \pm 2i\sqrt{3}$

5. $\dfrac{2 \pm i\sqrt{8}}{-6} = -\dfrac{1}{3} \pm \dfrac{i\sqrt{2}}{3}$

Section 10.5 (p. 375)

1. 0.4226; 0.9063

3. 0.7660; −0.6428

5. −0.5; −0.8660

7. −0.9397; 0.3430

9. −1; 0

11. −0.7660; −0.6428

13. 0.5; −0.8660

15. −0.9397; 0.3420

Section 10.6 (p. 378)

1. $1.9284 + 2.2980i$

3. $5.6568 + 5.6568i$

5. $-2.1213 + 2.1213i$

7. $3i$

9. $\sqrt{2}(\cos 45° + i \sin 45°)$

11. $\sqrt{2}(\cos 315° + i \sin 315°)$

13. $2(\cos 60° + i \sin 60°)$

15. $16(\cos 0° + i \sin 0°)$

Section 10.7 (p. 380)

1. $12(\cos 45° + i\sin 45°)$

3. $3(\cos 20° + i\sin 20°)$

5. $32(\cos 150° + i\sin 150°)$

7. $4(\cos 180° + i\sin 180°)$

9. $10(\cos 284° + i\sin 284°)$

11. $243(\cos 90° + i\sin 90°) = 243i$

Section 10.8 (p. 383)

1. $1, -\dfrac{1}{2} + \dfrac{\sqrt{3}}{2}i, -\dfrac{1}{2} - \dfrac{\sqrt{3}}{2}i$

3. $1.732 + i, -1.732 + i, -2i$

5. $2, 2(\cos 72° + i\sin 72°), 2(\cos 144° + i\sin 144°),$
   $2(\cos 216° + i\sin 216°), 2(\cos 288° + i\sin 288°)$

7. $1, \dfrac{1}{2} + \dfrac{\sqrt{3}}{2}i, -\dfrac{1}{2} + \dfrac{\sqrt{3}}{2}i, -1, -\dfrac{1}{2} - \dfrac{\sqrt{3}}{2}i, \dfrac{1}{2} - \dfrac{\sqrt{3}}{2}i$

Section 10.9 (p. 385)

1. Yes, by direct observation we see that G is closed with respect to addition and multiplication, G contains additive and multiplicative identity values (0 and 1), each value has an additive inverse (-1 for 1 and vice versa, 0 for itself), and each nonzero value has a multiplicative inverse (1 for 1, -1 for -1). The commutative, associative, and distributive conditions can be verified directly or by simply observing that these conditions are satisfied by virtue of G being a subcollection of the real-numbers.

Section 10.10

Self-Test 1 (p. 385)

1. (a) $1+2i$ (b) $-5+2i$ (c) $-3+11i$

   (d) $-\dfrac{9}{10}+\dfrac{7i}{10}$ (e) $2-3i$ (f) $\dfrac{3}{10}+\dfrac{i}{10}$

2. $2i$ and $-2i$ since they lead to division by 0, which is not defined.

3. No, since non-real complex-numbers cannot be ordered.

4. (a) $-2i, 2i$ (b) $\dfrac{1\pm i\sqrt{15}}{4}$

5. (a) $16(\cos 180° + i\sin 180°)$

   (b) $\sqrt{5}(\cos 154° + i\sin 154°)$

6. $z^8 = 16(\cos 0° + i\sin 0°) = 16$

7. $2, 2i, -2, -2i$

8. No; closure with respect to multiplication is not satisfied. For example, $i^2 = -1$, which is not of the form $0+yi$.

Self-Test 2 (p. 386)

1. (a) $-1+3i$ (b) $3-7i$ (c) $8+9i$

   (d) $\dfrac{-12-i}{29}$ (e) $-1+2i$ (f) $\dfrac{-2-5i}{29}$

2. No, since there is no complex-number z such that $z \cdot 0 = 0 \cdot z = 1$.

3. (a) $|z_1| = 2, |z_2| = \sqrt{13}$

   (b) No, the non-real-complex-numbers cannot be ordered. Geometrically speaking, $|z_1| < |z_2|$ means that the point $P(1, 1)$ representing z_1, is closer to the origin 0 than the point $Q(-2, 3)$ representing z_2.

4. $\dfrac{-1\pm i\sqrt{11}}{6}$

5. (a) $3(\cos 90° + i\sin 90°)$ (b) $4(\cos 120° + i\sin 120°)$

6. (a) $1 - 1.732i$ (b) $-3.5355 + 3.5355i$

7. $4(\cos 180° + i\sin 180°) = -4$

8. $\sqrt{2} + i\sqrt{2}$, $2(\cos 105° + i\sin 105°)$, $2(\cos 165° + i\sin 165°)$,
   $-\sqrt{2} - i\sqrt{2}$, $2(\cos 285° + i\sin 285°)$, $2(\cos 345° + i\sin 345°)$

9. No; consider $1 + i$, for example. Its multiplicative inverse is $\dfrac{1}{2} - \dfrac{1}{2}i$, which is not a number in G.

10. The product theorem for quotients introduced in section 1.6, $\sqrt[n]{ab} = \sqrt[n]{a} \cdot \sqrt[n]{b}$, requires that a and b be **positive**. It does not carry over to the situation at hand.

Section 11.1 (p. 390)

1. $A + B = \begin{bmatrix} 6 & 1 & 2 \\ 6 & 4 & 6 \end{bmatrix}$,

   $A - B = \begin{bmatrix} -2 & -3 & 4 \\ -4 & 0 & -6 \end{bmatrix}$,

   $B - A = \begin{bmatrix} 2 & 3 & -4 \\ 4 & 0 & 6 \end{bmatrix}$

3. $A + B = \begin{bmatrix} 10 & -8 \\ 9 & -6 \end{bmatrix}$,

   $A - B = \begin{bmatrix} -6 & 6 \\ 3 & 2 \end{bmatrix}$

5. $\begin{bmatrix} -2 & -2 & 1 \\ -4 & -5 & -2 \\ -3 & 0 & 0 \end{bmatrix}$

7. $A + B = \begin{bmatrix} 5 & 1 & -1 \end{bmatrix}$
   $A - B = \begin{bmatrix} 1 & 3 & 9 \end{bmatrix}$

9. (a) 4,　　　(b) −6,　　　(c) 17,　　　(d) −6

Section 11.1 (p. 396)

11. $AB = \begin{bmatrix} 5 & 12 & -5 \\ -2 & 0 & -1 \\ -3 & 4 & -4 \end{bmatrix}$

$A^2 = \begin{bmatrix} -6 & 5 & 11 \\ -3 & -2 & 1 \\ -9 & -3 & 6 \end{bmatrix}$

$BA = \begin{bmatrix} 3 & 5 & 2 \\ -2 & 3 & 5 \\ 3 & -2 & -5 \end{bmatrix}$

$B^2 = \begin{bmatrix} 5 & 4 & 0 \\ 5 & 10 & -1 \\ 1 & 1 & 4 \end{bmatrix}$

13. $\begin{bmatrix} 0 & 0 \\ 0 & 0 \end{bmatrix}$

15. $\begin{bmatrix} 0 & 0 & 0 \\ 0 & 0 & 0 \\ 0 & 0 & 0 \end{bmatrix}$

17. $\begin{bmatrix} a & b & c \\ d & e & f \\ g & h & i \end{bmatrix}$

19. (a) $A(BC) = (AB)C = \begin{bmatrix} 10 & -5 \\ -5 & 5 \end{bmatrix}$

(b) $A(B+C) = AB + AC = \begin{bmatrix} 16 & -7 \\ 9 & -8 \end{bmatrix}$

(c) $(B+C)A = BA + BC = \begin{bmatrix} -1 & 7 \\ 8 & 9 \end{bmatrix}$

21. (a) $BC = \begin{bmatrix} ej + fm & ek + fn \\ gj + hm & gk + hn \end{bmatrix}$,

$AB = \begin{bmatrix} ae + bg & af + bh \\ ce + dg & cf + dh \end{bmatrix}$,

$AC = \begin{bmatrix} aj + bm & ak + bn \\ cj + dm & ck + dn \end{bmatrix}$

(b) $A(BC) = \begin{bmatrix} a(ej + fm) + b(gj + hm) & a(ek + fn) + b(gk + hn) \\ c(ej + fm) + d(gj + hm) & c(ek + fn) + d(gk + hn) \end{bmatrix}$

$(AB)C = \begin{bmatrix} (ae + bg)j + (af + bh)m & (ae + bg)k + (af + bh)n \\ (ce + dg)j + (cf + dh)m & (ce + dg)k + (cf + dh)n \end{bmatrix}$

(c) $A(B+C) = \begin{bmatrix} a(e + j) + b(g + m) & a(f + k) + b(h + n) \\ c(e + j) + d(g + m) & c(f + k) + d(h + n) \end{bmatrix}$

$AB + AC = \begin{bmatrix} ae + bg + aj + bm & af + bh + ak + bn \\ ce + dg + cj + dm & cf + dh + ck + dn \end{bmatrix}$

$$
\text{23.} \quad BE = \begin{bmatrix} 13,300 & 13,900 \\ 17,200 & 18,350 \end{bmatrix} \quad \begin{matrix} \text{CMM steel} \\ \text{CST steel} \end{matrix}
$$

plant P_1 plant P_2

The entries of BE describe the cost of pollution control in making the two types of steel at the two plants: \$13,300 is the cost of pollution control in producing CMM steel at plant P_1; \$13,900 is the cost of pollution control in producing CMM steel at plant P_2; etc.

Section 11.2 (p. 404)

$$
\text{1.} \quad \overline{0} = \begin{bmatrix} 0 & 0 \\ 0 & 0 \\ 0 & 0 \\ 0 & 0 \end{bmatrix}
$$

$$
\text{3.} \quad I_4 = \begin{bmatrix} 1 & 0 & 0 & 0 \\ 0 & 1 & 0 & 0 \\ 0 & 0 & 1 & 0 \\ 0 & 0 & 0 & 1 \end{bmatrix}
$$

$$
\text{5. no, since} \quad \begin{bmatrix} 2 & \frac{1}{2} \\ \frac{1}{3} & -1 \end{bmatrix} \cdot \begin{bmatrix} 4 & 1 \\ 2 & 3 \end{bmatrix} = \begin{bmatrix} 9 & \frac{7}{2} \\ -\frac{2}{3} & -\frac{8}{3} \end{bmatrix} \neq \begin{bmatrix} 1 & 0 \\ 0 & 1 \end{bmatrix}
$$

Section 11.3 (p. 410)

$$
\text{1.} \quad \begin{bmatrix} \dfrac{1}{13} & \dfrac{3}{13} \\ \dfrac{4}{13} & -\dfrac{1}{13} \end{bmatrix}
$$

3. $\begin{bmatrix} -\dfrac{1}{10} & \dfrac{3}{10} \\ \dfrac{4}{10} & -\dfrac{2}{10} \end{bmatrix}$

5. $\begin{bmatrix} 1 & 0 & 0 \\ -1 & \dfrac{1}{2} & 0 \\ 0 & -\dfrac{1}{2} & \dfrac{1}{5} \end{bmatrix}$

7. no inverse

9. $\begin{bmatrix} -\dfrac{3}{4} & -\dfrac{1}{4} & \dfrac{1}{2} \\ \dfrac{5}{4} & \dfrac{5}{4} & -1 \\ -\dfrac{1}{4} & -\dfrac{3}{4} & \dfrac{1}{2} \end{bmatrix}$

11. $\begin{bmatrix} \dfrac{7}{42} & \dfrac{7}{42} & \dfrac{1}{6} \\ \dfrac{11}{42} & -\dfrac{1}{42} & -\dfrac{1}{6} \\ \dfrac{5}{42} & \dfrac{11}{42} & -\dfrac{1}{6} \end{bmatrix}$

Section 11.4 (p. 414)

1. $\left(\dfrac{16}{7}, \dfrac{22}{7}\right)$ 3. $(6, -2)$ 5. $(3, 2, 1)$

Section 11.5 (p. 418)

1. (a) $X = \begin{bmatrix} 25,900.89 \\ 22,199.01 \\ 84,710.89 \\ 16,809.99 \end{bmatrix}$ (b) $X = \begin{bmatrix} 28,378.37 \\ 24,496.31 \\ 99,665.84 \\ 19,287.46 \end{bmatrix}$

3. (a) In matrix terms we have $EX = B$, where

$$E = \begin{bmatrix} 1 & -0.6 & -0.6 \\ 0 & 1 & -0.3 \\ -0.1 & -0.1 & 1 \end{bmatrix}, \quad B = \begin{bmatrix} 80,000 \\ 60,000 \\ 50,000 \end{bmatrix}, \quad X = \begin{bmatrix} x \\ y \\ z \end{bmatrix}.$$

(b) $E^{-1} = \begin{bmatrix} 1.0874439 & 0.7399102 & 0.8744394 \\ 0.0336322 & 1.0538116 & 0.3363228 \\ 0.1121076 & 0.1793721 & 1.1210762 \end{bmatrix}$

From $X = E^{-1}B$, we obtain $X = \begin{bmatrix} 175,112 \\ 82,735 \\ 75,784 \end{bmatrix}$.

(c) For $B = \begin{bmatrix} 90,000 \\ 70,000 \\ 70,000 \end{bmatrix}$, $X = E^{-1}B$ yields $X = \begin{bmatrix} 210,874 \\ 100,336 \\ 101,121 \end{bmatrix}$.

Section 11.6 (p. 425)

1. (a) $A = \begin{bmatrix} 0.2 & 0.3 \\ 0.3 & 0.1 \end{bmatrix}$, $D = \begin{bmatrix} d_1 \\ d_2 \end{bmatrix}$, and $X = \begin{bmatrix} x_1 \\ x_2 \end{bmatrix}$ are the input-coefficient,

final-demand, and output matrices, respectively. We have:

$$I_2 - A = \begin{bmatrix} 0.8 & -0.3 \\ -0.3 & 0.9 \end{bmatrix}, \quad (I_2 - A)^{-1} = \begin{bmatrix} \dfrac{90}{63} & \dfrac{30}{63} \\ \dfrac{30}{63} & \dfrac{80}{63} \end{bmatrix}$$

$$X = (I_2 - A)^{-1} D = \begin{bmatrix} \dfrac{90}{63} d_1 & \dfrac{30}{63} d_2 \\ \dfrac{30}{63} d_1 & \dfrac{80}{63} d_2 \end{bmatrix}$$

(b) $5100 worth of commodity 1 and $5200 worth of commodity 2; $1020 worth of commodity 1 is needed to produce commodity 1, and $1560 worth of commodity 1 is needed to produce commodity 2; $1530 worth of commodity 2 is needed to produce commodity 1, and $520 worth of commodity 2 is needed to produce commodity 2.

(c) $6750 worth of commodity 1 and $6450 worth of commodity 2; $1350 worth of commodity 1 is needed to produce commodity 1, and $1935 worth of commodity 1 is needed to produce commodity 2; $2025 worth of commodity 2 is needed to produce commodity 1, and $645 worth of commodity 2 is needed to produce commodity 2.

Section 11.7

Self-Test 1 (p. 425)

1. $E + F = \begin{bmatrix} 2 & 3 & 4 \\ 12 & -6 & 4 \end{bmatrix}, \quad E - F = \begin{bmatrix} 6 & 1 & -2 \\ 0 & -2 & 2 \end{bmatrix}$

2. $AB = \begin{bmatrix} -10 & -2 \\ 8 & 17 \end{bmatrix}, \quad BA = \begin{bmatrix} -2 & 8 \\ 17 & 9 \end{bmatrix}, \quad A^2 = \begin{bmatrix} 14 & -14 \\ 7 & 7 \end{bmatrix},$

$B^2 = \begin{bmatrix} 7 & 8 \\ 12 & 31 \end{bmatrix}$

3. $AB = \begin{bmatrix} 23 & -3 \\ -8 & 5 \end{bmatrix}$, $BA = \begin{bmatrix} 9 & 5 & 0 \\ 14 & 14 & 8 \\ 11 & 10 & 5 \end{bmatrix}$

4. $-A = \begin{bmatrix} -5 & -3 \\ 4 & -1 \end{bmatrix}$ since $A + (-A) = (-A) + A = \bar{0}$

5. yes, since $AB = BA = I_2$

6. no multiplicative inverse

7. $F^{-1} = -\begin{bmatrix} \dfrac{5}{16} & \dfrac{7}{16} & \dfrac{3}{16} \\ \dfrac{1}{16} & \dfrac{5}{16} & \dfrac{9}{16} \\ \dfrac{3}{16} & \dfrac{1}{16} & \dfrac{5}{16} \end{bmatrix}$

8. $(1, 2)$

9. $(29, -18, -3)$

10. False; see Problem 2 of this Self-Test.

11. False; $AB = \bar{0}$ for $A = \begin{bmatrix} 1 & 2 \\ 2 & 4 \end{bmatrix}$, $B = \begin{bmatrix} 2 & 4 \\ -1 & -2 \end{bmatrix}$.

12. False; $AX = AY$ for $A = \begin{bmatrix} 2 & -1 \\ 4 & -2 \end{bmatrix}$, $X = \begin{bmatrix} 0 & 1 \\ 3 & 2 \end{bmatrix}$, $Y = \begin{bmatrix} -1 & -1 \\ 1 & -2 \end{bmatrix}$, but $X \neq Y$.

13. True

14. True

15. False; $A = \begin{bmatrix} 1 & 2 \\ 2 & 4 \end{bmatrix}$ has no multiplicative inverse; see section 10.3, Example 3.

Self-Test 2 (p. 426)

1. $G+H = \begin{bmatrix} 7 & 0 \\ 2 & 2 \end{bmatrix}$ $G-H = \begin{bmatrix} -1 & -2 \\ 6 & 2 \end{bmatrix}$

2. $AB = \begin{bmatrix} -5 \end{bmatrix}$ $BA = \begin{bmatrix} 3 & 9 & -6 \\ -2 & -6 & 4 \\ 1 & 3 & -2 \end{bmatrix}$ 3. $AX = AY = \begin{bmatrix} 6 & 0 \\ 3 & 0 \end{bmatrix}$

4. Yes, since $AB = BA = I_2$, is the identity matrix for the 2 by 2 matrices.

5. $A^{-1} = \begin{bmatrix} 2 & -5 \\ -1 & 3 \end{bmatrix}$

6. $E^{-1} = \begin{bmatrix} \dfrac{3}{7} & -1 & \dfrac{2}{7} \\ 0 & 1 & 0 \\ \dfrac{2}{7} & -1 & -\dfrac{1}{7} \end{bmatrix}$

7. Let $E = \begin{bmatrix} 1 & 2 \\ -2 & 1 \end{bmatrix}$, $B = \begin{bmatrix} 8 \\ 4 \end{bmatrix}$, and $X = \begin{bmatrix} x_1 \\ x_2 \end{bmatrix}$. Then $E^{-1} = \begin{bmatrix} \dfrac{1}{5} & -\dfrac{2}{5} \\ \dfrac{2}{5} & -\dfrac{1}{5} \end{bmatrix}$

and $X = E^{-1}B = \begin{bmatrix} 0 \\ 4 \end{bmatrix}$.

8. Let $E = \begin{bmatrix} 2 & 1 & -1 \\ 3 & 2 & 5 \\ 1 & 1 & 3 \end{bmatrix}$, $B = \begin{bmatrix} 1 \\ -2 \\ 3 \end{bmatrix}$, and $X = \begin{bmatrix} x_1 \\ x_2 \\ x_3 \end{bmatrix}$. Then

$E^{-1} = \begin{bmatrix} -5 & 6 & -7 \\ 10 & -11 & 13 \\ -1 & 1 & -1 \end{bmatrix}$ and $X = E^{-1}B = \begin{bmatrix} -38 \\ 71 \\ -6 \end{bmatrix}$.

9. (a) $A = \begin{bmatrix} 0.5 & 0.2 \\ 0.2 & 0.5 \end{bmatrix}$, $D = \begin{bmatrix} d_1 \\ d_2 \end{bmatrix}$, $X = \begin{bmatrix} x_1 \\ x_2 \end{bmatrix}$ are the input-coefficient,

final-demand, and output matrices, respectively. We have:

$$I_2 - A = \begin{bmatrix} 0.5 & -0.2 \\ -0.2 & 0.5 \end{bmatrix}, \quad (I_2 - A)^{-1} = \begin{bmatrix} \dfrac{50}{21} & \dfrac{20}{21} \\ \dfrac{20}{21} & \dfrac{50}{21} \end{bmatrix}$$

Thus:

$$X = (I_2 - A)^{-1} D = \begin{bmatrix} \dfrac{50}{21} d_1 + \dfrac{20}{21} d_2 \\ \dfrac{20}{21} d_1 + \dfrac{50}{21} d_2 \end{bmatrix}$$

INDEX

COMPANION TO ALGEBRA

William J. Adams

Pace University

CONTENTS

PREFACE

The objective of *Companion to Algebra* is to help optimize the use of *Algebra with Applications: Technique and THOUGHT* as a tool for teaching and studying algebra. It provides detailed solutions to, discussions of, and answers to even-numbered exercises.

W. J. A.

Answers, Solutions, and Discussion of Even-Numbered Exercises

Chapter 1
The World of Real-Numbers

Section 1.2 (p. 30)

2. 1

4. −6

6. −5

8. −2

10. 0

12. 6

14. $9xy$

16. $9x^2$

18. 7ab

20. $6y + 9a + 12z$

22. $-4w + (-2)v + (-3)z$

24. $(-2)a+(-4)b+(-6)c$

26. $-x+(-2)y+1$

28. $x^2+(-1)y$

30. $9(n+m)$

32. $3a(b+2c)$

34. $3x^2(y+2z)$

36. $2a(b+2c+4d)$

38. $2mn(p+2v+5w)$

40. $m(3xy+2y+5x)$

42. $2w^2(2v+x+4)$

44. $3rs(t+2+4k)$

46. No; since their sum is not 0.

48. 0 is the additive inverse of 0 because $0+0=0$

50. $-6, 1, -\dfrac{1}{4}, -\sqrt{2}, -\pi$, respectively.

Section 1.4 (p. 38)

2. -2

4. -7

6. 5

8. 1

10. 10

12. -5

14. 10

16. −3

18. −9

20. $\dfrac{1}{6}$

22. $-\dfrac{1}{8}$

24. 0

26. $-6a + 8b - 6c$

28. $-3a - 4b + 2c$

30. $-8x - 5y$

32. $7x + 10y + 7z$

34. $-7xy$

36. $-5ab + 2a$

38. $x(y - 4z - 5)$

40. $mn(4 - 5n + 3p)$

42. $\dfrac{1}{2}a(1 - 4b)$ or $a(\dfrac{1}{2} - 2b)$

44. $\dfrac{1}{3}xy(1 + 18z)$ or $xy(\dfrac{1}{3} + 6z)$

46. No; he cannot substitute 0 for x because it leads to division by 0 which is not defined.

48. No; $\dfrac{0}{0}$ is undefined and it is nonsensical to cancel 0's and obtain 1.

Section 1.5 (p. 44)

2. No

4. Yes

6. No

8. No

10. $\dfrac{1}{3}$

12. $\dfrac{2x}{3y}$

14. $\dfrac{3x+1}{x+1}$

16. $\dfrac{3}{x+3}$

18. $\dfrac{3a}{4x}, x \neq -y$

20. $2, a \neq -2b$

22. $\dfrac{x+y+1}{y+1}$

24. $1, x \neq 5$

26. $2, x \neq -5,$

28. $-2, a \neq 5$

30. $\dfrac{2+3b}{2(2+3ab)}, a \neq 0$

32. $3+4z, x \neq 0, y \neq 0$

34. $-x,\ y \neq -\dfrac{1}{3}$

36. $\dfrac{1}{3},\ yz \neq 3ys$

38. $\dfrac{5}{x},\ x \neq y$

40. For 23, $x=3$; for 24, $x=5$; for 25, $x=0$; for 26, $x=-5$, for 27, $x=2$.

Section 1.5 (p. 50)

42. No

44. Yes

46. No

48. $\dfrac{2}{7}$

50. $\dfrac{10a}{9c},\ b \neq 0$

52. $\dfrac{x^2}{z},\ y \neq 0$

54. $\dfrac{x}{3},\ x \neq 1$

56. $\dfrac{x}{x+1},\ x \neq 1$

58. $\dfrac{1}{x-5} \cdot \dfrac{5-x}{3} = \dfrac{-(\overset{1}{\cancel{-5+x}})}{3(\underset{1}{\cancel{x-5}})} = \dfrac{1}{3},\ x \neq 5$

60. $2a,\ a \neq 1$

62. $\dfrac{2a}{b^2},\ a \neq -\dfrac{1}{4}$

64. $\dfrac{3}{v-3} \cdot \dfrac{vx-3x}{x} = \dfrac{3 \overset{1}{\cancel{x}} (\overset{1}{\cancel{v-3}})}{(\cancel{v-3})\underset{1}{\cancel{x}}} = 3,\ x \neq 0,\ v \neq 3$

65. $-\dfrac{1}{4}$

68. 6

70. $-\dfrac{1}{9}$

72. $\dfrac{3(y-1)}{y},\ x \neq 0,\ y \neq 1$

74. $\dfrac{(x-1)^2}{6y},\ x \neq 1$

76. $\dfrac{a}{a+b},\ a \neq b$

78. $4,\ a \neq -4$

80. $\dfrac{n^2}{2m},\ m \neq -n$

82. $\dfrac{y}{2y-z} \div \dfrac{3z}{6y-3z} = \dfrac{y}{\underset{1}{\cancel{2y-z}}} \cdot \dfrac{3(\overset{1}{\cancel{2y-z}})}{3z} = \dfrac{y}{z},\ z \neq 2y$

Section 1.5 (p. 54)

84. $\dfrac{13}{12}$

86. $\dfrac{7}{6}$

88. $\dfrac{23}{6}$

90. $\dfrac{13}{10}$

92. $\dfrac{5}{x}$

94. $\dfrac{6x}{5}$

96. $\dfrac{2a-2}{2b}$

98. $4,\ x \neq 1$

100. $\dfrac{2b}{a-b}$

102. $\dfrac{7}{6x}$

104. $\dfrac{10x-7}{21}$

106. $\dfrac{m+4}{2}+\dfrac{2m-3}{3}=\dfrac{3(m+4)}{6}+\dfrac{2(2m-3)}{6}=\dfrac{7m+6}{6}$

108. $\dfrac{7x+3}{6x}$

110. $\dfrac{1}{m-n}+\dfrac{2}{m+n}=\dfrac{1(m+n)+2(m-n)}{(m-n)(m+n)}=\dfrac{3m-n}{(m-n)(m+n)}$

112. $\dfrac{5}{x-1}+6=\dfrac{5}{x-1}+\dfrac{6(x-1)}{x-1}=\dfrac{6x-1}{x-1}$

114. $\dfrac{3t+2x}{xyt}$

116. $\dfrac{4x+1}{(x-2)(x+1)}$

118. $\dfrac{x}{1-x}+\dfrac{2x}{x-1}=\dfrac{-x}{x-1}+\dfrac{2x}{x-1}=\dfrac{x}{x-1}$

120. $\dfrac{6x^2-2x+12}{6x}$

Section 1.5 (p. 57)

122. $\dfrac{1}{6}$

124. $\dfrac{7}{15}$

126. $\dfrac{11}{6}$

128. $-\dfrac{a}{3}$

130. $\dfrac{3x^2-4}{4x}$

132. $\dfrac{1}{a-1}-\dfrac{2}{b}=\dfrac{b}{b(a-1)}-\dfrac{2(a-1)}{b(a-1)}=\dfrac{b-2a+2}{b(a-1)}$. Some people obtain

$\dfrac{b-2a-2}{b(a-1)}$ as their conclusion. Where did they go wrong?

134. $\dfrac{-2x-3}{x}$

136. $\dfrac{7x+3}{2(x-1)}$

138. $\dfrac{1}{2x(2x-1)}$, or is it $\dfrac{-1}{2x(2x-1)}$? Why?

140. $\dfrac{9x-3}{2x}$

142. $\dfrac{-2n}{(m+n)(m-n)}$

Section 1.6 (p. 63)

2. $\dfrac{1}{9}$

4. 1

6. $\dfrac{1}{9}$

8. $\dfrac{8}{81}$

10. $\dfrac{(-2)^{-4}}{5^{-2}} = \dfrac{5^2}{(-2)^4} = \dfrac{25}{16}$

12. $[(-2)^{-3}]^{-2} = (-2)^6 = 64$

14. $\dfrac{-1}{432}$

16. $\dfrac{3y}{x^2}$

18. $\dfrac{3b^2}{a}$

20. $\dfrac{y^6}{x^2}$

22. $(1+a)^4$

24. $(m+2n)^2$

26. $\dfrac{1}{2x^3 y}$

28. $6x^6 + 8$

30. $\dfrac{2y^6}{3z}$

32. $(4a^2b^3)^{-2} = \dfrac{1}{(4a^2b^3)^2} = \dfrac{1}{4^2(a^2)^2(b^3)^2} = \dfrac{1}{16a^4b^6}$

34. $72x^7y$

36. $\dfrac{x^4}{y^8}$

38. $\dfrac{(2x^3)^{-3}}{(4x^2)^{-2}} = \dfrac{2^{-3}x^{-9}}{4^{-2}x^{-4}} = \dfrac{4^2}{2^3}x^{-9+4} = \dfrac{16x^{-5}}{8} = \dfrac{2}{x^5}$

40. $\dfrac{15}{x} + 10x^7$

42. $2x^5 + x^6$

44. $\dfrac{1}{a^8b}$

46. $8a^8$

48. $\dfrac{(z^2)^{-3} + (z^4)^{-2}}{z^3} = \dfrac{z^{-6} + z^{-8}}{z^3} = \dfrac{z^{-6}}{z^3} + \dfrac{z^{-8}}{z^3} = \dfrac{1}{z^9} + \dfrac{1}{z^{11}}$

Section 1.6 (p. 65)

50. -3

52. 7

54. 2

56. -5

60. $\dfrac{1}{(\sqrt{25})^{-3}} = \dfrac{1}{5^{-3}} = 5^3 = 125$

62. $\dfrac{1}{3}$

64. No; $\sqrt{(-2)^2}$ expresses the principal square root of $(-2)^2 = 4$,which is (positive) 2.

Section 1.6 (p. 68)

66. $6\sqrt{2}$

68. $10\sqrt{3}$

70. $20\sqrt{2}$

72. $3\sqrt{5}$

74. $2\sqrt{\dfrac{3}{5}}$

76. $7\sqrt{\dfrac{2}{5}}$

78. $6\sqrt{3}$

80. $5\sqrt[3]{2}$

82. $3\sqrt[3]{4}$

84. $3a^2\sqrt{c}$

86. $\sqrt{x^2 - 4}$

88. $x+1, \ x \geq -1$

Section 1.6 (p. 71)

90. 7

92. $\dfrac{1}{8}$

94. $\dfrac{1}{1000}$

96. $\dfrac{1}{6}$

98. $8^{-4/3} = \dfrac{1}{8^{4/3}} = \dfrac{1}{(\sqrt[3]{8})^4} = \dfrac{1}{2^4} = \dfrac{1}{16}$

100. $\dfrac{1}{4}$

102. $\dfrac{1}{\sqrt{2}}$

104. $32^{1/6}$

106. 3

108. x^{10}

110. $x^{1/2}$

112. $\left(x^{-1/4}y^{1/2}\right)^4 = \left(x^{-1/4}\right)^4\left(y^{1/2}\right)^4 = x^{-1}y^2 = \dfrac{y^2}{x}$

114. $\dfrac{\left(\sqrt[3]{x}\right)^2}{\sqrt[3]{x^5}} = \dfrac{\left(x^{1/3}\right)^2}{\left(x^5\right)^{1/3}} = \dfrac{x^{2/3}}{x^{5/3}} = \dfrac{1}{x}$

Chapter 2
Equations and Inequalities:
A First Look

Section 2.1 (p. 78)

2. $x = 2$

4. $x = 27$

6. $x = 6$

8. $x = 3$

10. $x = 2$

12. $x = -\dfrac{5}{2}$

14. $x = 6$

16. $(0,5)$, $(2,0)$, $(4,-5)$

18. $(0,-4)$, $(6,0)$, $(3,-2)$

22. $(0,0,4)$, $(2,0,0)$, $(0,4,0)$

24. $(0,0,4)$, $(5,0,0)$, $(0,-20,0)$

26. $y = \dfrac{12 - 2x}{-3}$

28. $z = \dfrac{20 - 4x + y}{5}$

30. $z = \dfrac{4 - 2y}{3x}$

32. $z = \dfrac{10 - x}{2xy}$

34. $w = \dfrac{p - 2x}{2}$

36. $r = \sqrt{\dfrac{A}{\pi}}$

Section 2.2 (p. 88)

The graphs of the equations stated in the even-numbered exercises follow.

4.

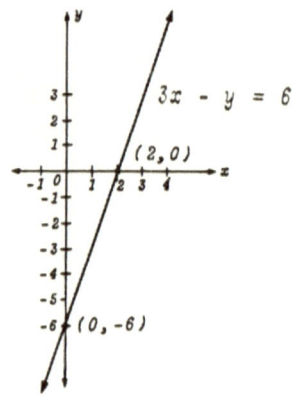

Figure 2.1

6.

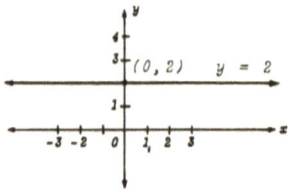

Figure 2.2

8.

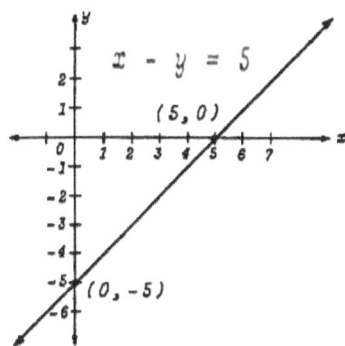

Figure 2.3

10.

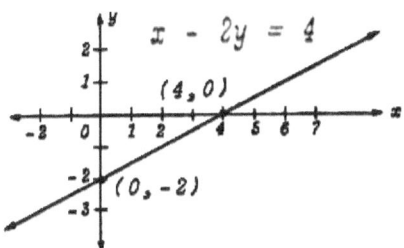

Figure 2.4

12.

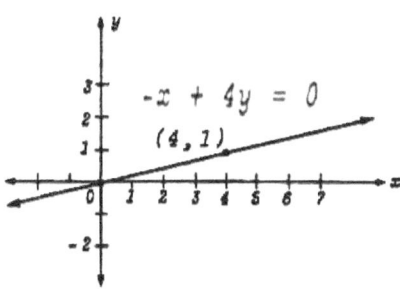

Figure 2.5

14.

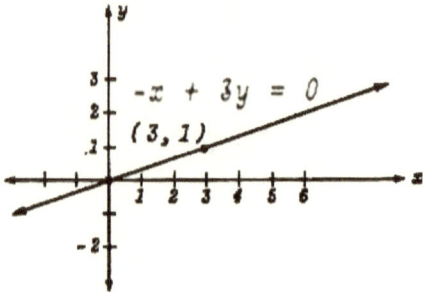

Figure 2.6

16.

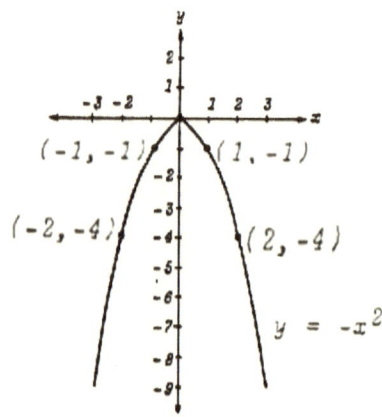

Figure 2.7

18.

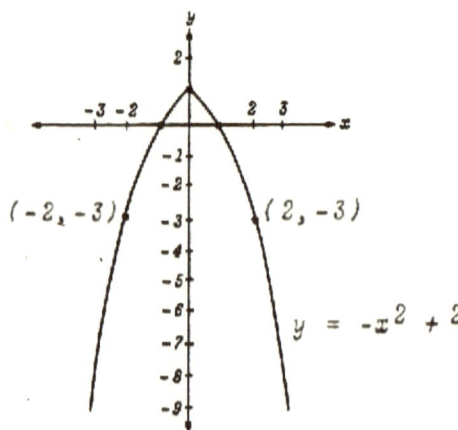

Figure 2.8

20.

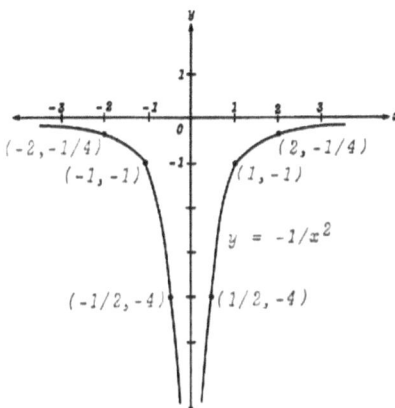

Figure 2.9

Section 2.3 (pp. 91 and 97)

2.
$$I = \$1200$$
$$A = \$6200$$

4. Total Interest: $500
   Total amount: $8500

6. (a) $2846.62
   (b) $2851.52
   (c) $2856.50

8. $2768.40

10. The present value of $200 under the given interest conditions is $A = 200(1.03)^{-3} = 183.02$ or $183.02 (the exponent -3 is $-mx = -4(9/12) = -3$). Thus the selling price of the appliance was $100 + \$183.02 = \$283.02$.

12. (a) $1920
    (b) 8.70% per annum

14. If A is the amount borrowed at simple discount rate d, the amount received now is $A - dA$, which equals $A(1-d)$, and the amount repaid 1 year from now is A. The interest on the loan is dA and the principal is $A(1-d)$. If r is the interest rate, we have:

$$rA(1-d) = dA$$

Solving for r yields:

$$r = \frac{dA}{A(1-d)} = \frac{d}{1-d}$$

Starting with $rA(1-d) = dA$, canceling A to obtain $r(1-d) = d$, and solving for d yields:

$$r - rd = d$$
$$r = d + rd$$
$$r = d(1+r)$$

$$\frac{r}{1+r} = d$$

16. 12.55%; 12.55% compounded annually yields the same interest at the end of the year as 12% per annum compounded 4 times a year.

Section 2.4 (p. 100)

2. $(0, 2)$, $(3, -2)$, $(-4, 0)$, $(\frac{1}{2}, 5)$, $(2, 3)$

4. $(0, 2)$, $(-4, 0)$, $(\frac{1}{2}, 5)$, $(2, 3)$

6. $(0, 2)$, $(\frac{1}{2}, 5)$, $(2, 3)$

8. $(3, -1, 4)$, $(-2, 3, 1)$

10. $(3, -1, 4)$, $(\frac{1}{2}, \frac{1}{3}, -3)$

12. $(2, 3, 1)$, $(1, 3, -1)$, $(3, -1, 4)$, $(\frac{1}{2}, \frac{1}{3}, -3)$, $(-2, 3, 1)$

Section 2.4 (p. 104)

14. Starting with $300 - y \geq 0$ and adding y to both sides yields $300 \geq y$, with is equivalent to $y \leq 300$.

    Starting with $y \leq 300$ and multiplying both sides by -1 yields:

    $$-y \geq -300$$

    Adding 300 to both sides yields $300 - y \geq 0$.

22. $x \leq -\dfrac{5}{2}$

24. $y < 3$

26. $y > -5$

28. $x > -\dfrac{11}{12}$

30. $x \leq 1$

32. $y \geq 7$

Section 2.5 (p. 110)

The graphs of the inequalities stated in the even-numbered exercises follow.

2.

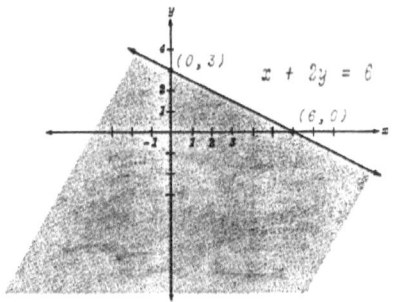

Figure 2.10

4.

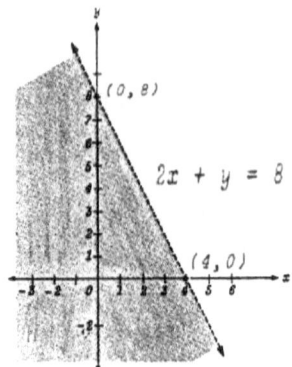

Figure 2.11

6.

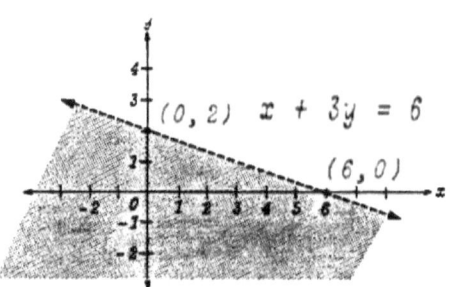

Figure 2.12

8.

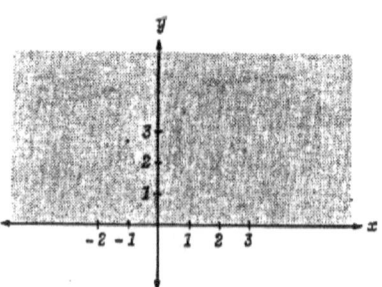

Figure 2.13

10.

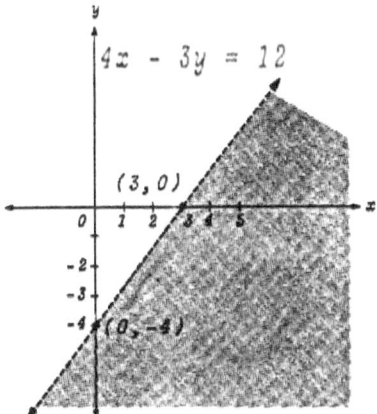

Figure 2.14

12.

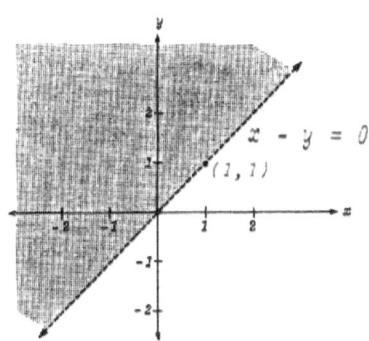

Figure 2.15

14.

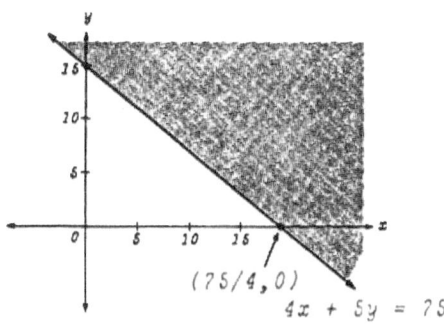

Figure 2.16

16.

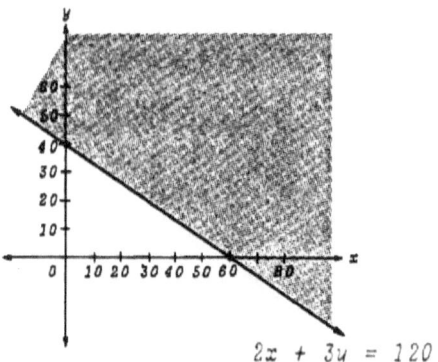

$2x + 3y = 120$

Figure 2.17

18.

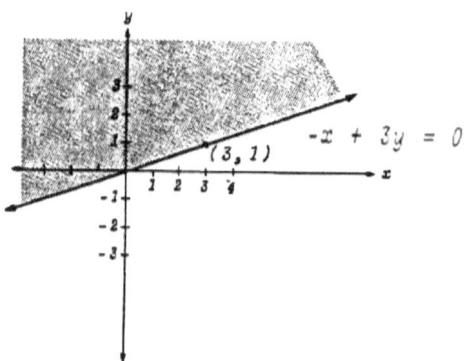

$-x + 3y = 0$

$(3, 1)$

Figure 2.18

20.

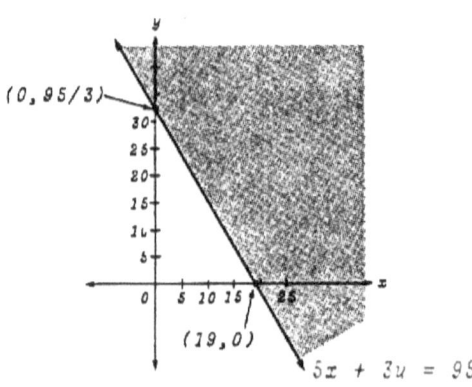

$(0, 95/3)$

$(19, 0)$

$5x + 3y = 95$

Figure 2.19

22.

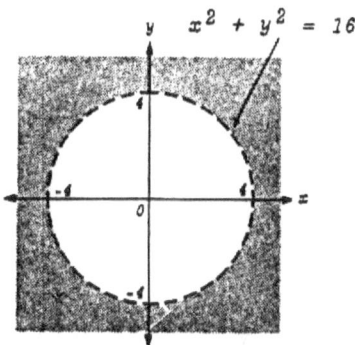

Figure 2.20

24.

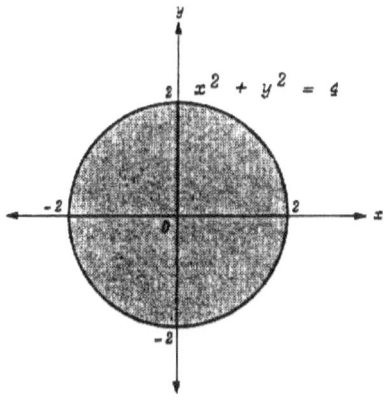

Figure 2.21

Chapter 3
Functions

Section 3.1 (p. 114)

2. $f(0)=1$, $f(1)=3$, $f(2)=15$, $f(3)=43$, $f(-1)=3$, $f(-2)=3$

4. $x \geq -1$; the square root of $x+1$ is not defined within the real-number system for $x < -1$. $f(-2)$ is not defined; $f(-1)=0$, $f(3)=2$, $f(2)=\sqrt{3}$.

6. All real-numbers except 2; 2 cannot be substituted for x because this leads to division by zero, which is not defined.

8. $f(1)=-1$, $f(2)=1$, $f(\frac{1}{2})=-1$, $f(0)=-1$, $f(-1)=-1$

Section 3.2 (p. 121)

2.

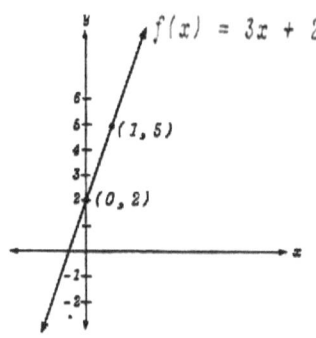

Figure 3.1

4.

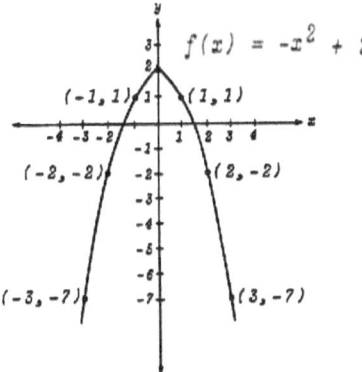

Figure 3.2

6.

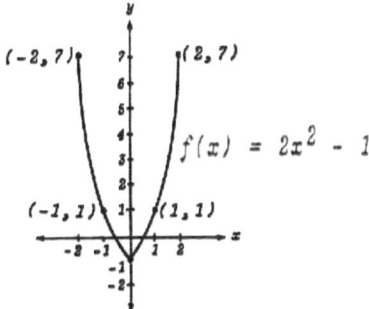

Figure 3.3

8.

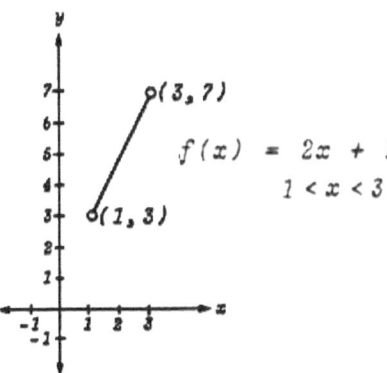

Figure 3.4

10.

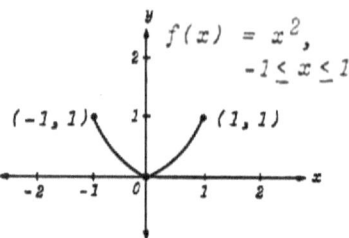

Figure 3.5

12.

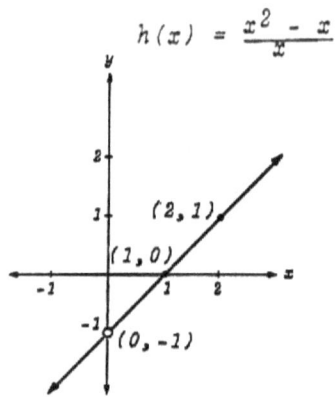

Figure 3.6

14.

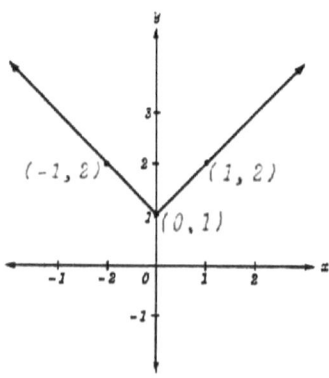

Figure 3.7

16.

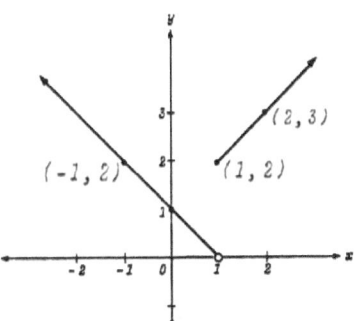

Figure 3.8

18.

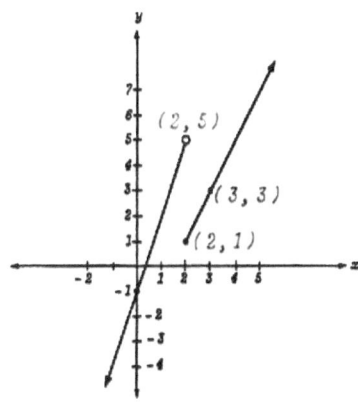

Figure 3.9

20.

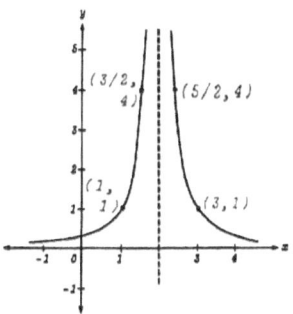

Figure 3.10

22.

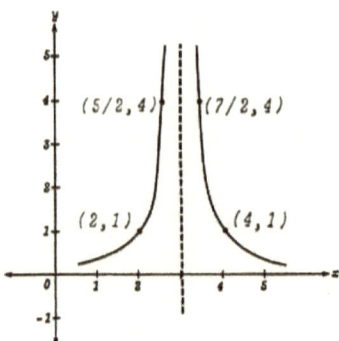

Figure 3.11

24.

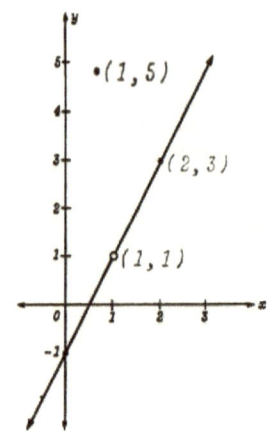

Figure 3.12

26.

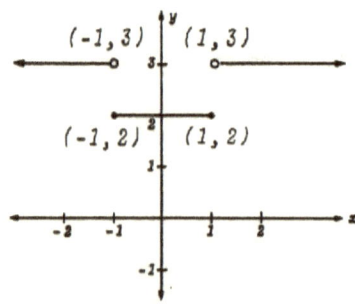

Figure 3.13

Section 3.3 (p. 129)

2. $f(t) = \begin{cases} 500, & \text{for } 0 \le t < \dfrac{1}{2} \\[2mm] 520, & \text{for } \dfrac{1}{2} \le t < 1 \\[2mm] 540.8, & \text{for } 1 \le t < \dfrac{3}{2} \\[2mm] 562.43, & \text{for } \dfrac{3}{2} \le t < 2 \\[2mm] 584.93, & \text{for } 2 \le t < \dfrac{5}{2} \\[2mm] 608.33, & \text{for } \dfrac{5}{2} \le t < 3 \\[2mm] 632.66, & \text{for } 3 \le t < \dfrac{7}{2} \\[2mm] 657.97, & \text{for } \dfrac{7}{2} \le t < 4 \\[2mm] 684.29, & \text{for } t = 4 \end{cases}$

4. $I(x) = \begin{cases} 1.50x, & \text{for } x = 0,1,\dots,5000 \\ 7500 + 0.2(x-5000), & \text{for } x = 5001,\dots, \end{cases}$ where x is the number of copies sold.

6. $T(x) = \begin{cases} 0, & \text{for } 0 \le x \le 5000 \\ 20 + 0.01(x-5000), & \text{for } 5000 < x \le 25{,}000 \\ 220 + 0.03(x-25{,}000), & \text{for } x > 25{,}000, \text{ where} \end{cases}$ x is taxable income in dollars.

8. $V(t) = 2000 - 200t$, for $0 \le t \le 10$, where t is time in years. $V(3) = 1400$;

$V(t) = C - Crt$, for $0 \le t \le \dfrac{1}{r}$

Section 3.5 (p. 133)

2. $\log_{10} 1000 = 3$

4. $\log_2 8 = 3$

6. $\log_{10} 0.01 = -2$

8. $\log_{32}\left(\dfrac{1}{2}\right) = -\dfrac{1}{5}$

10. $\log_5 1 = 0$

12. $\log_{10} 0.001 = -3$

14. $3^0 = 1$

16. $12^2 = 144$

18. $8^{2/3} = 4$

20. $10^4 = 10,000$

22. 3

24. -3

26. $\dfrac{3}{2}$

28. 3

30. -1

Section 3.5 (p. 137)

32. 1.5441

34. 1.4771

36. $9.4771 - 10$

38. $\dfrac{1}{0.4771} = 2.0960$

40. 3.3804

42. $9.8451 - 10$

44. $\log 10^2 = 2(\log_5 10) = 2.8612$

46. $\dfrac{1}{3}[\log_{10} 5 + \log_{10} 7] = 0.5147$

48. 0.2940

50. 0.0731

Section 3.6 (p. 144)

2.

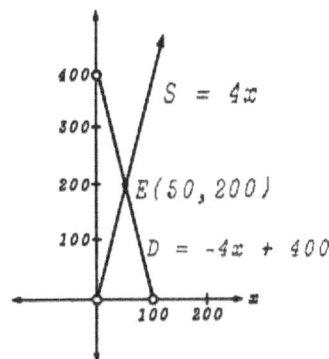

Figure 3.14

(c) No; its math validity is established, but its real world accuracy hinges on the real-world accuracy of the demand and supply functions we are given.

Section 3.7 (p. 148)

2. $f(3,2) = 4$, $f(-3,1) = -5$, $f(4,0)$ and $f(-4,1)$ are not defined.

Chapter 4
Linear Structures

Section 4.1 (p. 160)

2. $-3x + y = 6$

4. $x + 2y = 2$

6. $-x + 2y = 5$

8. $y = 1$

10. $-2x + y = 1$

Section 4.2 (p. 167)

2. $(20, 5)$

4. $(-1, -4)$

6. The system has infinitely many solutions, namely, all ordered pairs of numbers that satisfy $x - 2y = -4$. Both equations in the system are multiples of this equation.

8. $(4, 3)$

10. $(42, 12)$

12. $(6, 5)$

14. $(20, 80)$

Section 4.3 (p. 177)

2.

x_i	y_i	x_i^2	$x_i y_i$
5	24	25	120
10	36	100	360
13	28	169	364
15	36	225	540
18	36	324	648
21	54	441	1134
27	52	729	1404
31	76	961	2356
37	76	1369	2812
41	100	1681	4100
218	518	6024	13,838

$$m = \frac{10(13,838) - 218(518)}{10(6024) - (218)^2} = \frac{25,456}{12,716} = 2.002$$

$$b = \frac{518 - 218(2.002)}{10} = 8.16$$

Thus the sample linear regression equation is:

$$\hat{y} = 2.002x + 8.16$$

For $x = 50$, $\hat{y} = 108.26$; for an equipment maintenance expenditure of $50 thousand the predicted **average** profit before taxes is $108.26 thousand.

4.

x_i	y_i	x_i^2	$x_i y_i$
20	50	400	1000
25	50	625	1250
30	6	900	1800

32	64	1024	2048
40	70	1600	2800
50	75	2500	3750
60	80	3600	4800
60	85	3600	5100
65	90	4225	5850
70	90	4900	6300
452	714	23,374	34,698

$$m = \frac{10(13,698) - 452(714)}{10(23,374) - (452)^2} = \frac{24,252}{29,438} = 0.82$$

$$b = \frac{714 - 452(0.82)}{10} = 34.12$$

Thus the sample linear regression equation is:

$$\hat{y} = 0.82x + 34.12$$

For $x = 55$, $\hat{y} = 5$; for students who obtain a score of 55 on the mathematics proficiency exam in question, the predicted **average** final exam grade in the calculus course is 79.

Section 4.4 (p. 190)

2. $x = 20$, $y = 5$; a sequence of tableaus leading to this solution is shown in Figure 4.1.

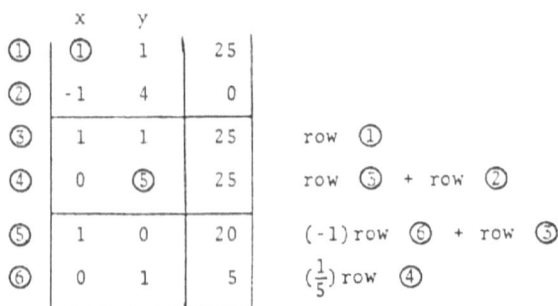

Figure 4.1

4. $(-1,-4)$

6. The system has infinitely many solutions, namely, all ordered pairs of numbers obtained from $x = 6 + 2y$, where y is arbitrary. A sequence of tableaus leading to this result is shown in Figure 4.2.

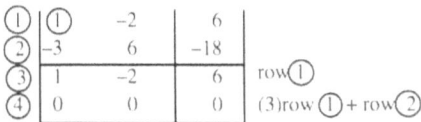

Figure 4.2

8. $(124, 260)$

10. $(3, 3)$

Section 4.5 (p. 195)

1. $x = 2$, $y = 1$, $z = -3$; a sequence of tableaus leading to this result is shown in Figure 4.3.

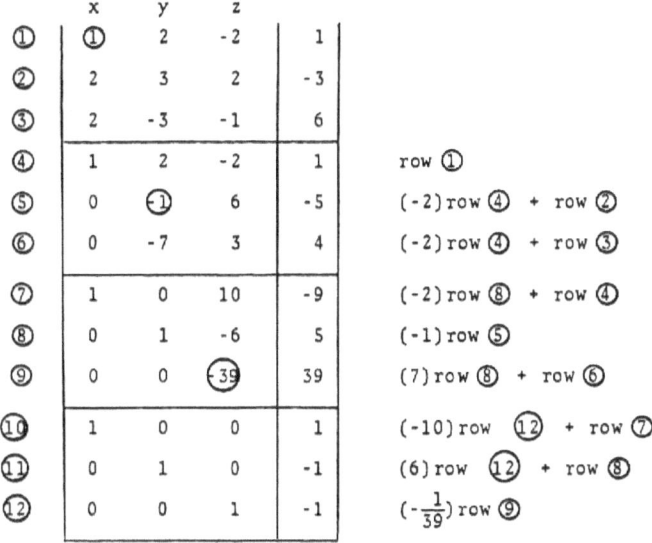

Figure 4.3

2. $x=1$, $y=-1$, $z=-1$; a sequence of tableaus leading to this solution is shown in Figure 4.4.

	x	y	z		
①	①	3	2	-1	
②	2	3	3	-2	
③	-2	2	-3	7	
④	1	3	2	-1	row ①
⑤	0	-3	-1	0	(-2) row ④ + row ②
⑥	0	8	①	5	(2) row ④ + row ③
⑦	1	-13	0	-11	(-2) row ⑨ + row ④
⑧	0	⑤	0	5	row ⑨ + row ⑤
⑨	0	8	1	5	row ⑥
⑩	1	0	0	2	(13) row ⑪ + row ⑦
⑪	0	1	0	1	$(\frac{1}{5})$ row ⑧
⑫	0	0	1	-3	(-8) row ⑪ + row ⑨

Figure 4.4

4. $u=-2$, $v=3$, $w=-1$; a sequence of tableaus leading to this solution is shown in Figure 4.5.

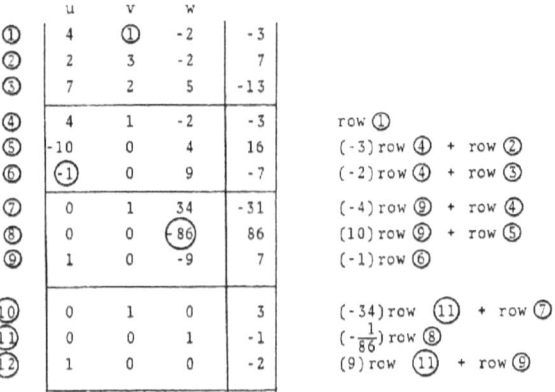

	u	v	w		
①	4	①	-2	-3	
②	2	3	-2	7	
③	7	2	5	-13	
④	4	1	-2	-3	row ①
⑤	-10	0	4	16	(-3) row ④ + row ②
⑥	⊙1	0	9	-7	(-2) row ④ + row ③
⑦	0	1	34	-31	(-4) row ⑨ + row ④
⑧	0	0	-86	86	(10) row ⑨ + row ⑤
⑨	1	0	-9	7	(-1) row ⑥
⑩	0	1	0	3	(-34) row ⑪ + row ⑦
⑪	0	0	1	-1	$(-\frac{1}{86})$ row ⑧
⑫	1	0	0	-2	(9) row ⑪ + row ⑨

Figure 4.5

6. $s=-1/2$, $t=-1$, $u=2$, $v=2$; a sequence of tableaus leading to this solution is shown in Figure 4.6.

	s	t	u	v		
①	8	①	-1	0	-7	
②	4	3	0	1	-3	
③	2	1	4	4	14	
④	4	2	-3	-1	-12	
⑤	8	1	-1	0	-7	row ①
⑥	-20	0	3	①	18	(-3) row ⑤ + row ②
⑦	-6	0	5	4	21	(-1) row ⑤ + row ③
⑧	-12	0	-1	-1	2	(-2) row ⑤ + row ④
⑨	8	1	-1	0	-7	row ⑤
⑩	-20	0	3	1	18	row ⑥
⑪	74	0	-7	0	-51	(-4) row ⑩ + row ⑦
⑫	-32	0	②	0	20	row ⑩ + row ⑧
⑬	-8	1	0	0	3	row ⑯ + row ⑨
⑭	28	0	0	1	-12	(-3) row ⑯ + row ⑩
⑮	(-38)	0	0	0	19	(7) row ⑯ + row ⑪
⑯	-16	0	1	0	10	$(\frac{1}{2})$ row ⑫
⑰	0	1	0	0	-1	(8) row ⑲ + row ⑬
⑱	0	0	0	1	2	(-28) row ⑲ + row ⑭
⑲	1	0	0	0	$-\frac{1}{2}$	$(-\frac{1}{38})$ row ⑮
⑳	0	0	1	0	2	(16) row ⑲ + row ⑯

Figure 4.6

7. $s = 4$, $t = 0$, $u = 0$; a sequence of tableaus leading to this result is shown in Figure 4.7.

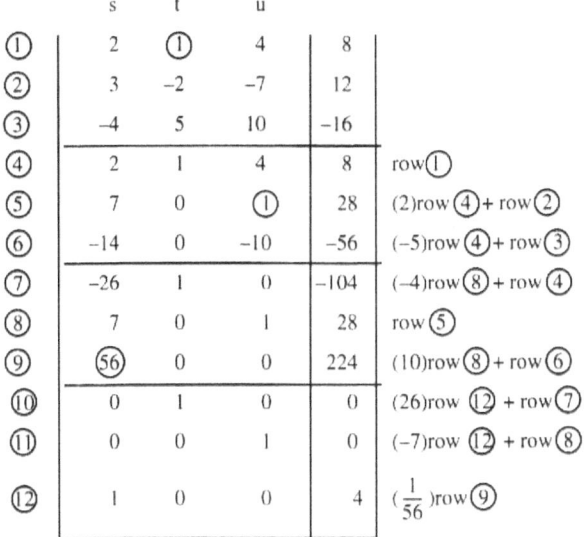

	s	t	u		
①	2	①	4	8	
②	3	-2	-7	12	
③	-4	5	10	-16	
④	2	1	4	8	row①
⑤	7	0	①	28	(2)row④+ row②
⑥	-14	0	-10	-56	(-5)row④+ row③
⑦	-26	1	0	-104	(-4)row⑧+ row④
⑧	7	0	1	28	row⑤
⑨	(56)	0	0	224	(10)row⑧+ row⑥
⑩	0	1	0	0	(26)row⑫ + row⑦
⑪	0	0	1	0	(-7)row⑫ + row⑧
⑫	1	0	0	4	$(\frac{1}{56})$row⑨

Figure 4.7

Section 4.6 (p. 201)

1. Let x, y, and z denote the total costs of service departments S_1, S_2, and S_3, respectively, for March. The condition that the total cost of each service department must equal the overhead of the department (for March) plus the charges for services provided by the other service departments leads to the following relations:

$$x = 40,000 + 0.1y + 0.2z$$
$$y = 30,000 + 0.1x + 0.05z$$
$$z = 20,000 + 0.05x + 0.1y$$

By transposing terms we obtain the following system of equations:

$$x - 0.1y - 0.2z = 40,000$$
$$-0.1x + y - 0.05z = 30,000$$
$$-0.05x - 0.1y + z = 20,000$$

A sequence of tableaus leading to the solution of this system is given in Figure 4.8. From the last tableau in the sequence we have, $x = 48,831$, $y = 36,186$, $z = 26,060$.

x	y	z	
①	-0.1	-0.2	40,000
-0.1	1	-0.05	30,000
-0.05	-0.1	1	20,000
1	-0.1	-0.2	40,000
0	(0.99)	-0.07	34,000
0	-0.105	0.99	22,000
1	0	-0.2070707	43,434.34
0	1	-0.070707	34,343.43
0	0	(0.9825758)	25,606.06
1	0	0	48,830.63
0	1	0	36,186.06
0	0	1	26,060.14

Figure 4.8

The allocation of the service departments' costs to the production departments is determined as follows:

P_1 : $0.40(48,831) + 0.40(36,186) + 0.35(26,060) = 43,128$ or $43,128

P_2 : $0.45(48,831) + 0.40(36,186) + 0.40(26,060) = 46,872$ or $46,872

2. Let x, y, and z denote the net incomes of the Ramunė, Algis, and Charles companies on a consolidated basis. From the affiliation diagram we have the following relations:

$$x = 80,000 + 0.6y + 0.6z$$
$$y = 60,000 + 0.3z$$
$$z = 50,000 + 0.01y + 0.1x$$

By transposing and rearranging terms we obtain the following system of equations:

$$x - 0.6y - 0.6z = 80,000$$
$$y - 0.03z = 60,000$$
$$-0.1x - 0.1y + z = 50,000$$

A sequence of tableaus leading to the solution of this system is given in Figure 4.9. From the last tableau in the sequence we have, $x = 175,112$, $y = 82,735$, $z = 75,785$.

x	y	z	
①	-0.6	-0.6	80,000
0	1	-0.3	60,000
-0.1	-0.1	1	50,000
1	-0.6	-0.6	80,000
0	①	-0.3	60,000
0	-0.16	0.94	58,000
1	0	-0.78	116,000
0	1	-0.3	60,000
0	0	⟨0.892⟩	67,600
1	0	0	175,112.100
0	1	0	82,735.425
0	0	1	75,784.753

Figure 4.9

3. By multiplying, transposing terms, and combining like terms, so that the unknowns x, y and z appear on one side of the equation and the constant appears on the other side, we obtain the following system of equations:

$$x + 0.095y \qquad\qquad = 2,589.415$$
$$0.01x + 1.010y + 0.01z = 1,529.7$$
$$0.48x + 0.480y + \qquad z = 12,253.24$$

A sequence of tableaus leading to the solution of this system is shown in Figure 4.10. From the solution obtained we have that the state income tax owed is $2,457.63, the state capital stock tax owed is $1,387.18, and the federal income tax owed is $10,407.73.

x	y	z	
1	0.095	0	2,589.415
0.01	1.01	0.01	1,529.7
0.48	0.48	①	12,253.24
①	0.095	0	2,589.415
0.0052	1.0052	0	1,407.1676
0.48	0.48	1	12,253.24
1	0.095	0	2,589.415
0	(1.004706)	0	1,393.7027
0	0.4344	1	11,010.321
1	0	0	2,457.6335
0	1	0	1,387.1746
0	0	1	10,407.733

x = 2,457.63,   y = 1,387.18,   z = 10,407.73

Figure 4.10

4. Let x, y, and z denote the income-tax due, the total in bonuses, and profit-sharing contribution, respectively. The stated relationships lead to the following conditions:

$$x = 0.5(2,000,000 - y - z)$$
$$y = 0.1(2,000,000 - x - z)$$
$$z = 0.05(2,000,000 - x - y)$$

By multiplying and transposing terms we obtain the following system of equations:

$$x + 0.5y + 0.5z = 1,000,000$$
$$0.1x + y + 0.1z = 200,000$$
$$0.05x + 0.05y + z = 100,000$$

A sequence of tableaus leading to a solution of this system is given in Figure 4.11. From the last tableau in the system, we have, $x = 924,324$, $y = 102,703$, $z = 48,649$.

x	y	z	
①	0.5	0.5	1,000,000
0.1	1	0.1	200,000
0.05	0.05	1	100,000
1	0.5	0.5	1,000,000
0	⟨0.95⟩	0.05	100,000
0	0.025	0.975	50,000
1	0	0.4736843	947,368.5
0	1	0.0526315	105,263.15
0	0	⟨0.9736843⟩	47,368.42
1	0	0	924,324.4
0	1	0	102,702.7
0	0	1	48,648.65

Figure 4.11

Section 4.7 (p. 210)

2. $t = -70 + 19s$, $u = -39 + 11s$, s is arbitrary; a sequence of tableaus leading to this result is shown in Figure 4.12.

	s	t	u	
①	3	①	-2	8
②	2	-3	5	15
③	-1	7	-12	-22
④	3	1	-2	8
⑤	11	0	(-1)	39
⑥	-22	0	2	-78
⑦	-19	1	0	-70
⑧	-11	0	1	-39
⑨	0	0	0	0

Figure 4.12

4. $x = -z$, $y = 3 - z$, z is arbitrary; a sequence of tableaus leading to this
    solution is shown in Figure 4.13.

x	y	z	
①	2	3	6
-2	1	-1	3
3	1	4	3
4	3	7	9
1	2	3	6
0	⑤	5	15
0	-5	-5	-15
0	-5	-5	-15
1	0	1	0
0	1	1	3
0	0	0	0
0	0	0	0

Figure 4.13

6. $s = 7 - 7u - w$, $t = -11 + 12u + w$, u and w are arbitrary; a sequence
    of tableaus leading to this result is shown in Figure 4.14.

s	t	u	w	
2	①	2	1	3
5	3	-1	2	2
2	1	2	1	3
⊝1	0	-7	-1	-7
0	1	-12	-1	-11
1	0	7	1	7

Figure 4.14

7. To begin, we introduce variables s, t, u, v, w and x as shown in the network
    diagram in Figure 4.15. Variable x denotes the number of cars passing

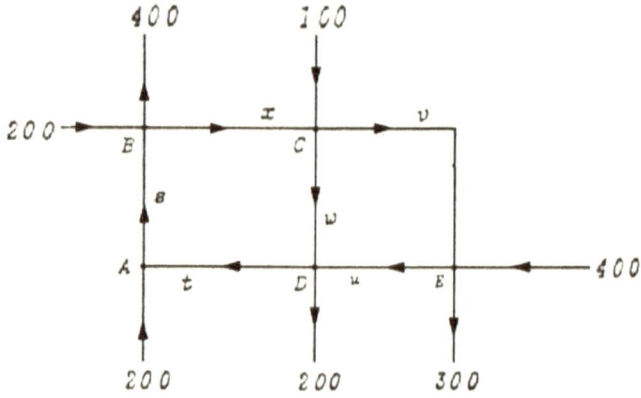

Figure 4.15

between points B and C per hour, v the number of cars passing between
points C and E per hour, and so on. For equilibrium at the points A, B,
C, D and E, the number of cars entering each of these points per hour
must equal the number of cars leaving each of these points per hour. This
leads to the following conditions:

$$A: t + 200 = s$$
$$B: s + 200 = x + 400$$
$$C: x + 100 = v + w$$
$$D: u + w = t + 200$$
$$E: v + 400 = u + 300$$

Transposing and rewriting terms yields the following system of equations:

$$s - t \qquad\qquad\qquad = 200$$
$$s \qquad\qquad -x = 200$$
$$v + w - x = 100$$
$$-t + u + \qquad w \qquad = 200$$
$$u - v \qquad\qquad = 100$$

A sequence of solution tableaus is shown in Figure 4.16. From the final tableau in the sequence we obtain the following description of solutions:

s	t	u	v	w	x	
①	-1	0	0	0	0	200
1	0	0	0	0	-1	200
0	0	0	1	1	-1	100
0	-1	1	0	1	0	200
0	0	1	-1	0	0	100
1	-1	0	0	0	0	200
0	①	0	0	0	-1	0
0	0	0	1	1	-1	100
0	-1	1	0	1	0	200
0	0	1	-1	0	0	100
1	0	0	0	0	-1	200
0	1	0	0	0	-1	0
0	0	0	1	1	-1	100
0	0	①	0	1	-1	200
0	0	1	-1	0	0	100
1	0	0	0	0	-1	200
0	1	0	0	0	-1	0
0	0	0	①	1	-1	100
0	0	1	0	1	-1	200
0	0	0	-1	-1	1	-100
1	0	0	0	0	-1	200
0	1	0	0	0	-1	0
0	0	0	1	1	-1	100
0	0	1	0	1	-1	200
0	0	0	0	0	0	0

Figure 4.16

$$s = x + 200$$
$$t = x$$
$$u = -w + x + 200$$
$$v = -w + x + 100$$
w and x are arbitrary

For this solution to make sense in terms of the traffic background in question, the values given to the underlying variables must, at a minimum, be restricted to nonnegative integers. From $s = x + 200$, we obtain:

$$x = s - 200$$

$x \geq 0$ yields

$$s - 200 \geq 0$$
$$s \geq 200$$

Thus traffic equilibrium cannot be maintained if fewer than 200 cars per hour pass between points A and B, which is clear from the network structure.

The condition $t = x$ tells us that traffic equilibrium cannot be maintained if the number of cars passing between D and A per hour is not equal to the number of cars passing between B and C. From $u \geq 0$ and $u = -w + x + 200$, we have:

$$-w + x + 200 \geq 0$$
$$x + 200 \geq w$$
$$w \leq x + 200$$

This tells us that traffic equilibrium cannot be maintained if the number of cars passing between C and D per hour exceeds 200 plus the number of cars passing between B and C per hour. Replacing x in $w \leq x + 200$ by $s - 200$ yields

$$w \leq s,$$

which tells us that traffic equilibrium cannot be maintained if the number of cars passing between C and D per hour exceeds the number of cars passing between A and B per hour.

Section 4.8 (p. 216)

2. $(8, 2), (8, 17), (20, 5), (9, 12), (14, 7)$
4. $(1, 1, 2), (0, 1, 3), (2, 1, 1), (4, 1, 0), (2, 1, 2)$
6.

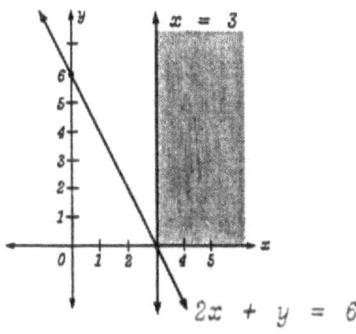

Figure 4.17

8.

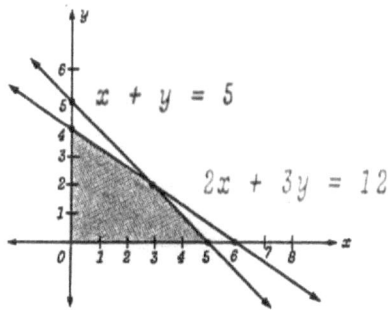

Figure 4.18

10.

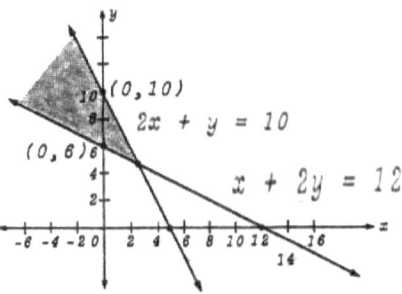

Figure 4.19

12.

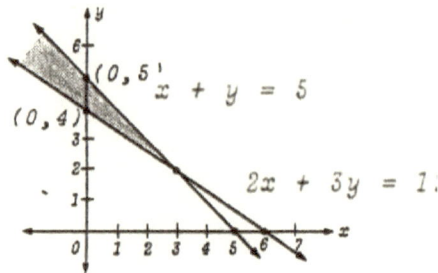

Figure 4.20

14.

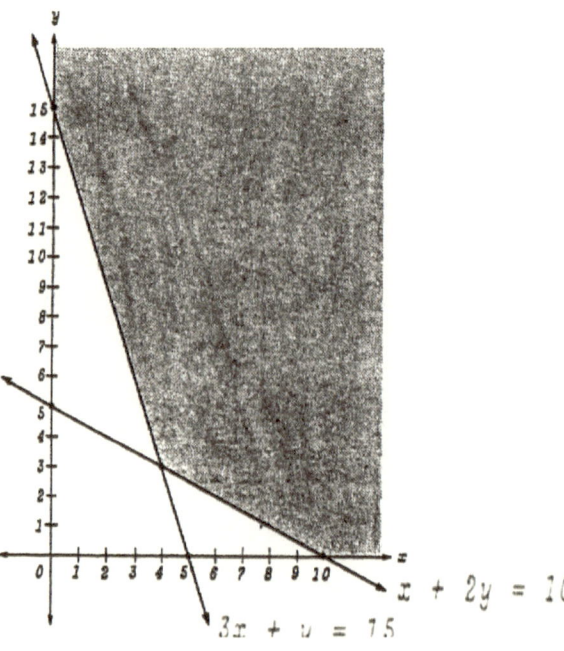

Figure 4.21

16.

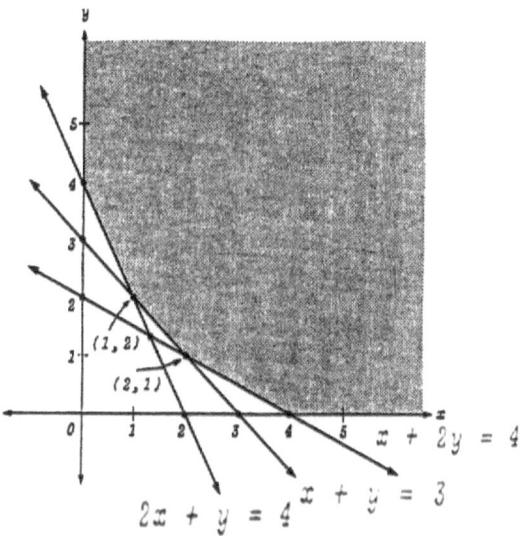

Figure 4.22

18.

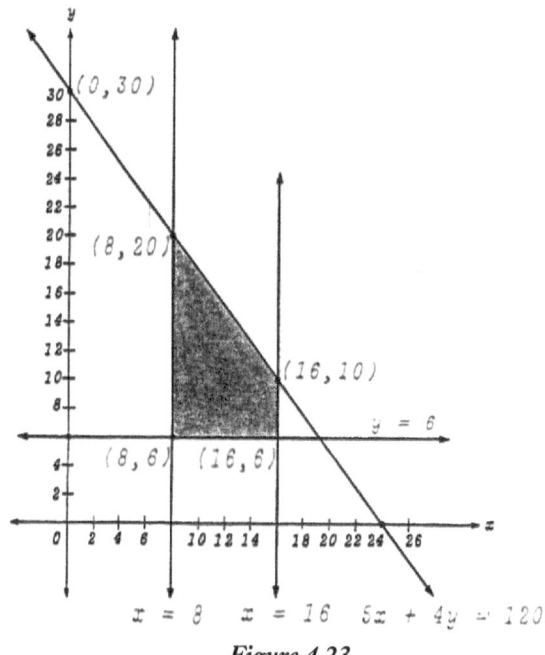

Figure 4.23

Section 4.9 (p. 219)

2. $F(x, y, z) = 3x + 4y + 2xz$ is not a linear function because of the product of variables in the 2xz term.

4. $F(x, y, z) = 3x + 2y + z^2$ is not a linear function because of the z^2 term.

6. $C(0, 2000) = 156,000$, $C(0, 12000) = 186,000$, $C(3000, 12000) = 192,000$, $C(15000, 0)$ is not defined, $C(10000, 5000) = 185,000$

Chapter 5
More on
Mathematical Modeling

Section 5.1 (p. 226)

2(a) T1: Cost of Cigarettes; $\$5.00(2.5)(365)(50) = \$228,125$

T2: Cost of Insurance: $\$400(30) = \$12,000$

T3: Cost of Health Maintenance: $\$1000(50) = \$50,000$

T4: Total Cost of Smoking: $290,125

(b) No; it's irrelevant to the validity issue. The validity of the theorems of Jacobs' model for the cost of smoking is established by the mathematical analysis given in (a), period.

(c) No; as in answer to (b), this is irrelevant to the validity issue.

(d) Not necessarily; the fact that Warner's and Wright's models took these factors into account is not enough. The question is, how realistic was the way they took these factors into account.

(e) Yes; it raises a question about the realism of postulates P2, and P2a, of these models.

Section 5.2 (p. 231)

2(a). Andy would have to take the trip to Kennebunkport along the routes
envisioned in both models and note how "close" the actual trip time is
to 6 hours for his model and 7 hours for Henry's model.

   If his actual trip time is close to 6 hours for the one and close to 7
hours for the other, this would establish that both of the theorems from
these models are realistic. And then there are the other possibilities:
"close" to 6 hours for the one, "far" from 7 hours for the other, etc.

4. No; the shorter of the trip times from the two models is inviting to focus
on, but it would be a mistake to do so. It's a valid conclusion of Andy's
model, but this does not guarantee that it's realistic.

Chapter 6
Linear Program Models

Section 6.3 (p. 257)

2. Corner points: $(\frac{3}{2},\frac{3}{2})$, $(3, 0)$, $(4, 0)$; $(3, 0)$ yields the min. value 30.

4. Corner points: (0, 250000), (0, 400000), (150000, 100000); (150000, 100000) yields the minimum value 2,400,000.

6. The feasible points are shown in Figure 6.1. This linear program does not have a solution because $F(x, y) = 3x + 2y$ does not have a largest value. No matter what value is given, we can make $3x + 2y$ exceed that value by choosing a suitable feasible point.

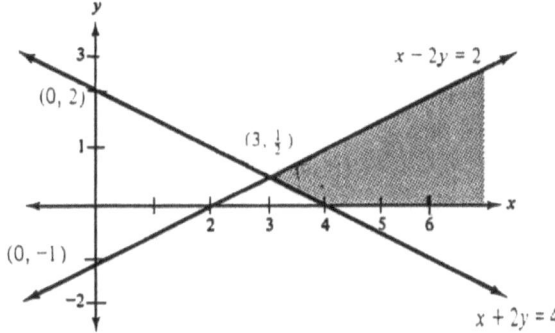

Figure 6.1

8. The linear program has no solution because its constraints are incompatible.

10. The graph of the feasible points is shown in Figure 6.2. (60, 15) yields the maximum value 13,050.

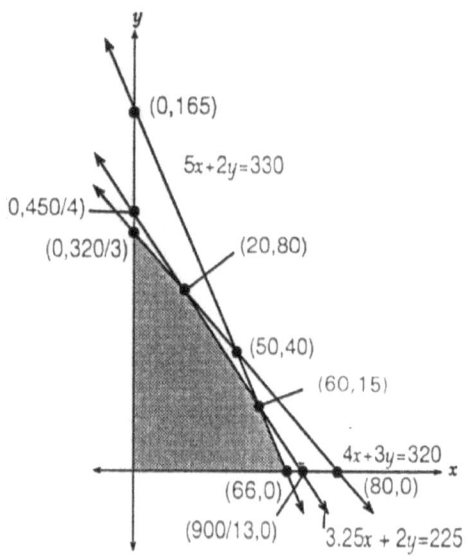

Figure 6.2

Section 6.4 (p. 262)

2. (a) + (b) Mathematics is **precise** in the sense that it yields **valid conclusions** with respect to an underlying starting point called a hypothesis. If there are two underlying hypotheses, then two quite different valid conclusions—solutions to linear program models in this case—might arise, each valid with respect to its own starting point.

   (c) What assumptions underlie M1 and M2? On what basis were they made? Convince me that they are realistic.

   (d) I would adopt that model based on what I believe to be **realistic assumptions**. If I were not convinced that the assumptions for either of M1 and M2 were realistic, I would not adopt either model. I would call for additional studies.

Section 6.6 (p. 269)

2. No; the corner point method ensures that the profit function is maximized. Whether profit is maximized depends on how close to reality's mark the profit function is.

4. I would recommend that he go over the underlying assumptions/ postulates with a fine-tooth comb and that if he were satisfied that they were realistic, implement (30, 20). If he were not satisfied about their realism, he should not implement (30, 20), but seek the judgment of another operations research firm about what assumptions/postulates would be appropriate (realistic).

6. Min. $C = 3x + 2y$ subject to
   $x \geq 0, y \geq 0$
   $2x + 3y \geq 120$
   $3x + 2y \geq 150$
   $x + y \geq 55,$

where x and y are the number of ounces of orange juice and apricot juice concentrate to be used. The graph of the feasible points is shown in Figure 6.3 (0, 75) and (40, 15) yield the minimum value 150.

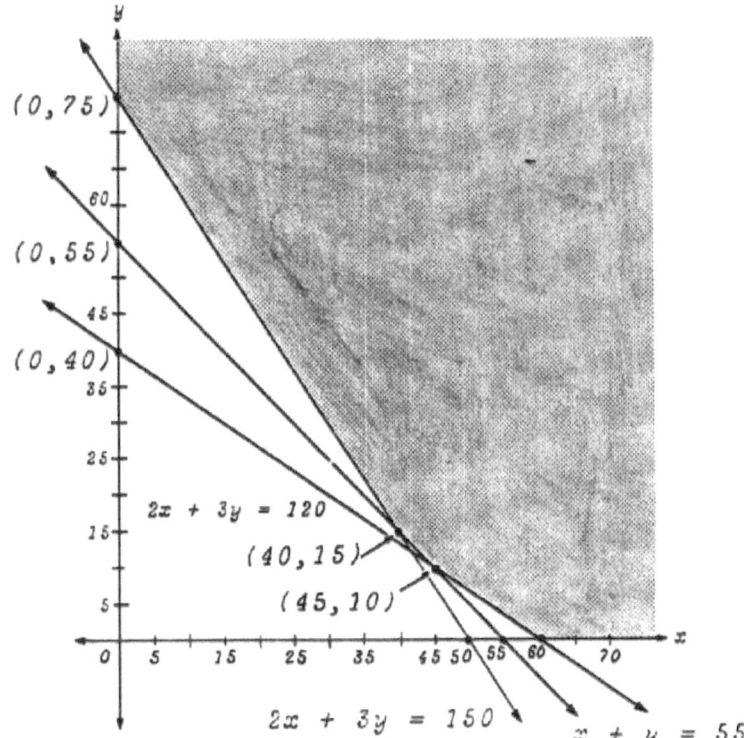

Figure 6.3

8. Max. $P = 50x + 30y$ subject to
$$x \geq 0, y \geq 0$$
$$5x + 2y \leq 220$$
$$3x + 2y \leq 180,$$

where x and y are the number of refrigerators and air conditioners, respectively, to be produced weekly. Corner points: (0, 0), (0, 90), (20, 60), (44, 0); (20, 60) yields the max. value 2800.

10. Max. $P = 1.5x + 2y$ subject to
$$x \geq 0, y \geq 0$$
$$4x + 6y \leq 2200$$
$$5x + 3y \leq 1400,$$

where x and y are the number of units of toy 1 and toy 2, respectively, to be bought. The graph of the feasible points is shown in Figure 6.4 (100, 300) yields the maximum value 750.

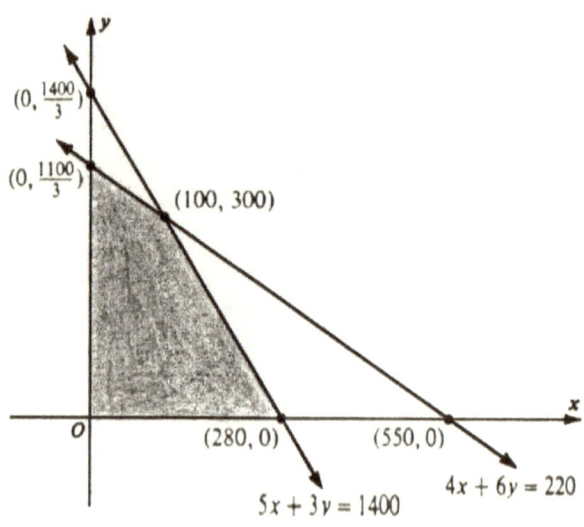

$$(0, \tfrac{1400}{3})$$

$$(0, \tfrac{1100}{3})$$

$(100, 300)$

$(280, 0)$

$(550, 0)$

$5x + 3y = 1400$

$4x + 6y = 220$

Figure 6.4

10. Min. $C = 500x + 700y$ subject to
$$x \geq 0, y \geq 0$$
$$x + 2y \geq 40$$
$$4x + 3y \geq 100$$
$$x \leq 24, y \leq 24,$$

where x and y denote the number of hours per day that the Brooks and Darius mines, respectively, are operated. Corner points: (16, 12), (7, 24), (24, 24), (24, 8); (16, 12) yields the minimum value 16,400.

Chapter 7
Further Food for Thought

Section 7.1 (p. 284)

2 (a) Max. $L(x,y) = 1.5x + 2.3y$
subject to
$$x \geq 0, y \geq 0$$
$$x \geq 12$$
$$y \geq 10$$
$$y \leq 20$$
$$10,000x + 20,000y \leq 100,000$$
or equivalently,
$$x + 2y \leq 10$$

x and y are the number of minutes of late afternoon time and the number of minutes of early evening time, respectively, to be purchased.

(b) From the points of view expressed by Bass, Lonsdale, and Kotler this L.P. model is a crude one. Whether or not it is sufficient for the needs of Chuck Associates is a decision their management will have to make.

(c) The linear program of this L.P. model has no solution; its constraints are not compatible. If, for example, we take $x = 12$ and $y = 10$ (to satisfy $x \geq 12$ and $y \geq 10$, $x + 2y \leq 10$ is not satisfied.

(d) Increase the advertising budget, adjust $x \geq 12$ and $y \geq 10$ to satisfy $x + 2y \leq 10$, or adjust $x \geq 12$ and $y \geq 10$ to satisfy $10,000x + 20,000y \leq K$, where K is the adjusted advertising budget.

Section 7.2 (p. 292)

2. The hypothesis of the corner point method is that the linear program it is being applied to has a solution.

4. The graph of the feasible points of this integer program is shown in Figure 7.1 from which we see that the

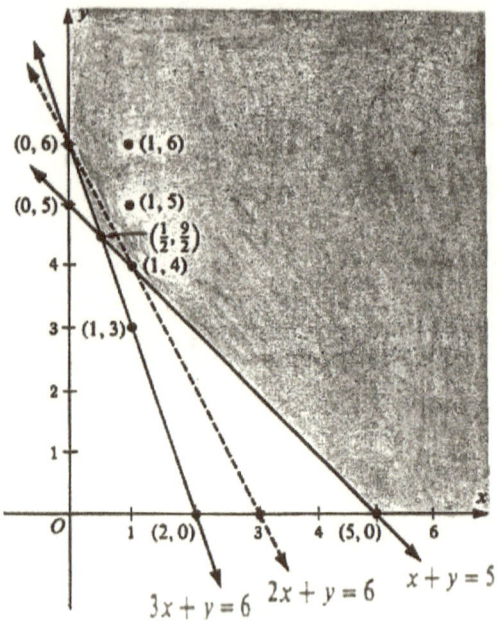

Figure 7.1

corner points are $(0, 6)$, $(\frac{1}{2}, \frac{9}{2})$ and $(5, 0)$. From Table 7.1 we have that $(\frac{1}{2}, \frac{9}{2})$,

Table 7.1

Corner Point	$F(x, y) = 3x + 2y$
$(0, 6)$	12
$(\frac{1}{2}, \frac{9}{2})$	10.5
$(5, 0)$	15

a non-integer corner point yields the minimum value 10.5. Thus we must introduce new constraints so as to,

1. eliminate $(\frac{1}{2},\frac{9}{2})$ as a corner point,

2. not eliminate any integer feasible points,

3. yield corner points in integers.

From Figure 7.1 we see that one additional constraint need be introduced, $2x+y \geq 6$, based on the boundary line $2x+y = 6$ passing through $(0,6)$ and $(1,4)$. Thus the modified linear program to be solved is the following.

Minimize $F(x,y) = 3x + 2y$
subject to
$$x \geq 0$$
$$y \geq 0$$
$$3x + y \geq 6$$
$$x + y \geq 5$$
$$2x + y \geq 6$$

The graph of the feasible points of this modified linear program is shown in Figure 7.2. The corner points are $(0, 6)$, $(1, 4)$ and $(5, 0)$; $(1, 4)$ is the solution and 11 is the minimum value.

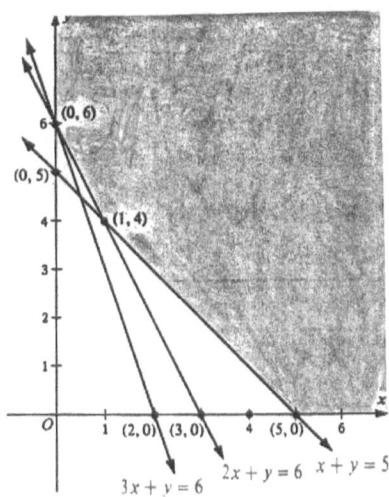

Figure 7.2

Chapter 8
Topics in Algebra with Consideration of Realism

Section 8.1 (p. 304)

2. $2x^4 + x^3 - x^2 + 9$

4. $2x^3 - x^3 h - 7x^2 h - xh^2 + 4h$

6. $2x^2 - 4x + 6$

8. $2x^2 h + 7xh - h + 7$

10. $x^2 h^2 + 3x^2 h + 5xh + 4$

Section 8.1 (p. 306)

12. $4x^3 + 2x$

14. $6x^3 - 6x^2 + 8x$

16. $-6x^4 - 8x^3 + 18x^2$

18. $15x^3 h - 9x^2 h^2 + 12xh$

20. $x^2 - 9$

22. $x^2 - 2hx + h^2$

24. $5x^2 - 14x - 3$

26. $10x^2 - 26x + 12$

28. $x^3 + 2x^2 + x - 4$

30. $2hx^3 + 4hx^2 - 3x + 2h^2x^2 + 4h^2x - 3h$

32. $8x^2 - 14x - 30$

34. $2hx^4 + 6hx^3 - 8x^2 + h^2x^2 + 3h^2x - 4h$

36. $4x^3 + 12x^2 - 40x$

38. $x^4 - 4hx^3 + 6h^2x^2 - 4h^3x + h^4$

40. $6x^3 + 38x^2 + 66x + 18$

42. $x^5 - 5hx^4 + 10h^2x^3 - 10h^3x^2 - 10h^3x^2 + 5h^4x - h^5$

Section 8.1 (p. 311)

44. $a(3a + 2b + 1)$

46. $3b(1 + 2b + 4b^2)$

48. $ab(2 - 3ab + a^2b)$

50. $(x + 3)(x + 1)$

52. $(x - 3)(x + 3)$

54. cannot be factored

56. $2(x - 8)(x - 1)$

58. $2y(y - 4)(y - 2)$

60. $(x - 12)(x - 6)$

62. $(x - 8)^2$

64. $(t + 15)(t - 4)$

66. cannot be factored

68. $2y(y - 9)(y + 3)$

70. $(2x-5)(2x-1)$

72. cannot be factored

74. $(10t-1)(t+5)$

Section 8.2 (p. 322)

2. $x=-6,\ x=1$

4. $x=0,\ x=-4$

6. $x=1,\ x=8$

8. $x=0,\ x=2,\ x=7$

10. $x=-\dfrac{3}{2},\ x=\dfrac{2}{3}$

12. $x=-5,\ x=11$

14. $x=\dfrac{1\pm\sqrt{21}}{2}$

16. $t=-5,\ t=\dfrac{1}{10}$

18. $x=-1,\ x=\dfrac{5}{3}$

20. $x=-1,\ x=\dfrac{8}{5}$

22. $(-2,-1)$ is the minimum point. The graph of $y=x^2+4x+3$ is shown in Figure 8.1.

24. $(2,-7)$ is a minimum point.

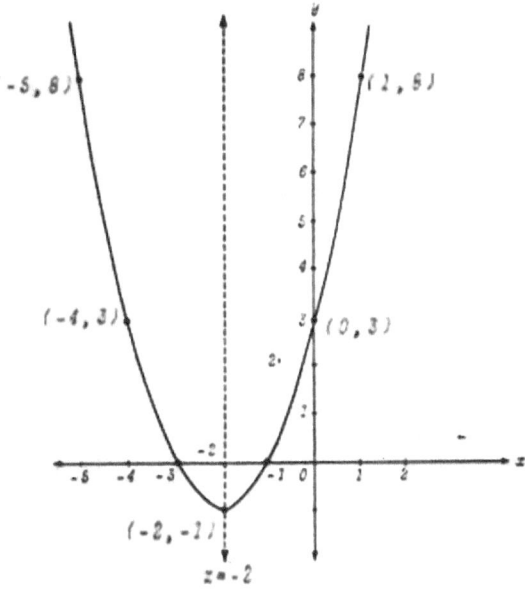

Figure 8.1

26. $(0.3, -8.55)$ is the minimum point.

28. (a) To determine the break-even points we set R(x) equal to c(x) and solve for x. This yields

$$310x - 2x^2 = \frac{1}{2}x^2 + 10x + 1500$$

$$-\frac{5}{2}x^2 + 300x - 1500 = 0$$

Multiplying both sides by $-\frac{2}{5}$ yields the more convenient form:

$$x^2 - 120x + 600 = 0$$

Solving this quadratic equation by means of the quadratic formula yields the break-even levels $x = 5.2$ and $x = 114.8$.

(b) No; this analysis establishes the **mathematical validity** of the break-even values with respect to the given cost and revenue functions. The **real world accuracy** of the break-even output levels hinges on the real world accuracy of the given cost and revenue functions.

(c) The graphs of the cost and revenue functions are shown in Figure 8.2.

(d) Let P(x) denote the profit function of the firm. Then

$$P(x) = R(x) - c(x)$$

$$= 310x - 2x^2 - (\frac{1}{2}x^2 + 10x + 1500)$$

$$-\frac{5}{2}x^2 + 300x - 1500$$

This function has its maximum value at $x = 60$.

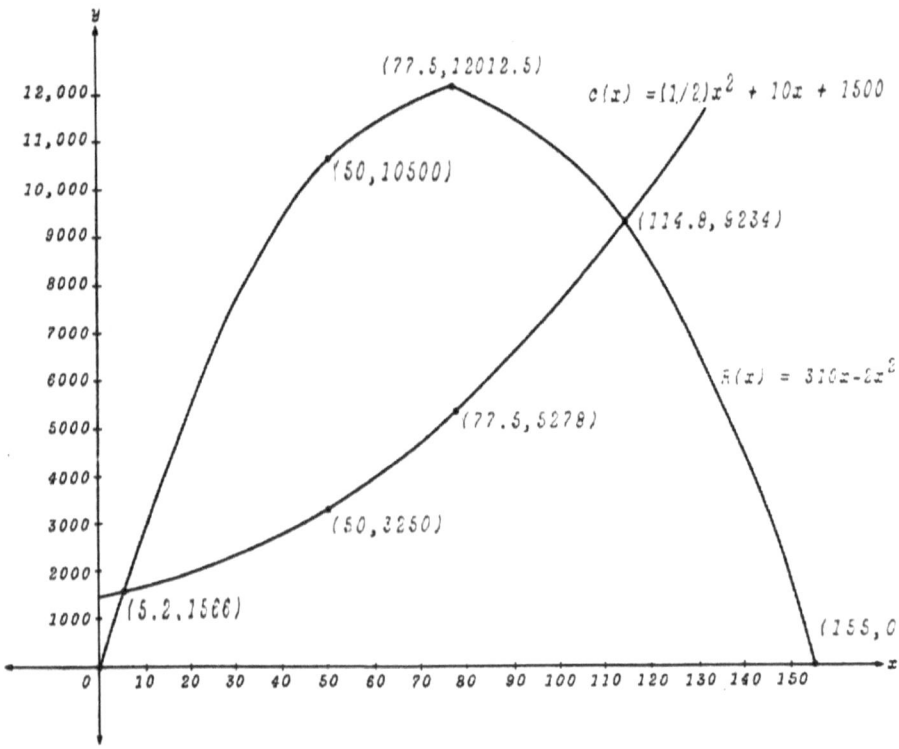

Figure 8.2

(e) No; **mathematical validity** is established, but **real world accuracy** hinges on the real world accuracy of the profit function.

Section 8.3 (p. 326)

2. (a) $y = 8/x$

(b) $y = 8/5$

4. (a) $r = \dfrac{4t^3}{3s^2}$

(b) $r = 500/27$

6. (a) $S = 200p^2$

(b) $S = 1800$

8. (a) $C = 50wd$

(b) $C = 2400$

Chapter 9
Math Modeling for Money

Section 9.1 (p. 334)

2. $l = 95$, $S_{25} = 875$

4. $S_{200} = 40,200$

6. $r = 3$, $l = 243$, $S_6 = 364$

8. $500(2)^5 = 16,000$

10. $S_{52} = \dfrac{1-2^{52}}{1-52}$, which is greater than $7 \cdot 10^{13}$.

12. (a) $6000,   (b) $4800   (c) $26,779

Section 9.2 (p. 341)

2. $R = \$5000$, $n = 30$, and $i = 0.05$. Thus
$$F = 5000 \cdot s_{\overline{30}|0.05} = 5000(66.43885) = \$332,194.25$$

4. (a) $F = \$2500$, $n = 9$, and $i = 0.05$. Thus
$$R = F \cdot \frac{1}{s_{\overline{9}|0.05}} = 2500(0.09069) = 226.73 \text{ or } \$227.$$

(b) $F = \$2500$, $n = 18$, and $i = 0.02$. Thus
$$R = F \cdot \frac{1}{s_{\overline{18}|0.02}} = 2500(0.04670) = 116.75 \text{ or } \$117.$$

(c) $F = \$2500$, $n = 36$, and $i = 0.01$. Thus

$$R = F \cdot \frac{1}{s_{\overline{36}|0.01}} = 2500(0.02321) = 58.03 \text{ or } \$58.$$

Section 9.2 (p. 347)

6. $R = \$400$, $n = 40$, and $i = 0.03$. Thus
$$P = R \cdot a_{\overline{40}|0.03} = 400(23.11477) = 9245.91 \text{ or } \$9246.$$

8. $R = \$100$, $n = 36$, and $i = 0.01$. Thus
$$P = R \cdot a_{\overline{36}|0.01} = 100(30.10751) = \$3010.75.$$

9. Set A costs $55 plus the present value P of an annuity with $R = \$10$,

    $n = 12$ and $i = 0.01$. Thus $P = R \cdot a_{\overline{12}|0.01} = 10(11.25508) = \$112.55$,

    and the cost of set A is $\$55 + \$112.55 = \$167.55$. Set B costs $25 plus

    the present value P of an annuity with $R = \$10$, $n = 10$, and $i = 0.01$.

    Thus $P = R \cdot a_{\overline{15}|0.01} = 10(13.86505) = 138.65$, and the cost of set B is

    $\$25 + \$138.65 = \$163.65 \cdot$

    Thus set B is cheaper by $\$167.55 - \$163.65 = \$3.90$.

10. The cost of the car, $850, is equal to the downpayment D plus the
    present value P of an annuity with $R = \$40$, $n = 12$, and $i = 0.01$.
    Thus $\quad P = R \cdot a_{\overline{12}|0.01} = 40(11.25508) = \$450.20$. Therefore

    $850 = D + 450.20$, from which we obtain $D = \$399.80$.

12. (a) $P = \$6000$, $n = 5$ and $i = 0.04$. Thus

$$R = P \cdot \frac{1}{a_{\overline{5}|0.04}} = 6000(0.22463) = 1347.78 \text{ or } \$1348.$$

(b) $P = \$6000$, $n = 10$ and $i = 0.02$. Thus

$$R = P \cdot \frac{1}{a_{\overline{10}|0.02}} = 6000(0.11133) = 667.98 \text{ or } \$668.$$

(c) $P = \$6000$, $n = 20$ and $i = 0.01$. Thus

$$R = P \cdot \frac{1}{a_{\overline{20}|0.01}} = 6000(0.05542) = 332.52 \text{ or } \$333.$$

13. We first determine the present value P of an annuity that pays \$1500 on the girl's 18th, 19th, 20th, and 21st birthdays.

$R = \$1500$, $n = 4$, and $i = 0.04$. Therefore

$P = R \cdot a_{\overline{4}|0.04} = 1500(3.62990) = 4444.85$ or \$4445.

Thus \$4444.85 invested under the given interest conditions on Amy's 17th birthday will yield the desired annuity. The amount A to be invested on Amy's 14th birthday to grow to \$4444.85 on her 17th birthday is

$A = 4444.85(1.04)^{-3} = 3951$ or \$3951.

14. (a) $F = \$10,000$, $n = 20$ and $i = 0.03$. Thus

$$R = F \cdot \frac{1}{s_{\overline{20}|0.03}} = 10,000(0.03722) = 372.20 \text{ or } \$372$$

Thus \$372.20 must be deposited into the fund at the end of each 6-month period to accumulate to \$10,000 by the time Mr. Baxter is 65 years old.

(b) $P = \$10,000$, $n = 20$ and $i = 0.03$. Thus

$$R = P \cdot \frac{1}{a_{\overline{20}|0.03}} = 10,000(0.06722) = 672.20 \text{ or } \$672.20 \text{ at the}$$

end of each 6-month period.

15. $R = \$200$, $n = 10$, and $i = 0.05$. Thus, at the end of 10 years, Susan will have $F = R \cdot s_{\overline{10}|0.05} = 200(12.57789) = 2515.58$ or $\$2515.58$.

This amount left on deposit for an additional 5 years will grow to $2516(1.05)^5 = 2516(1.27628) = 3211$ or $\$3211$. Thus $\$3211$ is withdrawn at the end of 15 years.

16. We first find the rent that Karen is paying. $P = \$2000$, $n = 24$, and $i = 0.01$. Thus $R = P \cdot \dfrac{1}{a_{\overline{24}|0.01}} = 2000(0.04707) = 94.14$ or $\$94.14$.

Therefore at the end of 15 months she will have paid $\$94.14(15) = \$1412.25$ in principal and interest.

We now determine the interest she will have paid after 15 payments. $\$2000$ at $\$12\%$ per annum compounded monthly grows to $2000(1.01)^{15} = \$2321.94$. Thus the interest accumulated is $\$2321.94 - \$2000 = \$321.94$. At the end of 15 months Karen will have paid $\$321.94$ in interest. Thus the principal paid after 15 payments is $\$1412.25 - \$321.94 = \$1090.31$.

Therefore the amount of principal remaining to be paid after 15 months is $\$2000 - \$1090.31 = \$909.69$.

17. Let us observe that after 20 payments $\$1000$ will have been paid in principal and interest. To determine how much of this $\$1000$ is interest we must determine the original price P, which is the present value of the annuity with $R = \$50$, $n = 24$, $i = 0.01$. $P = R \cdot a_{\overline{24}|0.01} = 1062$ or $\$1062$

Under the given interest conditions the amount that $\$1062$ grows to after 20 months is $1062(1.01)^{20} = 1296$ or $\$1296$. Thus the interest accumulated is $\$1296 - \$1062 = \$234$. At the end of the 20th payment she will have paid $\$234$ interest.

The principal that will have been paid after the 20th payment is, therefore, $\$1000 - \$234 = \$766$. The amount of principal remaining to be paid after 20 payments is thus $\$1062 - \$766 = \$296$.

18. If payments are made at the beginning of each payment period, then by following in the footsteps of the analysis given on page 338 of the text we obtain as the future value F of the annuity

$$F = R(1+i) + R(1+i)^2 + \ldots + R(1+i)^n \qquad (1)$$

If payments are made at the end of each payment period, then the future value of the annuity is expressed by (see pp. 342-344):

$$R + R(1+i) + \ldots + R(1+i)^{n-1} = R \cdot s_{\overline{n}|i} \qquad (2)$$

By multiplying both sides of (2) by $(1+i)$ we obtain:

$$R(1+i) + R(1+i)^2 + \ldots + R(1+i)^n = (1+i)R \cdot s_{\overline{n}|i} \qquad (3)$$

From (1) let us observe that the left side of (3) is F. Thus we have:

$$F = (1+i)R \cdot s_{\overline{n}|i}$$

By means of a similar analysis it can be shown that the present value P of the annuity for which payments are made at the beginning of each payment period is expressed by:

$$P = (1+i)R \cdot a_{\overline{n}|i}$$

Chapter 10
The World of Complex-Numbers

Section 10.2 (p. 355)

2. $(-1, 7)$; $(-18, -1)$

4. $(\frac{9}{2}, 1)$; $(8, 11)$

6. (x, y); $(0, 0)$

8. $(0, 0)$; $(-x^2 + y^2, -2xy)$

10. $(\frac{28}{13}, \frac{36}{13})$; $(1, 0)$

12. $z, z_2 = (1, 0)$

14. $(0, 1 + y)$; $(0, -y)$

16. $(-11, 3)$

18. $(-27, 31)$

20. $(1, 8)$

Section 10.2 (p. 362)

22. No; their sum is not 0.

24. (a) $-7 + 2i$ (b) $3 - 4i$ (c) $-2 - 3i$ (d) $1 + 4i$

26. No; there is no complex-number z such that $0 \cdot z = z \cdot 0 = 1$.

28. $-1 - 3i$

30. $19 + 9i$

32. $2 - 4i$

34. $15 + 2\sqrt{3} + (6 - 5\sqrt{3})i$

36. $-\dfrac{2}{5} + \dfrac{7}{10}i$

38. $\dfrac{10}{13} + \dfrac{11}{13}i$

40. $z = i$ and $-i$; for these values $z^2 + 1 = 0$ and division by 0 is not defined.

Section 10.3 (p. 366)

2. $2 + 3i,\ -1 + 4i,\ -4 + 2i,\ 5 - 3i,\ 6 - 2i,\ -3 - 7i$

Section 10.4 (p. 367)

2. $\dfrac{3 \pm i\sqrt{15}}{2}$

4. $\dfrac{-1 \pm i}{2}$

6. $\dfrac{-2 \pm 6i}{-10}$

Section 10.5 (p. 375)

2. $0.8988;\ 0.4384$

4. $-0.9397;\ -0.3420$

6. $-0.9397;\ 0.3420$

8. $0;\ 1$

10. 1; 0

12. −0.7660; −0.6428

14. 0.8660; −0.5

Section 10.6 (p. 378)

2. $1+(1.7320)i$

4. $-4.929+(0.868)i$

6. $-2i$

8. $2.5-(4.33)i$

10. $\sqrt{2}(\cos 135°+i\sin 135°)$

12. $8(\cos 180°+i\sin 180°)$

14. $5(\cos 127°+i\sin 127°)$

16. $8(\cos 90°+i\sin 90°)$

Section 10.7 (p. 380)

2. $12(\cos 80°+i\sin 80°)$

4. $2(\cos 90°+i\sin 90°)$

6. $2(\cos 90°+i\sin 90°)$

8. $169(\cos 136°+i\sin 136°)$

10. $2^5(\cos 0°+i\sin 0°)=2^5$

Section 10.8 (p. 383)

2. $2(\cos 120°+i\sin 120°)$; $2(\cos 240°+i\sin 240°)$; 2

4. $\sqrt{2}+i\sqrt{2}$; $-\sqrt{2}-i\sqrt{2}$

6. $\cos 60° + i \sin 60°$; -1; $\cos 300° + i \sin 300°$

8. $\sqrt{3} + i$; $-1 + i\sqrt{3}$; $-\sqrt{3} - i$; $1 - i\sqrt{3}$

Section 10.9 (p. 385)

2.   (a) Yes. Rational-number addition and multiplication are closed. The commutative, associative properties, etc. are satisfied by virtue of the rationals being contained within the reals. 0 and 1 are rational.

   (b) Yes. Since the rationals are part of the real-number system, which is ordered, so are the rationals.

   (c) The real-number system is a complete ordered field, that is, for every set of real-numbers which is bounded above, there is a least upper bound which is contained in the reals. The rational-number field is not a complete ordered field. Not every set of rational numbers which is bounded above has a least upper bound which is a rational-number—the set of rational-numbers r such that $r^2 < 2$, for example. The least upper bound of this set of rational-numbers is $\sqrt{2}$, which is not a rational-number.

Chapter 11
Life in the Land of Matrices

Section 11.1 (p. 391)

2. $A + B = \begin{bmatrix} 1 & 3 \\ 8 & -1 \\ -2 & -2 \end{bmatrix}$, $A - B = \begin{bmatrix} 3 & -1 \\ 2 & -3 \\ -6 & 8 \end{bmatrix}$

4. $I_2 - A = \begin{bmatrix} -1 & -3 \\ -4 & -4 \end{bmatrix}$

6. $A + B = \begin{bmatrix} 0 \\ 5 \\ 2 \end{bmatrix}$, $B - A = \begin{bmatrix} -2 \\ 1 \\ -6 \end{bmatrix}$

8. (a) -3 (b) 0 (c) 14 (d) 6
    (e) -7 (f) 16 (g) 12 (h) -8

Section 11.1 (p. 396)

10. $GH = \begin{bmatrix} 0 & 10 \\ -7 & 9 \end{bmatrix}$, $HG = \begin{bmatrix} 10 & -8 \\ 10 & -1 \end{bmatrix}$, $G^2 = \begin{bmatrix} 7 & 0 \\ 0 & 7 \end{bmatrix}$

$H^2 = \begin{bmatrix} 8 & 4 \\ 1 & 13 \end{bmatrix}$, $G^2 H^2 = \begin{bmatrix} 56 & 28 \\ 7 & 91 \end{bmatrix}$

12. $AB = \begin{bmatrix} 2 & 4 & -7 \\ 4 & 8 & -14 \\ -2 & -4 & 7 \end{bmatrix}$, $BA = \begin{bmatrix} 17 \end{bmatrix}$

14. $AB = \begin{bmatrix} 0 & 0 \\ 0 & 0 \end{bmatrix}$

16. $AI_2 = I_2A = \begin{bmatrix} a & b \\ c & d \end{bmatrix}$. Thus $AI_2 = I_2A = A$.

18. $AX = AY = \begin{bmatrix} -3 & 0 \\ -6 & 0 \end{bmatrix}$. Thus. $AX = AY$, where $X \neq Y$.

20. (a) $BC = \begin{bmatrix} -2 & 4 & 3 \\ 11 & -1 & 3 \\ 1 & -1 & -1 \end{bmatrix}$, $AB = \begin{bmatrix} 4 & 5 & 6 \\ 3 & 0 & -3 \\ 2 & 2 & -1 \end{bmatrix}$, $AC = \begin{bmatrix} 2 & 0 & 0 \\ -1 & 5 & 5 \\ 2 & 2 & 3 \end{bmatrix}$

(b) $A(BC) = (AB)C = \begin{bmatrix} 19 & 3 & 10 \\ -12 & 6 & 0 \\ 0 & 2 & 1 \end{bmatrix}$. Thus $A(BC) = (AB)C$.

(c) $B + C = \begin{bmatrix} 0 & 2 & 2 \\ 4 & 2 & 2 \\ 2 & 1 & 0 \end{bmatrix}$, $A(B+C) = \begin{bmatrix} 6 & 5 & 6 \\ 2 & 5 & 2 \\ 4 & 4 & 2 \end{bmatrix}$

$AB + AC = \begin{bmatrix} 6 & 5 & 6 \\ 2 & 5 & 2 \\ 4 & 4 & 2 \end{bmatrix}$. Thus $A(B+C) = AB + AC$.

(d) $B + C = \begin{bmatrix} 0 & 2 & 2 \\ 4 & 2 & 2 \\ 2 & 1 & 0 \end{bmatrix}$, $(B+C)A = \begin{bmatrix} 6 & -2 & 10 \\ 10 & 6 & 6 \\ 4 & 3 & 1 \end{bmatrix}$

$BA = \begin{bmatrix} 3 & 4 & 0 \\ 9 & -1 & 12 \\ 1 & -1 & 1 \end{bmatrix}$, $CA = \begin{bmatrix} 3 & -6 & 10 \\ 1 & 7 & -6 \\ 3 & 4 & 0 \end{bmatrix}$

$$BA + CA = \begin{bmatrix} 6 & -2 & 10 \\ 10 & 6 & 6 \\ 4 & 3 & 1 \end{bmatrix}. \text{Thus } (B+A)C = BA + CA.$$

22. $PC = \begin{bmatrix} 30,040 \\ 37,780 \end{bmatrix}$ Service departments' costs allocated to P_1

Service departments' costs allocated to P_2

Section 11.2 (p. 404)

2. $\begin{bmatrix} -3 & -1 \\ 2 & -2 \\ -1 & 4 \\ 3 & -6 \end{bmatrix}$ since $\begin{bmatrix} 3 & 1 \\ -2 & 2 \\ 1 & -4 \\ -3 & 6 \end{bmatrix} + \begin{bmatrix} -3 & -1 \\ 2 & -2 \\ -1 & 4 \\ 3 & -6 \end{bmatrix} = \begin{bmatrix} 0 & 0 \\ 0 & 0 \\ 0 & 0 \\ 0 & 0 \end{bmatrix}.$

4. Each matrix is the multiplicative inverse of the other because their product in both directions is I_2, the identity matrix for the class of 2 by 2 matrices.

6. Each matrix is the multiplicative inverse of the other because their product in both directions is I_2, the identity matrix for the class of 3 by 3 matrices.

Section 11.3 (p. 410)

2. $\begin{bmatrix} \dfrac{3}{7} & \dfrac{2}{7} \\ -\dfrac{2}{7} & \dfrac{1}{7} \end{bmatrix}$

4. The sequence of inversion tableaus is displayed in Figure 11.1. The entry in row ⑥, column (2) is 0, and thus cannot be converted to 1 by pivoting. Therefore the given matrix does not have an inverse.

$$
\begin{array}{cc}
① \\
②
\end{array}
\left[\begin{array}{cc:cc}
2 & 4 & 1 & 0 \\
1 & 2 & 0 & 1
\end{array}\right]
$$

$$
\begin{array}{cc}
③ \\
④
\end{array}
\left[\begin{array}{cc:cc}
① & 2 & 0 & 1 \\
2 & 4 & 1 & 0
\end{array}\right]
\begin{array}{l}
\text{row } ② \\
\text{row } ①
\end{array}
$$

$$
\begin{array}{cc}
⑤ \\
⑥
\end{array}
\left[\begin{array}{cc:cc}
1 & 2 & 0 & 1 \\
0 & 0 & 1 & -2
\end{array}\right]
\begin{array}{l}
\text{row } ③ \\
(-2)\,\text{row } ⑤ + \text{row } ④
\end{array}
$$

Figure 11.1

6. $\begin{bmatrix} \dfrac{2}{3} & 0 & -\dfrac{1}{3} \\[2mm] \dfrac{1}{3} & 0 & -\dfrac{2}{3} \\[2mm] -\dfrac{2}{3} & 1 & \dfrac{1}{3} \end{bmatrix}$ is the multiplicative inverse of $\begin{bmatrix} 2 & -1 & 0 \\ 1 & 0 & 1 \\ 1 & -2 & 0 \end{bmatrix}$. The

inversion tableaus leading to this result are shown in Figure 11.2.

$$
\begin{array}{c}
① \\
② \\
③
\end{array}
\left[\begin{array}{ccc:ccc}
2 & -1 & 0 & 1 & 0 & 0 \\
1 & 0 & 1 & 0 & 1 & 0 \\
1 & -2 & 0 & 0 & 0 & 1
\end{array}\right]
$$

$$
\begin{array}{c}
④ \\
⑤ \\
⑥
\end{array}
\left[\begin{array}{ccc:ccc}
① & 0 & 1 & 0 & 1 & 0 \\
2 & -1 & 0 & 1 & 0 & 0 \\
1 & -2 & 0 & 0 & 0 & 1
\end{array}\right]
\begin{array}{l}
\text{row } ② \\
\text{row } ① \\
\text{row } ③
\end{array}
$$

$$
\begin{array}{c}
⑦ \\
⑧ \\
⑨
\end{array}
\left[\begin{array}{ccc:ccc}
1 & 0 & 1 & 0 & 1 & 0 \\
0 & ①\!\!-1 & -2 & 1 & -2 & 0 \\
0 & -2 & -1 & 0 & -1 & 1
\end{array}\right]
\begin{array}{l}
\text{row } ④ \\
(-2)\,\text{row } ⑦ + \text{row } ⑤ \\
(-1)\,\text{row } ⑦ + \text{row } ⑥
\end{array}
$$

$$
\begin{array}{c}
⑩ \\
⑪ \\
⑫
\end{array}
\left[\begin{array}{ccc:ccc}
1 & 0 & 1 & 0 & 1 & 0 \\
0 & 1 & 2 & -1 & 2 & 0 \\
0 & 0 & ③ & -2 & 3 & 1
\end{array}\right]
\begin{array}{l}
\text{row } ⑦ \\
(-1)\,\text{row } ⑧ \\
(2)\,\text{row } ⑪ + \text{row } ⑨
\end{array}
$$

$$
\begin{array}{c}
⑬ \\[4mm]
⑭ \\[4mm]
⑮
\end{array}
\left[\begin{array}{ccc:ccc}
1 & 0 & 0 & \dfrac{2}{3} & 0 & -\dfrac{1}{3} \\[2mm]
0 & 1 & 0 & \dfrac{1}{3} & 0 & -\dfrac{2}{3} \\[2mm]
0 & 0 & 1 & -\dfrac{2}{3} & 1 & \dfrac{1}{3}
\end{array}\right]
\begin{array}{l}
(-1)\,\text{row } ⑮ + \text{row } ⑩ \\[2mm]
(-2)\,\text{row } ⑮ + \text{row } ⑪ \\[2mm]
(\tfrac{1}{3})\,\text{row } ⑫
\end{array}
$$

Figure 11.2

8. $\begin{bmatrix} 0 & -\dfrac{1}{4} & \dfrac{1}{2} \\[2mm] 1 & -\dfrac{5}{4} & \dfrac{1}{2} \\[2mm] 0 & \dfrac{3}{4} & -\dfrac{1}{2} \end{bmatrix}$ is the multiplicative inverse of $\begin{bmatrix} 1 & 1 & 2 \\ 2 & 0 & 2 \\ 3 & 0 & 1 \end{bmatrix}$. The

inversion tableaus leading to this result are shown in Figure 11.3.

①	①	1	2	1	0	0	
②	2	0	2	0	1	0	
③	3	0	1	0	0	1	
④	1	1	2	1	0	0	row ①
⑤	0	-2	-2	-2	1	0	(-2) row ④ + row ②
⑥	0	-3	-5	-3	0	1	(-3) row ④ + row ③
⑦	1	0	1	0	$\frac{1}{2}$	0	(-1) row ⑧ + row ④
⑧	0	1	1	1	$-\frac{1}{2}$	0	$(-\frac{1}{2})$ row ⑤
⑨	0	0	-2	0	$-\frac{3}{2}$	1	(3) row ⑧ + row ⑥
⑩	1	0	0	0	$-\frac{1}{4}$	$\frac{1}{2}$	(-1) row ⑫ + row ⑦
⑪	0	1	0	1	$-\frac{5}{4}$	$\frac{1}{2}$	(-1) row ⑫ + row ⑧
⑫	0	0	1	0	$\frac{3}{4}$	$-\frac{1}{2}$	$(-\frac{1}{2})$ row ⑨

Figure 11.3

10. $\begin{bmatrix} \dfrac{1}{2} & -\dfrac{1}{2} & -\dfrac{1}{2} \\[2mm] \dfrac{3}{10} & -\dfrac{7}{10} & -\dfrac{1}{10} \\[2mm] \dfrac{1}{10} & \dfrac{1}{10} & \dfrac{3}{10} \end{bmatrix}$ is the multiplicative inverse of $BC = \begin{bmatrix} 2 & -1 & 3 \\ 1 & -2 & 1 \\ -1 & 1 & 2 \end{bmatrix}$.

The inversion tableaus leading to this result are shown in Figure 11.4.

$$\left[\begin{array}{ccc:ccc} 2 & -1 & 3 & 1 & 0 & 0 \\ 1 & -2 & 1 & 0 & 1 & 0 \\ -1 & 1 & 2 & 0 & 0 & 1 \end{array}\right]$$

$$\left[\begin{array}{ccc:ccc} ① & -2 & 1 & 0 & 1 & 0 \\ 2 & -1 & 3 & 1 & 0 & 0 \\ -1 & 1 & 2 & 0 & 0 & 1 \end{array}\right]$$

$$\left[\begin{array}{ccc:ccc} 1 & -2 & 1 & 0 & 1 & 0 \\ 0 & 3 & 1 & 1 & -2 & 0 \\ 0 & -1 & 3 & 0 & 1 & 1 \end{array}\right]$$

$$\left[\begin{array}{ccc:ccc} 1 & -2 & 1 & 0 & 1 & 0 \\ 0 & -1 & 3 & 0 & 1 & 1 \\ 0 & 3 & ① & 1 & -2 & 0 \end{array}\right]$$

$$\left[\begin{array}{ccc:ccc} 1 & -5 & 0 & -1 & 3 & 0 \\ 0 & ⑩ & 0 & -3 & 7 & 1 \\ 0 & 3 & 1 & 1 & -2 & 0 \end{array}\right]$$

$$\left[\begin{array}{ccc:ccc} 1 & 0 & 0 & \frac{1}{2} & -\frac{1}{2} & -\frac{1}{2} \\ 0 & 1 & 0 & \frac{3}{10} & -\frac{7}{10} & -\frac{1}{10} \\ 0 & 0 & 1 & \frac{1}{10} & \frac{1}{10} & \frac{3}{10} \end{array}\right]$$

Figure 11.4

12. The given matrix does not have a multiplicative inverse. As we see from the last tableau in the sequence shown in Figure 11.5, the given matrix cannot be converted to I_3.

$$\left[\begin{array}{ccc:ccc} ① & -3 & 2 & 1 & 0 & 0 \\ 3 & -4 & 1 & 0 & 1 & 0 \\ 2 & -1 & -1 & 0 & 0 & 1 \end{array}\right]$$

$$\left[\begin{array}{ccc:ccc} 1 & -3 & 2 & 1 & 0 & 0 \\ 0 & ⑤ & -5 & -3 & 1 & 0 \\ 0 & 5 & -5 & -2 & 0 & 1 \end{array}\right]$$

$$\left[\begin{array}{ccc:ccc} 1 & 0 & -1 & -\frac{4}{5} & \frac{3}{5} & 0 \\ 0 & 1 & -1 & -\frac{3}{5} & \frac{1}{5} & 0 \\ 0 & 0 & 0 & 1 & -1 & 1 \end{array}\right]$$

Figure 11.5

Section 11.4 (p. 414)

2. Let $E = \begin{bmatrix} 1 & -2 \\ 2 & 3 \end{bmatrix}$, $B = \begin{bmatrix} 6 \\ 9 \end{bmatrix}$, and $X = \begin{bmatrix} x_1 \\ x_2 \end{bmatrix}$. Then $E^{-1} = \begin{bmatrix} \dfrac{3}{7} & \dfrac{2}{7} \\ -\dfrac{2}{7} & \dfrac{1}{7} \end{bmatrix}$

and $E^{-1}B = \begin{bmatrix} \dfrac{36}{7} \\ -\dfrac{3}{7} \end{bmatrix}$.

3. Let $E = \begin{bmatrix} 1 & 1 & 2 \\ 2 & 0 & 2 \\ 3 & 0 & 1 \end{bmatrix}$, $B = \begin{bmatrix} 3 \\ 4 \\ 6 \end{bmatrix}$, and $X = \begin{bmatrix} x_1 \\ x_2 \\ x_3 \end{bmatrix}$. From Exercise 8

discussed on page 411, $E^{-1} = \begin{bmatrix} 0 & -\dfrac{1}{4} & \dfrac{1}{2} \\ 1 & -\dfrac{5}{4} & \dfrac{1}{2} \\ 0 & \dfrac{3}{4} & -\dfrac{1}{2} \end{bmatrix}$, and $X = E^{-1}B = \begin{bmatrix} 2 \\ 1 \\ 0 \end{bmatrix}$.

6. Let $E = \begin{bmatrix} 1 & -1 & -1 \\ 3 & -5 & 5 \\ 2 & 3 & 4 \end{bmatrix}$, $B = \begin{bmatrix} -4 \\ 6 \\ 7 \end{bmatrix}$, and $X = \begin{bmatrix} x_1 \\ x_2 \\ x_3 \end{bmatrix}$. Then

$E^{-1} = \begin{bmatrix} \dfrac{35}{52} & -\dfrac{1}{52} & \dfrac{10}{52} \\ \dfrac{2}{52} & -\dfrac{6}{52} & \dfrac{8}{52} \\ -\dfrac{19}{52} & \dfrac{5}{52} & \dfrac{2}{52} \end{bmatrix}$, and $X = E^{-1}B = \begin{bmatrix} -\dfrac{19}{13} \\ \dfrac{3}{13} \\ \dfrac{30}{13} \end{bmatrix}$.

Section 11.5 (p. 418)

2. (a) In matrix terms, we have $EX = B$, where

$$E = \begin{bmatrix} 1 & -0.1 & -0.2 \\ -0.1 & 1 & -0.05 \\ -0.05 & -0.1 & 1 \end{bmatrix}, B = \begin{bmatrix} 40,000 \\ 30,000 \\ 20,000 \end{bmatrix}, X = \begin{bmatrix} x \\ y \\ z \end{bmatrix}.$$

(b) $E^{-1} = \begin{bmatrix} 1.0228732 & 0.1233614 & 0.2107427 \\ 0.1053713 & 1.0177332 & 0.0719608 \\ 0.0616807 & 0.1079413 & 1.0177331 \end{bmatrix}$

The inversion tableaus leading to E^{-1} are shown in Figure 11.6.

$$\begin{bmatrix} ① & -0.1 & -0.2 & \vdots & 1 & 0 & 0 \\ -0.1 & 1 & -0.05 & \vdots & 0 & 1 & 0 \\ -0.05 & -0.1 & 1 & \vdots & 0 & 0 & 1 \end{bmatrix}$$

$$\begin{bmatrix} 1 & -0.1 & -0.2 & \vdots & 1 & 0 & 0 \\ 0 & \boxed{0.99} & -0.07 & \vdots & 0.1 & 1 & 0 \\ 0 & -0.015 & 0.99 & \vdots & 0.05 & 0 & 1 \end{bmatrix}$$

$$\begin{bmatrix} 1 & 0 & -0.2070707 & \vdots & 1.010101 & 0.101010 & 0 \\ 0 & 1 & -0.0707070 & \vdots & 0.1010101 & 1.0101010 & 0 \\ 0 & 0 & \boxed{0.9825758} & \vdots & 0.060606 & 0.1060606 & 1 \end{bmatrix}$$

$$\begin{bmatrix} 1 & 0 & 0 & \vdots & 1.0228732 & 0.1233614 & 0.2107427 \\ 0 & 1 & 0 & \vdots & 0.1053713 & 1.0177332 & 0.0719608 \\ 0 & 0 & 1 & \vdots & 0.0616807 & 0.1079413 & 1.0177331 \end{bmatrix}$$

Figure 11.6

$$\text{For } B = \begin{bmatrix} 40,000 \\ 30,000 \\ 20,000 \end{bmatrix}, \ X = E^{-1}B \text{ yields } X = \begin{bmatrix} 48,831 \\ 36,186 \\ 26,060 \end{bmatrix}.$$

(c) $\ B = \begin{bmatrix} 42,000 \\ 36,000 \\ 24,000 \end{bmatrix}, \ X = E^{-1}B \text{ yields } X = \begin{bmatrix} 52,460 \\ 42,791 \\ 30,902 \end{bmatrix}.$

(d) If $\ C = \begin{bmatrix} 0.40 & 0.40 & 0.35 \\ 0.45 & 0.40 & 0.40 \end{bmatrix}$ and X, obtained from $E^{-1}B$, is

the total cost matrix of the service departments, CX describes the allocation of the total costs of the service departments to the production departments.

3. (a) In matrix terms we have $EX = B$, where

$$E = \begin{bmatrix} 1 & -0.6 & -0.6 \\ 0 & 1 & -0.3 \\ -0.01 & -0.1 & 1 \end{bmatrix}, \ B = \begin{bmatrix} 80,000 \\ 60,000 \\ 50,000 \end{bmatrix}, \ X = \begin{bmatrix} x \\ y \\ z \end{bmatrix}.$$

(b) $\ E^{-1} = \begin{bmatrix} 1.0874439 & 0.7399102 & 0.8744394 \\ 0.0336322 & 1.0538116 & 0.3363228 \\ 0.1121076 & 0.1793721 & 1.1210762 \end{bmatrix}$

The sequence of inversion tableaus leading to E^{-1} is shown in Figure 11.7.

$$\begin{bmatrix} ① & -0.6 & -0.6 & \vdots & 1 & 0 & 0 \\ 0 & 1 & -0.3 & \vdots & 0 & 1 & 0 \\ -0.1 & -0.1 & 1 & \vdots & 0 & 0 & 1 \end{bmatrix}$$

$$\begin{bmatrix} 1 & -0.6 & -0.6 & \vdots & 1 & 0 & 0 \\ 0 & ① & -0.3 & \vdots & 0 & 1 & 0 \\ 0 & -0.16 & 0.94 & \vdots & 0.1 & 0 & 1 \end{bmatrix}$$

$$\begin{bmatrix} 1 & 0 & -0.78 & \vdots & 1 & 0.6 & 0 \\ 0 & 1 & -0.3 & \vdots & 0 & 1 & 0 \\ 0 & 0 & \boxed{0.892} & \vdots & 0.1 & 0.16 & 1 \end{bmatrix}$$

$$\begin{bmatrix} 1 & 0 & 0 & \vdots & 1.0874439 & 0.7399102 & 0.8744394 \\ 0 & 1 & 0 & \vdots & 0.0336322 & 1.0538116 & 0.3363228 \\ 0 & 0 & 1 & \vdots & 0.1121076 & 0.1793721 & 1.1210762 \end{bmatrix}$$

Figure 11.7

From $X = E^{-1}B$, we obtain $X = \begin{bmatrix} 175,112 \\ 82,735 \\ 75,784 \end{bmatrix}$.

(c) For $B = \begin{bmatrix} 90,000 \\ 70,000 \\ 70,000 \end{bmatrix}$, $X = E^{-1}B$ yields $X = \begin{bmatrix} 210,874 \\ 100,336 \\ 101,121 \end{bmatrix}$.

Section 11.6 (p. 425)

1. $A = \begin{bmatrix} 0.2 & -0.3 \\ 0.3 & 0.1 \end{bmatrix}$, $D = \begin{bmatrix} d_1 \\ d_2 \end{bmatrix}$, and $X = \begin{bmatrix} x_1 \\ x_2 \end{bmatrix}$ are the input-coefficient,

final-demand, and output matrices, respectively. We have:

$$I_2 - A = \begin{bmatrix} 0.8 & -0.3 \\ -0.3 & 0.9 \end{bmatrix}, \ (I_2 - A)^{-1} = \begin{bmatrix} \dfrac{90}{63} & \dfrac{30}{63} \\ \dfrac{30}{63} & \dfrac{80}{63} \end{bmatrix}$$

Thus

$$X = (I_2 - A)^{-1} D = \begin{bmatrix} \dfrac{90}{63} d_1 + \dfrac{30}{63} d_2 \\ \dfrac{30}{63} d_1 + \dfrac{80}{63} d_2 \end{bmatrix}$$

The inversion tableaus that lead to $(I_2 - A)^{-1}$ are shown in Figure 11.8.

$$\left[\begin{array}{cc|cc} \dfrac{8}{10} & -\dfrac{3}{10} & 1 & 0 \\ -\dfrac{3}{10} & \dfrac{9}{10} & 0 & 1 \end{array} \right]$$

$$\left[\begin{array}{cc|cc} ⑧ & -3 & 10 & 0 \\ -3 & 9 & 0 & 10 \end{array} \right]$$

$$\left[\begin{array}{cc|cc} 1 & -\dfrac{3}{8} & \dfrac{10}{8} & 0 \\ 0 & \left(\dfrac{63}{8}\right) & \dfrac{30}{8} & 10 \end{array} \right]$$

$$\left[\begin{array}{cc|cc} 1 & 0 & \dfrac{90}{63} & \dfrac{30}{63} \\ 0 & 1 & \dfrac{30}{63} & \dfrac{80}{63} \end{array} \right]$$

Figure 11.8